The Ocean
of Life

The Ocean
of Life

THE FATE OF MAN
AND THE SEA

Callum Roberts

VIKING

VIKING
Published by the Penguin Group
Penguin Group (USA) Inc., 375 Hudson Street, New York, New York 10014, U.S.A.
Penguin Group (Canada), 90 Eglinton Avenue East, Suite 700, Toronto, Ontario, Canada M4P 2Y3
(a division of Pearson Penguin Canada Inc.)
Penguin Books Ltd, 80 Strand, London WC2R 0RL, England
Penguin Ireland, 25 St. Stephen's Green, Dublin 2, Ireland (a division of Penguin Books Ltd)
Penguin Books Australia Ltd, 250 Camberwell Road, Camberwell, Victoria 3124, Australia
(a division of Pearson Australia Group Pty Ltd)
Penguin Books India Pvt Ltd, 11 Community Centre, Panchsheel Park, New Delhi – 110 017, India
Penguin Group (NZ), 67 Apollo Drive, Rosedale, Auckland 0632, New Zealand
(a division of Pearson New Zealand Ltd)
Penguin Books (South Africa) (Pty) Ltd, 24 Sturdee Avenue, Rosebank,
Johannesburg 2196, South Africa

Penguin Books Ltd, Registered Offices: 80 Strand, London WC2R 0RL, England

First published in 2012 by Viking Penguin, a member of Penguin Group (USA) Inc.

1 3 5 7 9 10 8 6 4 2

LIBRARY OF CONGRESS CATALOGING IN PUBLICATION DATA
Roberts, Callum.
The ocean of life / The Fate of Man and the Sea / Callum Roberts.
p. cm.
Includes bibliographical references and index.
ISBN 978-0-670-02354-7
1. Ocean and civilization.   2. Ocean—History.   I. Title.
CB465.R62 2012
551.46—dc23   2012000252

Printed in the United States of America

ALWAYS LEARNING   PEARSON

To my daughters, who love the sea

Roll on, thou deep and dark blue Ocean, roll!
Ten thousand fleets sweep over thee in vain;
Man marks the earth with ruin, his control
Stops with the shore . . .

—Lord Byron, "Childe Harold's Pilgrimage"

Contents

PART 2: Changing Course

Prologue

T he water felt chilly as I waded out to fetch the battered skiff from its mooring. It slid easily over the glassy lagoon to the beach, where Julie waited with our diving gear. We had been married a month, and in lieu of a honeymoon I had persuaded her to accept two months of fieldwork studying fish behavior on this remote patch of Australia's Great Barrier Reef. It was June 1987, winter in Australia and summer back home. Two herons picked their way along the shoreline looking for breakfast. They flapped off as the outboard engine coughed to life, and we set course for a spot a mile away, across a maze of coral so complex that it would have baffled the most capable navigator.

We anchored on a rubble ridge that separated the lagoon from the open sea. This was our first ocean dive here, and the thrill of expectation was tempered by a frisson of fear. Ahead of us the homely greens and browns of the shallow reef gave way to the dark indigo of the deep. Huge buttresses of coral plunged down hundreds of feet in parallel walls separated by deep channels. Vivid purple finger-corals vied for space with yellow lettuce-coral, while great mounds of blue and green polyps rose from the bottom.

The reef was a confusing swirl of movement, like Grand Central Station at rush hour. Stocky surgeonfish with electric blue stripes raced around purposefully while loose groups of gaudy parrotfish dodged them. Clouds of damselfish and anthias hovered above, delicately plucking invisible food from the water. Along the edge of the reef I counted eight loggerhead turtles, one for each buttress in view.

A gray reef shark headed our way but barely acknowledged us as he passed by. The whole scene felt timeless and primordial. I was transfixed.

I wonder what I would have thought if on return to shore we had been greeted by a curmudgeonly sage who prophesied that in one hundred years this magnificent reef would be a crumbling ruin, that its bright coral escarpments would be replaced by green fuzz, and that the ranks of fish would have thinned and given way to swarms of jellyfish and gelatinous plankton. I would probably have thought him mad. Nothing could have seemed less likely. And yet less than twenty-five years later most serious marine scientists predict just such a fate. We can already see it happening.

Eleven years after our honeymoon dive, in a foretaste of a warmer world to come, the oceans heated so dramatically that a quarter of the world's coral died. In much of the Indian Ocean 70 percent to more than 90 percent of all corals died, taking with them countless creatures who used coral for living space or food. If three quarters of our forests had withered and died that year, people would have demanded to know why, and aggressive plans would have been drawn up for their recovery. Yet this global catastrophe passed largely unseen and unremarked outside the world of marine science.

The world's oceans have been very stable for most of civilized history. Since the seas leveled off six or seven thousand years ago, after the last ice age, they have for the most part been predictable. Yes, coasts have retreated and advanced under the relentless influence of wave and tide, but the oceans themselves seemed changeless. Their constancy contrasted with the world above water, where landscapes underwent dramatic alterations as first pastoralists and agriculture spread, and then later cities and industry. Today it is the turn of the oceans.

This is a book about the sea change unfolding around our planet. In the last half century human dominion over nature has finally reached the oceans. The speed and extent of these changes has caught us unprepared. The sea is becoming more hostile to life, and not just for the creatures that swim, scuttle, or crawl beneath the waves, but for us too. Only in the last decade or so have we begun to recognize how our activities are reshaping the oceans, and what that means for our own well-being.

We have long known how humankind has changed the land, modifying landscapes to suit our needs and affecting its wildlife across thousands of years. Tens of thousands of years, if you consider how Australian aboriginals and Native Americans used fire to clear vegetation and facilitate hunting and gathering. But we persist in believing, as to paraphrase Byron in the epigraph to this book, that humanity's dominion stops at the sea. Yet even in Byron's time human impacts on the sea were significant. The great auk stood two decades from extinction and the Atlantic gray whale had vanished forever. Fishing had begun to deplete stocks close to coasts and damage habitats with trawls and dredges. People were throwing up coastal defenses, and large areas of marsh and estuary had been reclaimed for ports and agriculture. Rivers in populous regions filled coastal estuaries and bays with mud washed from land exposed by the plow.

Our influence has grown exponentially since then. The last two hundred years have seen marine habitats wiped out or transformed beyond recognition. And with an ever accelerating tide of human impact, the oceans have changed more in the last thirty years than in all of human history before. In most places the oceans have lost upward of 75 percent of their megafauna—large animals such as whales, dolphins, sharks, rays, and turtles—as fishing and hunting spread in waves across the face of the planet. For some species, numbers are down as much as 99 percent as is the case for oceanic whitetip sharks of the high seas, American sawfishes, and the "common" skate of Northern Europe. By the end of the twentieth century, almost nowhere shallower than three thousand feet remained untouched by commercial fishing and some places are now fished to ten thousand feet down.

The oceans have been our conduit for commerce for thousands of years. Today they are the highways of a globalized world, and the roar of engines can be heard in every corner of the sea, even beneath polar ice. Increasingly they also provide us with oil and gas, and growing scarcity has driven us to venture deeper and farther offshore. We are within a decade or so of the onset of deep-sea mining. Riches beckon in the blackness of the abyss, miles below. Dark nuggets of precious metals and rare earths lie scattered over the bottom, seamounts are crusted in cobalt, and deposits of gold, silver, and manganese spewed forth from superheated springs are nearly within reach of miners.

Why, in the face of widespread evidence of human impact, do so many people persist in thinking that the oceans remain wild and beyond our influence? The answer lies in part in the creeping rate of change. Each generation forms its own view of the state of the environment. Younger people generally fail to perceive changes experienced by the old, and so knowledge of past conditions fades with time. Younger generations are often dismissive of the tales of old-timers, rejecting their stories in favor of things they have experienced themselves. The result is a phenomenon known as "shifting baseline syndrome": we take for granted things that two generations ago would have seemed inconceivable.

Loren McClenachan unearthed a telling example of shifting baselines in the archives of the Monroe County Library in Florida when she was doing research there as a doctoral student at Scripps Institution of Oceanography.[1] She found a series of photographs of fish catches landed in Key West by one recreational fishing charter company between the 1950s and the 1980s. She extended the series into the twenty-first century with her own pictures at the same dock. In the 1950s, huge goliath groupers and sharks dominated catches, many of them bigger and fatter than the anglers. As the years pass by the fish shrink, and groupers and sharks give way to smaller snappers and grunts. But the grins on the anglers' faces are just as broad today as they were in the 1950s. Modern-day tourists have no idea that anything has changed.

The oceans are changing faster than at almost any time in Earth's history, and we are the agents of that transformation. Many of these changes will test the ability of its denizens to survive into the future. These alterations are also reshaping our own relationship with the sea and threaten many of the things that we most cherish and take for granted. Our failure to notice creeping environmental degradation has compromized more than our quality of life. In extreme circumstances, it threatens human welfare. History offers many examples of civilizations that have been destroyed by environmental catastrophes that they have unwittingly brought on themselves. Easter Islanders cut down all their trees to raise statues to their gods and then starved when their soil dried up. Mesopotamians invented sophisticated irrigation agriculture, but the technique eventually left fields so salty they could no longer grow crops. Mayan hillslope farming practices stripped the region of its topsoil, precipitating the collapse of this extraordinary civilization during prolonged drought. In these and many similar cases the adverse effects were local to an island or region. Today our influences are global, and we will have to act globally to reverse the impact of what we have already done.

I began my career studying coral reef fish. From the moment I first dived a Red Sea reef, I was hooked. (My wife, Julie, was similarly hooked by our honeymoon trip to the Great Barrier Reef, and she also became a marine biologist.) Thirty years on fish are still at the heart of my research, but my outlook has expanded to a much wider interest: the relations between people and the sea over the course of history. Nonetheless, when I began the research for this book there were huge stretches of ocean science of which I was only dimly aware. Scientists are specialists and devote their lives to research within narrow fields that become further constricted as time passes. Each pores over a fragment of the world, turning it over and over in his or her mind like a chip of some mosaic. Management of pollution is segregated from that of fisheries, which in turn are rarely considered in the same place as shipping or climate change. This means that impacts are discussed in isolation at different meetings and by different people, who never quite see the overall picture. I decided to write this book because I felt there was an urgent need to bring all these separate fiefdoms together. What I found along the way has been a revelation.

We fear change and resist it. Perhaps that resistance is hardwired into our genes: the familiar seems safer than the unknown. Many animals go to strenuous lengths to return to the place where they were born to breed, probably because past success there gives greater assurance of future success. This is a dynamic world, and change sometimes brings good things, but some changes, notably those which erode resilience, are harmful, for when resilience is worn away all bets for the future are off. The path we are on today, as I will show, is pushing ocean ecosystems to the edge of viability. We are emptying the seas of fish and filling them with pollution heedless of consequences, and our unplanned experiment with greenhouse gases is gradually infiltrating the deep sea.

Although a few of the human influences that I will describe in this book have been underway for centuries, others have really only taken hold in the last fifty years. In this sense our impacts on the oceans have been sudden—instantaneous, almost—given that they have taken less than one thousandth of the hundred and fifty thousand years or so for which modern humans have existed. The response demanded to counter these impacts will also have to be sudden and global in scale. Few people yet grasp the gravity of our predicament.

In this book I will take you on a voyage beneath the waves to reveal the seas as few know them. I will show how human activities have for centuries been unpicking the fabric of marine life. We have been able to ignore much of the harm done by our heedless use of the sea—until recently. But as the scale and intensity of human influence have grown, the rate of change has accelerated, and we must now confront the consequences.

To understand the present we must first know the past. I begin at the beginning, when the world began, before picking up our own story, when humankind first makes its appearance on Earth. For tens of thousands of years our only real impact on the oceans was the removal of fish and shellfish, so I start with a brief history of hunting and fishing, and how they have evolved through time. The Industrial Revolution heralded the emergence of people as agents of planetary change, and I move on to describe how the use of fossil fuels, and their impact on currents and climate, is transforming the sea in ways not seen for hundreds of thousands or millions of years. Sea levels are rising faster than the highest rates predicted only twenty years ago. They now threaten dozens of the world's great metropolises, and within fifty years could inundate vast tracts of our best agricultural land, imperiling food security. Within the seas themselves, in one of the least known but potentially most damaging effects of greenhouse gas emissions, acidity is going up in step with carbon dioxide. The outcome could be catastrophic. Shell-forming animals, including many that sustain ocean food webs and thereby our own fisheries, will find their lives increasingly difficult. Not for fifty-five million years has there been a disruption of comparable severity to the calamity that lies just a hundred years ahead if we fail to curtail emissions fast.

By absorbing heat the oceans have so far spared us the worst of global warming. But warming seas have set life on the move, so in the years to come the fortunes of some fishing nations will wane as others rise. But warming will have far more grave effects on productivity, driving it to excess in some places while turning others into oceanic deserts. If the only threats to life in the sea were climate change and fishing, matters would be bad enough. But ocean life faces other pressures, like pollution. I survey the effects of pollutants ranging from toxic chemicals and the now ubiquitous plastic to sewage and fertilizers, and to less familiar pollutants, such as noise and invasive species. Pollution problems have grown in severity with time, so much so that in many regions we see the emergence and spread of dead zones as oxygen is sucked from the water by decaying plankton. The cocktail of human stresses is mixed differently from place to place, but the result is the same: their effects in combination are far worse than in isolation. We are transforming life in the sea, and with it undermining our own existence.

This book is not a catalog of unavoidable disasters ahead. There is much we can do to change course, if we take the opportunity now, but time is of the essence. The longer we ignore the problems, or prevaricate, the less leeway we have to avoid the direst of our possible futures. I devote the rest of the book to how we can plot a new course to safeguard the oceans and ourselves. What we need, I will argue, is an ambitious plan to reverse long-term trends of depletion and degradation, recapitalize the value of the oceans to people and wildlife, and improve the quality of everyone's lives, especially of generations yet to come. We don't have to look on helplessly as all that we love about the sea is sullied. Change for good is within our reach.

PART I

Changing Seas

CHAPTER I

Four and a Half Billion Years

In a remote part of Western Australia there is a range of hills so smoothed by time that little more than undulations are left. The summer sun bakes the red earth, and sparse trees dig deep to quench their thirst. Close up, jumbled slabs reveal beds of smoothed pebbles brought here by ancient floods—very ancient, for the Jack Hills' rocks were laid down three billion years ago. This landlocked desert may seem like a strange place to begin a book about the oceans, but these stones, as unremarkable as they appear, have rewritten the earliest history of our planet.

Zircons are pyramid-shaped crystals that coalesce from cooling magma. They are incredibly resilient and survive through repeated remelting and hardening, as the crust of our restless planet is recycled by plate tectonic movements. Zircons form gems of many hues. The name is thought to come from the Persian *zargun*, which means golden, probably in reference to the translucent yellow stones that were once traded from Sri Lanka through Persia to Europe and to China. Brilliant sky blue crystals were mined from river mud by the people who built the temple city of Angkor Wat in Cambodia.

Jack Hills' zircons are less showy. In fact, they are so small they can hardly be seen with the naked eye. It took over two hundred pounds of Jack Hills' rock to yield less than a thimble of zircons.[1] But it was worth the effort. Each crystal was found to contain traces of uranium, which decays over time into lead, making it a geological chronometer that ticks away the eons. You can date the time of crystal formation to within a few tens of millions of years by measuring the ratio of uranium to lead. The most ancient Jack Hills' crystal is 4.4 billion years old, which takes us to within a whisker of the formation of Earth (well, a 170-million-year whisker, but that is close!). It lets us glimpse a planet in its infancy.

We can never be sure of all the details of Earth's birth and childhood; no one was there to see it, and we can't rewind time to have a look. The best we can do is search the world for ancient rocks and fossils and probe them for their secrets. We spin theories that fit the evidence and rework them when new data find them wanting. These days, new ways of interrogating the past are developed with breathtaking speed, so the picture is coming into clearer focus. Although many uncertainties remain in the story I will tell in this chapter, the broad pattern of events is well established. The Earth coalesced from the rotating disk of dust and debris that would become our solar system about 4.57 billion years ago.[2] The Sun is around the same age, so our planet formed when the match was struck to light the solar system. It grew hotter as debris pounded the growing planet into shape, until the rocks melted. Seen from the darkness of space, our world would have glowed like a faint sun. A planet the size of Mars barreled into this new world sometime around 4.53 billion years ago and smashed off pieces that became the Moon. The impact was so enormous that it vaporized rock, and the primeval world was shrouded by a thick atmosphere of rock and other gases. As the planet cooled a time came when minerals condensed, and for two thousand years the skies rained droplets of molten rock onto an ocean of magma below. The atmosphere remained thick with other vapor and gases even after this thousand-degree deluge ended, and the atmospheric pressure at the surface of the flaming sea would have been hundreds of times higher than today.[3]

We owe a lot to this collision. It knocked Earth's axis of rotation askew, which gives us seasons.[4] Over the vast plain of geological time the Moon has slowed and stabilized the Earth's rotation, giving us longer days. A billion years ago a day was just eighteen hours long, and the year lasted 480 days. The Moon was much closer to the Earth early on, and would have loomed large in the sky. Its gravitational pull gives us much greater tides than the more distant Sun alone, and the flood and ebb of tides would have been violent as they rose higher and fell faster in the short days.[5]

The searing heat of fireball Earth is called the Hadean. For a long time we thought these hellish conditions lasted for the first half billion years of Earth history, but a zircon from Jack Hills has recently transformed that view. The 4.4 billion year old crystal and tiny particles of other materials trapped within it formed at temperatures characteristic of coalescing granite in contact with liquid water.[6] The continents are built from granite, so we can see within this insignificant speck the imprint of a cooler world that had land and water long before we suspected such a place existed. Instead of comprising seething oceans of fire, the world might have been more like a large, steamy sauna.

The Earth is unique within the solar system for its liquid oceans and seas. Where did all this water come from? There was water in the disk of dust and gas from which the solar system formed, but some scientists believe that the inner region of this disk, where the Earth formed, was too hot for water.[7] They contend that icy comets and asteroids delivered water from far reaches of the solar system long after Earth's formation. This barrage goes on today; every few seconds a "snowball" the size of a large truck melts into the outer atmosphere. But measurements of the isotopic composition of these snowballs suggest that they contributed only a few percent of the planet's water.[8] Meteorites were assumed to have delivered most of the rest, but recently a comet has been found with water having an isotopic composition very similar to Earth's oceans, so comets could have contributed a bigger share than once thought.[9] Another view is that water molecules drifting in space stuck to particles of dust, and therefore rock and water came together at the same time. Our world actually has enough water for five to ten oceans, perhaps more: most remains trapped within the bedrock. About half a percent of basalt rock is water, by weight, trapped within the mineral lattice.[10] As the early world heated up and rocks melted, water boiled off into the atmosphere. For over a hundred million years the oceans were in the sky as a dense shroud of vapor that churned above the glowing surface of our planet.

A new rain began to fall when the Earth cooled down a bit, this time of scalding water. The downpour lasted for thousands of years, and would be repeated several times over the next half billion years, as giant impacts from extraterrestrial debris boiled off the upper layers of the seas. The Earth continued to be struck by asteroids for a half billion years more, ones far bigger than those that killed off the dinosaurs sixty-five million years ago. Few traces of these impacts remain, for our crust is continuously reworked into the planet's interior, but we can read their fury from the Moon, whose cratered surface recorded the bombardment and has remained immobile since it cooled.

To begin with, oceans covered most of the world; their volume could have been twice that of the present seas.[11] Islands cropped up where blocks of the Earth's crust collided and volcanoes built ash and lava mountains, but there were no continents. Those came later and formed slowly. The world today is broken up into continents and sea because the crust is made up of two kinds of rock. It is made of basalt below the sea and is slightly denser than the granites that make up the continents. Repeated remelting in the furnace of our young world separated off the lighter granitic rocks that form today's continents. Both kinds of rock float on a sea of hot, viscous magma in the mantle below, but the continents float higher. As with icebergs, how much you see above water depends on how much lies unseen. Continents have deep roots, so they float higher in the mantle than oceanic crust. This is why the average height of land is 2,770 feet above sea level, while the average depth of the oceans is 12,140 feet; it is a consequence of the different densities of their rocks.[12]

The sticky magma of the Earth's mantle keeps the world's surface in constant motion. Hot magma rising from deep down creates new crust in some places. Since the world is not getting bigger, that creative force is matched by destruction somewhere else, as crust slides back into the mantle. The Earth's surface is thus divided into blocks, called plates, that each moves in a slow-motion geological dance through time. Here again the density difference between oceanic and continental crusts comes into play. Because oceanic crust is thin and dense (three to six miles thick, compared to the twenty to twenty-five miles of most continental granite), it is recycled quickly into the mantle, about ten times faster than that of the continents. (None of the oceanic crust is greater than 200 million years old, whereas about 7 percent of the land is older than 2.5 billion years.[13]) Think of the continents as the froth bobbing on a pool beneath a waterfall. The water pours on beneath, but the froth endures.

Continents began to form very early on. Crust was recycled quickly at the beginning on a hot Earth, but the rate slowed as the planet cooled, and intense meteorite bombardment came to an end four billion years ago. Continents grew over time and reached roughly their present landmass two and a half billion years ago. Since then they have been recycled at about the same rate as they are created. Today the plates creep slowly; the Atlantic Ocean is opening at an inch a year, a little slower than fingernails grow.

To us, the oceans are immense. They cover 140 million square miles and fill a volume of 324 million cubic miles. It is hard to imagine a cubic mile. If you flooded all of New York's Central Park with water to the height of a thirty-story building, the volume would be a little greater than one eighth of a cubic mile; it would take 2.66 billion Central Parks of water to fill the oceans, an almost unthinkable volume of water. Yet at the scale of the planet, the oceans form a layer only as thick as the skin of an apple.[14]

Zircons aside, the oldest rock in the world is the four-billion-year-old Acasta Gneiss in northern Canada. This formed deep underground, so it tells us little about what was going on at the surface. The oldest surface rocks are the highly metamorphosed Isua sediments of southern Greenland, which formed underwater and give us the first direct evidence of oceans.[15] Remarkably, these deposits suggest that life had already evolved by the time they were formed. There are no fossils in Isua sediments, but the chemical composition of carbon buried in these rocks is characteristic of the presence of life.

The first life-forms evolved in the early Archean eon—the billion-and-a-half-year eon that followed the Hadean. It was a very different world from the one we live in today. There was almost no free oxygen, for a start, and the sun burned 25 percent less bright.[16] Methane-producing microbes evolved sometime between 3.8 billion and 4.1 billion years ago, creating a greenhouse gas twenty-five times more powerful than carbon dioxide.[17] Methane levels rose and the planet warmed.[18] For over a billion years, until at least 2.5 billion years ago, the greenhouse shroud sustained these liquid seas as the Sun's fire warmed the Earth. The oceans would have frozen if it hadn't been for this dense blanket of gases in the atmosphere, and life might never have kindled, or it could have started only to be snuffed out early on.

There is not a trace or shadow of life for hundreds of millions of years after the first spark, beyond chemical alterations in the rocks.[19] But spectacular recent advances in genetics and computing enable us to grow the tree of life backward, from its leaves to the tip of the root from which it sprung. Every living thing, from the humblest virus to the greatest of the whales, shares a common heritage that is written in their genes. That similarity tells us, just as Charles Darwin predicted, that everything living today is descended from life's primordial spark. The genes that code for different metabolic functions can be placed in the sequence of their emergence on the tree of life, and show something of their timing. They tell us when life passed critical evolutionary milestones, and from them we can infer how the environment was changing. In some cases early life-forms had to respond to planetary upheavals, but often they themselves were responsible for changing the world around them.

Hot springs under intense pressure can be found in the deep sea, spewing forth water superheated to six or seven hundred degrees Fahrenheit and so laden with minerals that their plumes are opaque black or white. They deposit metal-rich compounds nearby that in the early oceans might have catalyzed important chemical reactions. Today these hot springs support rich communities based entirely on energy captured from chemical reactions undertaken by microbes.

Here, then, is a vestige of how the earliest microbes might have made their living, perhaps even of the place where life itself was forged.

The planet was ruled by singleton cells and microbial slime for more than three billion years. It is hard to grasp such a vast number. Think of it this way: If every one of those years lasted just a second it would take ninety-five years for three billion to pass by. Microbes evolve fast, because they have fleeting generations. Even taking into account the likelihood that Archean oceans were far less productive than our own, there was time for hundreds of billions of generations to come and go.[20] Every new generation offered the possibility of variations that are the raw material of evolutionary innovation. This was a time of extraordinary inventiveness, when the foundations were laid for almost everything life does today.

Microbes developed the ability to produce energy early on, by converting hydrogen sulfide—the gas that gives rotten eggs their smell and that pours forth from deep-sea hot springs—into sulfates.[21] This was an essential step. Some developed complex chemical machinery to draw energy from sunlight to accomplish this conversion in shallow water, and photosynthesis was born. Microbes are not easily fossilized, but they are petrified into flinty rocks, called chert. The earliest fossils are microscopic threads locked into 3.45 billion year old Australian cherts, although their interpretation is controversial.[22] They look like cyanobacteria, a group still common today. These were the creatures that would later on develop the capacity for photosynthesis, the method of generating energy from sunshine that dominates all primary production today. They use the sun's energy to create carbon compounds—food, in other words—from carbon dioxide and water. Their waste product is oxygen.

The first chemical traces of oxygen-producing photosynthesis were found in 2.7 billion year old shales rich in organic matter.[23] We must thank this innovation for the way our world works, because this kind of photosynthesis has produced essentially all of the free oxygen around us today. It took hundreds of millions of years for enough oxygen to be made for us to detect its traces. Dark shales from Mount McRae in Western Australia laid down 2.5 billion years ago give us the first whiff of oxygen.[24]

Soon after (at least, in geological terms, since it took another fifty million years) we begin to find evidence of oxygen in rocks all over the world. The next 150 million years is known to geologists as the Great Oxidation Event, because it heralds the first major step in the formation of the atmosphere we have today. But far from being a boon—that would come later—oxygen first plunged the Earth into a crisis. When oxygen and methane get together the result is carbon dioxide and water. Methane is twenty-five times more powerful a greenhouse gas than carbon dioxide, so the Earth's comfort blanket thinned, and the planet froze.

Some scientists think this ice age was so severe that the sea and the continents iced over all the way to the tropics.[25] When ice forms at latitudes lower than about thirty degrees, it reflects so much heat back into space that glaciation runs away with itself. This is because more of the sun's heat is absorbed by the Earth at the tropics than the poles, so more heat is reflected into space by low-latitude ice than high-latitude ice. Some simulations of such a world suggest that the oceans could have frozen to over three thousand feet deep. The ice would have retreated only after millions of years of volcanic activity had added enough extra carbon dioxide to the atmosphere to warm Earth enough to melt it. It is hard to see how life could have survived "Snowball Earth," so one suggestion is that the tilt of the planet in relation to the Sun must have been different at this time, so that the poles were warmer than the tropics and ice-free conditions persisted through the great freeze in some places.[26] Another possibility is that weather systems kept some areas of ocean ice-free.

You would not have wanted to swim in the oceans of the past. In Hadean and Archean times they are thought to have been rich in dissolved iron, and anoxic (meaning oxygen-free). The iron came from deep-sea hot springs and weathered rock. Iron dissolves in the absence of free oxygen and is easily washed to the sea, whereas if oxygen is present, iron oxides tend to stay put. Early microbes put this iron to work. They developed ways to use the power of sunshine to oxidize free iron and make food from carbon dioxide and water. Iron was precipitated to the bottom of these seas to form thick deposits known today as banded iron formations.[27]

A slice of two-and-a-half-billion-year-old seabed sits on my shelf at home. It is only a centimeter thick but surprisingly heavy. Wavy layers of rust brown, yellow, and orange silica alternate through the slab with dark black and gray stripes of magnetite. Thicker layers of shimmering tiger eye fill gaps where the rock was later twisted and deformed under pressure. It is beautiful. I find it amazing to run my finger across this fragment of our primordial ocean. Most banded iron formations have long since been recycled back into the Earth's mantle, but bits of ancient seabed lie stranded in rock formations in Australia, Canada, Russia, and elsewhere to form some of the richest sources of iron ore in the world. Banded iron formations are largely confined to rocks older than 2.4 billion years. They disappear from the record for the next 400 million years. For a long time it was thought that free oxygen produced by cyanobacteria had dissolved in the ocean and stripped it of iron, but based on the chemistry of rock deposits, another possibility now seems more likely. Donald Canfield, a polymath geobiologist based in Denmark, thinks that the sea had no oxygen below a thin surface layer for hundreds of millions of years after oxygen first began to rise in the atmosphere. His idea is that oxygen reacted with sulfides in terrestrial rocks and washed into the sea as sulfate. It was the sulfate that stripped the oceans of their iron, not the oxygen. The Great Oxidation Event would not penetrate to the deep sea until a billion years later.

Why did it take so long to ventilate the deep? Oxygen levels in the atmosphere were still very low, no more than about 1 percent of today's. We would have suffocated in no time. Oxygen is transported to the deep sea when surface waters plunge down, but it is used up to break down organic matter—dead microbes, in other words—that sinks from the surface. It would not take much sinking organic matter to strip oxygen from the deep faster than it could be replaced. Still, it is hard to explain how deep water could remain oxygen-free for such an immense stretch of time after the rise of oxygen-producing photosynthesis. Early oxygen producers lived in shallow, sunlit waters. Just like life today, they needed nutrients to fuel their growth as well as sunshine. Since most nutrients sink with the dead bodies of microbes, they have to be recycled by upward mixing of deep water. An intriguing new idea is that anoxic waters rich in hydrogen sulfide reached into the lower sunlit layer, where they encountered photosynthesizers that used up hydrogen sulfide but didn't make oxygen, and intercepted most of the nutrients before they could reach the oxygen-producing cells closer to the surface.[28] There is another twist in the nutrient tale. Very early on life incorporated the trace metals iron and molybdenum into enzymes that fix nitrogen, thus securing one of life's essential nutrients. Sulfidic seas would have little of either of these metals in solution, because both would have reacted with hydrogen sulfide to produce insoluble compounds that were deposited at the seabed. So it was that quirks of oceanography and limitation of nutrient supply appear to have held back planetary oxygenation for what seems like eternity.

There are still places today where sulfur-based photosynthesis continues. The Black Sea is an almost completely enclosed basin of water that is capped by a warm, sunlit surface layer that is less dense than the cool water underneath. Little of the oxygen in this warm surface layer mixes downward, so the Black Sea has not had any oxygen for thousands of years below about five hundred feet. If you were to somehow bring a sample to the surface it would stink of rotten eggs, just as the sulfidic ancient oceans would have. As this anoxic water nears the surface, green and purple sulfur bacteria use sunlight to produce food and energy, just as their predecessors did eons ago.

It is hard to know if all the world's oceans were once sulfidic like today's Black Sea. Most of the rocks that could tell us were long ago recycled back into the mantle, so there are few places to look for clues. What we do know is that eight hundred million to nine hundred million years ago conditions began to change. What paleontologists refer to as the "boring billion years" ended. Ocean chemistry shifted toward the composition of modern oceans during this period. A phase of intense mountain building around this time may have enhanced the delivery of trace nutrients to surface waters, enabling oxygen producers to flourish. With the amount of oxygen in the atmosphere rising, the sea became ventilated beyond the lower limit of the sunlit layers, ending the hegemony of the sulfur bacteria and freeing more nutrients for oxygen producers. This positive feedback led gradually to the irreversible oxygenation of our atmosphere. It set the scene for life's next great innovation: multicellularity.

When Charles Darwin wrote *On the Origin of Species*, there were fossils in abundance all the way back to the onset of the great Cambrian explosion, when life-forms with hard body parts which readily fossilized first came into being. Before that, nothing: The rocks appeared blank. It was as if some creator had conjured forth a menagerie of primitive life that would later diversify into all subsequent beings. It has taken a hundred and fifty years of intensive search to find the enigmatic shadows of earlier life, much of it microbial, but it is there. Darwin would have loved to know what we know today.

Oxygenation of the Earth's atmosphere was a key event in the history of life. A metabolism based on oxygen respiration produces sixteen times more energy than equivalent anoxic pathways. When free oxygen became available it created tremendous opportunities, which were soon realized. The first oxygen-using enzymes have been traced by following the evolution of proteins backward in time, using the latest genetic sequencing libraries and powerful computer techniques. Although there are precious few pre-Cambrian fossils, in a sense every one of us contains a fossil library of genes that goes all the way back to the Archean eon, billions of years ago. The history of changing protein shapes gives us a molecular clock from which we can read the time when new innovations first emerged. The first hesitant steps toward the use of free oxygen seem to have been taken 2.9 billion years ago, 400 million years before the first clear evidence of free oxygen appears in the rocks.[29] Proteins involved in oxygen metabolism multiply fast through the Great Oxidation Event. Then there is a second wave of evolutionary innovation in oxygen use that begins about 1.2 billion years ago and continues all the way to the Cambrian explosion of life. Atmospheric oxygen had increased to about 12 percent of today's level by the onset of the Cambrian period.

Oxygen is produced by photosynthesis when organic matter is created, and consumed both by respiration and by the decay of that organic matter. If the two sides of the equation were exactly equal, oxygen would not build up in the air. But some organic matter is buried at the bottom of the sea or in swamps and lake beds. During late pre-Cambrian times, when oxygen levels in the atmosphere rose, the rate of organic carbon burial must have increased. Free oxygen is also produced when sulfate combines with iron to produce iron sulfide (pyrite, or fool's gold). So increased rates of pyrite burial would also oxygenate the world. We don't know exactly what caused burial rates to increase, but the oxygenation of the oceans is believed to have played a key role.

Whatever the cause, there are many who think that rising oxygen levels paved the way for life's explosion. The two go hand in hand. Large-bodied aerobic animals need enough oxygen to ventilate their tissues. Their virtual absence in the previous three billion–plus years of life might just be because their size was impossible until oxygen levels rose high enough to sustain such bulk.

The Cambrian period was a time of amazing evolutionary creativity. We can read in the rocks, within 20 million years of its onset, 542 million years ago, the appearance of virtually every major animal group alive today. If periods are characterized by their most successful creatures, this was the age of the arthropods, which are distinguished by their external skeletons and jointed legs. Insects, crabs, lobsters, millipedes, and spiders and the like are all their living descendants. But the now extinct trilobites really define the Cambrian period better than any other group. Cambrian seas were full of a bewildering mix of these low-bodied, scuttling creatures, which wore body plates and helmets that were sometimes flamboyantly ornamented, with spikes and spines and compound eyes with crystalline lenses. Trilobites made good use of this articulated armor to defend themselves from the predators that were about. One explanation for the Cambrian explosion describes it as a runaway arms race in which some animals scrambled to escape or repel predators, while others developed ever better ways to catch prey. Grazing and predation predate the Cambrian period by hundreds of millions of years, but it was then that the ability to swallow large prey was firmly established. The foundations of modern food webs were laid in Cambrian seas, a period that came to a close 488 million years ago.

Although life was diversifying, a series of environmental crises lopped branches from the spreading crown of the tree of life. Just when the atmosphere had become more breathable, anoxia returned to the oceans and sulfidic seas reappeared.[30] What seems to have happened is that waters devoid of oxygen welled up onto shallow continental shelves, perhaps following rising seas, and they snuffed out many trilobites and other creatures. Paradoxically, these crises may have driven atmospheric oxygen levels up. Ocean anoxia is linked to higher surface productivity and greater organic carbon burial in Cambrian oceans. Sinking feces and dead plants and animals would descend faster out of the sunlit surface layers than the microbes of more ancient seas, so more carbon would be buried, especially in anoxic waters where breakdown is sluggish. Free oxygen would get a boost.

The next leap forward in the construction of modern ocean ecosystems has again been linked to rising oxygen. Perhaps surprisingly, we owe the existence of sharks to the evolution of vascular land plants.[31] These plants were the first to develop differentiated organs, such as leaves, roots, and stems, and a complex architecture of tubes to conduct water, nutrients, and food throughout their system. This group includes ferns, club mosses, conifers, and flowering plants, although the last would come much later. Vascular plants debuted about 420 million years ago, and within 30 million years they had sprouted roots to draw nutrients from the soil and stabilize bigger growth forms. By 370 million years ago plants had clothed the continents in a lush tangle of green.

How does this relate to sharks, you might ask? Roots and other adaptations that extracted nutrients speeded the rate at which phosphate, a key plant nutrient, weathered the soil and washed into the sea. Marine productivity boomed, enabling longer food chains and bigger predators. At the same time, land plants massively increased organic carbon burial in swamps and, through runoff, in marine sediments. This brought oxygen to levels similar to those of the present (21 percent of the air), where they have stayed, give or take 5 percent, for most of the last 350 million years. Freed from the constraint of low oxygen, the stage was set for active predators with big oxygen demands.

Animals a yard long were already present in late pre-Cambrian seas 580 million years ago. They were probably sedentary grazers or detritus feeders. Eighty million years on, there was more diversity, and hunters emerged who were the size of a small child. Fast-forward to 400 million years ago, and the seas were filled with primitive fish, including the heavily armored placoderms. Some had become terrifying beasts by 370 million years ago, as large as buses. The predatory megafauna had arrived. Over the next 120 million years they were joined by sharks and reptiles such as ichthyosaurs, plesiosaurs, and turtles. The predatory expansion reached its zenith recently, almost yesterday in geological terms. During the Miocene epoch, which spanned 23 million to 5 million years ago, giant sharks the size of great whales patrolled the sea, and whales ate whales. There were formidable sperm whales the same length as today's but with teeth three times as big. They have been named *Livyatan melvillei* after Herman Melville's mythic white whale.[32]

The presence of these huge predators suggests seas more productive than our own. Indeed, the oceans and their lives have had many ups and downs over the past 500 million years. The variety has gradually increased, but not in a steady rise, and not by the simple addition of new species to the existing inventory. The vast majority of species that have ever lived on Earth have disappeared. Five mass extinctions followed the crises that wiped out most of the trilobites during the Cambrian period, the ones possibly caused by re-emergence of anoxic oceans.[33] The mother of all extinctions came 251 million years ago, at the end of the Permian period, when the fossils of more than 90 percent of marine species and two thirds of terrestrial species simply stop appearing in the rocks. Before the Permian extinction seabed sediments were filled with lively burrowing animals, which churned them over and over. Afterward the sediments were nearly stilled.

What caused this cataclysm remains a source of heated argument, but most likely there was a half-million-year episode of volcanism that discharged half a million cubic miles of basalt across the surface of what today is Siberia.[34] The lava flooded an area of 620,000 square miles, to depths of close to 2 miles. It passed through layers of carbonate rock and coal and released massive quantities of carbon dioxide into the air. Carbon dioxide had declined more than tenfold in the approximately 240 million years after the Cambrian period, and the loss of this greenhouse gas had driven the world into a glaciation that lasted over sixty million years.

Now, at the onset of the Triassic, the period which followed the Permian, it got hot. As carbon dioxide climbed, the world warmed, and as it warmed the poles melted. Global temperatures spiked eleven degrees Fahrenheit higher than it is today. The oceans rose and the Arctic reached a balmy sixty degrees Fahrenheit to seventy degrees Fahrenheit.[35] For reasons I will return to, this temporarily slowed deep-ocean mixing, and the stagnant water warmed all the way down to the deep seabed. The warm conditions melted huge polar and undersea deposits of methane, causing runaway global warming. Land plants withered and soils washed away. The stagnant sea became anoxic in all but surface waters. Meanwhile, dissolved carbon dioxide turned the oceans more acidic, which, again for reasons I will come back to, was disastrous for chalky animals like corals, urchins, calcareous seaweeds, and sponges.[36] The mass extinction at the close of the Permian period almost wiped life's slate clean. But life rose again. Within 10 million years, the oceans had refilled with new species.

Sitting in my garden one day thinking about this book, my mind wandered back to the world's origin four and a half billion years ago. It is hard to believe that that world is the same as our own. Our planet, hurtling through the emptiness of space, so alien yet so familiar. I imagine hitting the fast-forward button to see continents ascend, meander, coalesce, and crumble; oceans rise and fall; seas, deserts, and ice caps come and go. Within the belly of these oceans creatures outlandish and familiar, bizarre and terrifying appear and disappear. Great reefs are built, destroyed, and remade. Life flourishes, is choked off and resurrects itself again and again. And through the strangeness and inconstancy swim turtles and sharks, nautilus and jellyfish, scarcely altered, like a connecting thread that reaches back through time. A moment before the tape ends, we appear. All of us carry this thread within ourselves. We are creatures of the sea with a lineage that stretches back to sponges and beyond, all the way through single-celled microbes to life's origin. The rest of this book tells the story of what happened to the oceans after we arrived.

CHAPTER 2

Food from the Sea

Anthropology and archeology have long been in thrall to an image of early humans as big-game hunters of the open plains. Game-hunter thinking has us evolving from tree dwellers into savannah dwellers who started to walk on two legs. Sharp wits and ingenuity were necessary to thrive in the open, where we had to fend off dangerous carnivores, and our bipedalism freed us up to hold tools and weapons. Our large brains later allowed us to develop language, which unlocked the possibility of technology and culture.

This view of human origins has a certain mythological ring to it, suggesting as it does that our plucky species succeeded in a heroic struggle against great odds.[1] But the story has holes. Baboons live on savannahs and have not become brilliant bipeds, and we have a mixture of adaptations that make little sense in the absence of water. Today we carry ten times more subcutaneous fat than other primates (we're about as fat as fin whales), which if this was true of our ancestors, wouldn't have been particularly helpful to endurance hunters running down their prey. But it would have insulated us from water. *Homo erectus* had dense bones more akin to those of diving mammals like manatees than fleet-footed plains hunters. We are prone to dehydration, an uncommon trait in savannah dwellers, and have an instinctual breath-hold reaction when we plunge into water. The only other primate that is regularly bipedal today is the proboscis monkey, which wades through swamp waters on its hind legs. Could our shift to bipedalism have been an aquatic adaptation developed by wading to gather shellfish?

Marc Verhaegen, a Belgian doctor, and his colleagues have recently revived the concept of man as an "aquatic ape."[2] Verhaegen believes that *Homo erectus* evolved at the waterside and put the fruit-and-nut-cracking techniques developed by ancestral forest species to a new use by smashing open shellfish, turtles, and crabs.

Shellfish are easy to find and gather, and although not especially calorific (it would take 150,000 cockles to match the calorie content of a large deer), they are rich in protein and other nutrients needed for brain development. The nervous system, which evolved in the oceans half a billion years ago, was built in part from omega-3 fatty acids made by algae and plankton. These compounds were in short supply for land dwellers, which helps explain why animals like the rhino, which weigh a ton, have brains two-thirds smaller than our own, while marine mammals like dolphins have large brains. Savannah models of human evolution see us getting our brain food from scavenged or hunted brains of terrestrial mammals, but waterside sources such as shellfish, waterbirds, eggs, and turtles would have been easier and more regular fare.

The idea of man as a big-game hunter has been hard to shake off. Museums freeze the thinking of the day into dioramas that endure for decades, impressing their stories into the minds of generations. I remember well from childhood the tableaux of stocky, hairy people wrapped in skins, triumphant in their slaughter of buffalo or antelope. Usually these scenes were set in open savannah, occasionally giving a nod to our dependence on water with a lake or river painted into the distant background. As Verhaegen points out, the savannah-hunter picture of human evolution predates almost all fossil finds of early hominins.[3] In his accounting, all *Homo* fossils discovered before or since have been associated with lakes, rivers, deltas, and coasts. The concept of the aquatic ape is widely contested in anthropology, but to me it seems more persuasive than the convoluted logic needed to explain our evolution on Africa's dusty plains.

The earliest remains of *Homo* species, dating from 2.5 to 2 million years ago, are from inland water bodies such as the Gona floodplain in

Ethiopia and freshwater springs near Lake Olduvai in Kenya. It was later on that we reached the coast. Around 2.8 million years ago, Africa, like the rest of the world, underwent major climatic upheaval, cycling between prolonged phases in which much of the continent was arid and inhospitable and wetter phases with more varied climate. Wet phases greened the Sahara desert, perhaps opening up routes out of Africa, and arid phases pushed them into the small pockets of land that remained hospitable. One such place was southern Africa, where humans first developed their predilection for seafood. Our genes point that way, tracing our ancestry all the way back to Angola and Namibia.[4]

Pinnacle Point Cave 13B—the name sounds like a condo address—is on the coast of South Africa that was occupied on and off for thirty thousand years. It has enormous significance, for this cave was home to some of the earliest modern humans, *Homo sapiens sapiens*.[5] Today it lies beneath the ninth hole of the luxury Pinnacle Point Golf Resort, and more modern condominiums line the cliffs. Deep inside the cave, thick deposits mark the passage of time and tell the story of our awakening humanity. At the very beginning, 164,000 years ago, fragments of bright red ochre signal the first use of pigment, probably for adornment. Scratch marks crisscrossed over one piece suggest the emergence of symbolic art and small, sharp stone tools. At Blombos Cave, farther west on the coast, archaeologists found jewelry made seventy-five thousand years ago from tiny seashells perforated by predatory snails.

The first occupants at Pinnacle Point gathered shellfish to eat at low tide, and much of the cave-floor deposits are drifts of shells. Shellfish was also consumed 140,000 years ago at Blombos and remained important for the entire time Blombos was used as a home.[6] (Having once spent a summer in the company of a student whose research involved rotting heaps of shells, I can assure you that the smell would have tested the sternest of modern constitutions.) Bones of black musselcracker and mullet, fish that follow the rising tides to catch their prey, appear 77,000 years ago in Blombos. Since no hooks have been found with these fish bones, people probably speared them or caught them by hand. Or they may have been lured into shallow water by a chum made of broken urchins or shells and then speared with bone-tipped spears. A slightly later intellectual leap made at Pinnacle Point at least 71,000 years ago was the discovery that treating rock with heat made it easier to shape into tools such as spear points.[7]

People have hunted food in large packages since our African ancestors first appeared long ago. The earliest evidence of consumption of aquatic foods comes from northern Kenya, where the remains of butchered fish, crocodiles, and turtles have been found alongside stone tools made by the predecessors of *Homo erectus*.[8] In South Africa's Blombos and Klasies River caves, alongside fish and shellfish, remains of penguins and seals complement catches of eland, antelope, and buffalo.

Caves in Gibraltar occupied by Neanderthals over thirty thousand years ago likewise contain the bones of monk seals and bottlenose dolphins.[9] The dolphins were probably scavenged carcasses of stranded animals, but the seals would have been hunted throughout the Mediterranean, relatively easy prey on their breeding beaches. Paleolithic paintings on the walls of Cosquer Cave, near Marseille on the south coast of France, show monk seals hunted with spears.[10] Today the entrance lies 120 feet below sea level, a testament to the rising sea after the last ice age, but the cave slopes upward to a large chamber that remains dry. It was discovered by a French diver, Henri Cosquer, in 1985, but the paintings were not found until 1991, when he and two colleagues penetrated far into the cave and spotted images of human hands stenciled on the rock face. Subsequent expeditions revealed 177 paintings of bison, aurochs, and giant elk, many of them exquisite artworks. Cosquer Cave is almost unique in its representation of marine species. There are images of the now extinct great auk, as well as of monk seals, and of what seem to be jellyfish. The paintings date from around nineteen thousand years ago and were created by the ancestors of modern Europeans.

Jon Erlandson, an archaeologist from the University of Oregon, believes that seafood use and other adaptations to life on the coast were pivotal to human migration out of Africa, as our ancestors followed the coast to Asia and later jumped over the Bering Strait to the Americas.[11] The richly productive coastal habitats found along the way, such as coral reefs, mangroves, and kelp forests, would have provided abundant year-round food. From Asia, people dispersed through Indonesia, where land bridges joined many places that today are islands. The world was ice-bound at the time of this great migration, and sea levels fluctuated one hundred to two hundred feet below those of today. People did face water barriers, the most daunting of which must have been the gap between Indonesia and Australia. Remarkably, they made the leap to Australia at least fifty thousand years ago. A separate hop was made to the Ryukyu Islands south of Japan some thirty-two thousand years ago. Colonization of the Bismarck Archipelago and Solomon Islands had occurred between thirty-five thousand and twenty-eight thousand years ago, requiring additional voyages of over fifty miles, some of them out of sight of land.

The peopling of Australia provides the first concrete evidence of boat use by modern humans.[12] No physical remains of boats survive from this time, nor is there much direct evidence of fishing gear. Things made from wood or plant fibers withered away long ago. Even more durable materials, such as shell, from which fish hooks were often made, can only survive under some conditions. In most cases they have simply dissolved or crumbled to dust.

The Earth was gripped by ice for much of the last 125,000 years. Frozen sheets heaped up on continents drew sea levels down by 390 feet. When the world began to warm again, around 20,000 years ago, it triggered a rise in the seas that continued until present sea levels were reached about 6,000 years ago. Much of the physical evidence of our coastal lives between 15,000 and 120,000 years ago simply washed away or was submerged.

How did we catch fish at the very beginning? Early efforts to attract fish into the shallows with bait could have developed into active construction of tidal traps from stones. Remnants of similar traps are well-known from the Cape coast of South Africa, often built to block off natural gullies, though remnants of these traps date back no more than a century or so.[13] Tidal traps made of brushwood and stakes are widely known from the last few thousand years, but time has erased all traces of the earliest ones.

Some remarkable early evidence of our fishing prowess comes from East Timor, an island just north of Australia.[14] Uplifted coral terraces at the east end of the island are riddled with caves and fissures and were occupied for thousands of years. Fish bone remains in the Jerimalai Rock Shelter date back forty-two thousand years and include inshore fish familiar from coral reefs, like parrotfish, groupers, and surgeonfish. But they also include fish that might have been caught offshore from boats, like tuna and shark. Although these animals could also have been caught from the shore, this interpretation fits with the idea that the ancestors of these people crossed the sea to Australia at least eight thousand years previously.

Some of the most detailed evidence for the development of fishing has come from caves and shell middens in California's Channel Islands. Their secrets have gradually been uncovered over the last two decades, in excavations by Jon Erlandson and his colleagues.[15] Drifts of abalone, oysters, and clams, half buried in blown sand, afford mute testimony to the Native American predilection for seafood. These mounds go back twelve thousand years. So important was seafood in their diet that Erlandson and his colleagues suggest that a ring of kelp forests stretching from Japan to Mexico provided a "kelp highway" that helped seafarers from the Old World colonize the New thousands of years ago. California middens and cave shelters have yielded stone tools, fish gorges, hooks, harpoon points, and woven sea-grass artifacts. Fishers also had nets and traps by the time of their first contact with Europeans.

Fish gorges provide the earliest trace of line fishing. A gorge is a stick, bone, or piece of shell that is sharpened at both ends, baited, and tied to a line in the middle. When the fish takes the bait the line is pulled and the gorge lodges at a right angle in its mouth or gut. Gorges are known from thirty thousand years ago in Europe,[16] and may have been invented even earlier. Remarkably, they were still used by fishers on the Pacific Island of Palau, and probably by many others, well into the twentieth century. Fishers there said that, even though they were less effective than a hook, gorges were used because they were so easy to make.[17] Early fishing lines were probably woven from animal hair or plant fibers. Palauan fishers made a strong twine from braided coconut fibers until the twentieth century, a method that had most probably been used for thousands of years.

We have no idea when nets were first invented. Fragments have been found on the Black Sea coast and in caves in South Africa dating back thousands of years. They are depicted in 5,000-year-old Sumerian wall reliefs and 4,500-year-old Egyptian wall paintings, but they had doubtless been used for much longer. Archaeologists believe the 42,000-year-old tuna catches in East Timor were made with nets. Nets were probably made of flax, hemp, or spun grass. The first evidence of domesticated flax comes from caves in the Czech Republic and dates to 30,000 years ago.[18] The first evidence of weaving, rather prosaically, comes from human bottom prints and bag impressions made on wet clay between 25,000 and 23,000 years ago, also in the Czech Republic.[19]

Most fishing methods were probably invented many times over. Single-piece fish hooks that pop up in archaeological finds all over the world were used several thousand years ago. The earliest found to date was a shell hook from Jerimalai Cave in East Timor that could have been made 23,000 years ago. They appeared 3,000 years ago in California[20] and in Australia about 1,200 years ago.[21]

Against the vast panorama of human history, commercial fishing is a relatively recent development. Its first stirrings can be found in the Mediterranean and Black seas, but for over a hundred thousand years, people fished to meet their own needs or those of their close kin. Early man developed and honed his fishing skills in the rivers and lakes of Mesopotamia and Egypt. Wall paintings and reliefs

show nets, traps, hook and line, and even the first representation of a
fishing rod.[22] Fishing appears to have developed as a specialist occu-
pation here several thousand years ago.

Evidence for commercial fishing gets stronger around 1000 BCE.
The city of Gades, or Gadir, today known as Cadiz, was probably the
foremost fishing port of the ancient world. It lies on the Andalusian
coast of Spain, just west of the Strait of Gibraltar. The town owed its
early fishing expertise to the Phoenicians, highly skilled seafarers
from the Levant who reputedly founded the city around the tenth or
eleventh century BCE. Gadir was one of the first Phoenician colonies,
and it became an important source of grain, silver, tin, textiles, dyes,
and salt fish. Over the next several centuries, Phoenician influence
spread widely on trade routes that crisscrossed the Mediterranean. By
800 BCE dozens of Phoenician colonies punctuated these routes at
places like Carthage in Africa and Genoa, Marseille, and Palermo
along the modern-day coasts of Italy and France.

Gadir was ideally placed for fishing. A great variety of fish thronged
local waters, and the town was close to the seasonal migration route of
the bluefin tuna, a six-foot giant well-known to the ancients that pops
up repeatedly in literature. One author much taken by fish was Oppian,
a poet from Corycus, a town in today's southeastern Turkey. His father
apparently displeased a visiting Roman dignitary and was banished to
Malta in the second century CE. There as a young man Oppian wrote
a thirty-five-hundred-line poem in Greek hexameters, *Halieuticks of the
Nature of Fishes and Fishing of the Ancients.* The poem was so admired
by the Roman emperor Marcus Aurelius that he paid him a gold piece
for every line and pardoned his father. Here is Oppian's description of
the migration of bluefin tuna into the Mediterranean:

> The breed of Tunnies comes from the spacious Ocean,
> and they travel into the regions of our sea when they lust
> after the frenzy of mating in the spring. First the Iberians
> who plume themselves upon their might capture them
> within the Iberian brine; next by the mouth of the Rhone

the Celts and the ancient inhabitants of Phocaea hunt them; and thirdly those who are dwellers in the Trinacrian isle and by the waves of the Tyrrhenian sea. Thence in the unmeasured deeps they scatter this way or that and travel over all the sea. Abundant and wondrous is the spoil for fishermen when the host of Tunnies set forth in spring. First of all the fishers mark a place in the sea which is neither too straitened under beetling banks nor too open to the winds, but has due measure of open sky and shady coverts. There first a skillful Tunny-watcher ascends a steep high hill, who remarks the various shoals, their kind and size, and informs his comrades. Then straightway all the nets are set forth in the waves like a city, and the net has its gate-warders and gates withal and inner courts. And swiftly the Tunnies speed on in line, like ranks of men marching tribe by tribe—these younger, those older, those in the mid season of their age. Without end they pour within the nets, so long as they desire and as the net can receive the throng of them; and rich and secret is the spoil.²³

Malta was then and remains to this day a major hub for tuna fishing. The accuracy of Oppian's description of the bluefin tuna's migration, breeding, and capture reveals just how well the ancients understood this fish. It also shows the great antiquity of the almadraba method of catching tuna, still in use today: an elaborate, chambered, net trap to intercept their coastal migration. Oppian died of plague at thirty, but his fame lives on for the remarkable insights he gave us into Mediterranean fisheries two thousand years ago.²⁴

It wasn't long before eastern Mediterranean cultures developed methods to preserve fish with salt. By at least the fifth century BCE, salt fish was traded across the Mediterranean and Black seas. The importance of bluefin tuna in this trade is attested by the number of coastal cities that depicted the fish on their money. Coins from Gadir show two tuna, while those from Abdera in Spain show tuna as the pillars of a temple.[25] Tuna was cut into pieces, cured with salt, packed into amphorae, and shipped to consumers hundreds, perhaps thousands of miles away. "Canned" tuna has been around a very long time! Several other fish were preserved and shipped this way, including sea bream and mullet, and huge catfish and sturgeon from rivers and estuaries.

Oppian writes that the different kinds of nets are "innumerable" and lists eight main varieties: hand casting, draw, drag, round bag, seine (he used the word "sagene"), cover, ground, ball, and (my favorite) hollow all-catching. The art of fishing was clearly well advanced. Oppian tells of a clever device used when the fish lie close to the sea bed:

> They have a stout log, not long but as thick as may be, about a cubit in length. On the end of it are put abundant lead and many three-pronged spears set close together; and about it runs a well-twisted cable exceeding long. Sailing up in a boat to where the gulf is deepest, mightily they launch into the murky deep the pine-log's stubborn strength. Straightway with swift rush, weighed down by lead and iron, it speeds to the nether foundations of the sea, where it strikes upon the weak Pelamyds [bonito tuna] huddling in the mud and kills and transfixes as many as it reaches of the hapless crowd. And the fishermen swiftly draw them up, impaled upon the bronze and struggling pitifully under the iron torture. Beholding them even a stone-hearted man would pity them for their unhappy capture and death.[26]

It is hard to imagine today that a plank loaded with spikes thrown off the bow of a boat would catch anything at all. It could only work in an ocean crowded with fish or in water shallow and clear enough to see the target.[27]

Polychrome mosaics were invented around the first and second centuries CE, and marine scenes soon became popular. There is a magnificent floor mosaic of sea fishing from the Catacomb of Hermes at Hadrumetum in Tunisia (another former Phoenician colony). Fishermen in boats ply their trade amid a sea crowded with a wondrous diversity of fish and lobster. Some use basket traps, others hook and line, another a harpoon; one casts a net and two men work a drift seine net buoyed by bobbing corks. The boats are small and powered by oars.

The Mediterranean had long been plied by large, seaworthy craft, but there was little point in fishing from a big boat. Only the very wealthy could afford ice (the Roman writer Galen mentions preserving fish in snow in the second century CE[28]), so fishermen had to work close to the coast to prevent their catches from going off on long journeys back to port.

Large-scale salting works were built around the shores of the Mediterranean in the fifth century BCE, especially in productive western waters. They made salt fish, but increasingly turned to the manufacture of fish sauces beloved by Greeks and Romans. Production varied over time but seems to have peaked between the fifth and fourth centuries BCE, serving mainly Greek markets, and again in the first to the second century CE, when most of the region was under Roman rule.[29] Some salting works contain enough vats to hold nearly forty thousand cubic feet of fish and sauce.

Like Marmite or blue cheese today, fish sauces weren't to everyone's taste. In the first century CE Pliny the Elder called fish sauce an "exquisite liquid," and the best grades of *garum* sold for a price equal to perfume,[30] while Seneca railed, "It's the overpriced guts of rotten fish!" at about the same time, and he wasn't far from the truth. Most fish-salting works were smelly, and therefore located outside city walls, well away from residential areas.

Recipes for fish sauce can be found in a tenth-century CE collection of the works of ancient writers.[31] One directs you to take the intestines, blood, and gills of a bluefin tuna, add salt, and ferment the ensuing goo in a vase for two months. Another gives the method for a lower-grade sauce made from a mixture of many kinds of fish, usually those good for little else. (Anchovies were a favorite.) To one measure of fish add two of salt and let stand overnight, instructs another recipe. Then place the contents in a clay vessel and leave uncovered

and exposed to the sun for two or three months, occasionally stirring
with a stick. It sounds unappetizing (all the more so when you consider that *allex*, another kind of fish sauce, was made of the residue
scraped from the bottom of the empty fermentation vessel).

Spanish food technologists have recently created a version of
garum using modern methods to mimic ancient processes.[32] The key
is the protein-digesting action of enzymes within the fish guts, which
break down the flesh to a slurry of amino acids, fats, and nutrients.
The resulting paste, rich in omega-3 fatty acids as well as many vitamins
and minerals,[33] was not dissimilar to Asian fish sauces of today. Fish
sauce was an early health food.

As the Roman Empire spread north through Europe they took
their tastes with them. Fish sauce was transported in large quantities
to their colonies—around a fifth of all amphorae found at one northern European town were for fish sauce[34]—and production sites were
established along the North Sea coast of Gaul. With the collapse of
the Roman Empire these factories fell into disrepair, and fish sauces
dropped off the menu. Sea fishing dwindled, except in the chilly
extremities of Scandinavia, where the climate was too harsh for most
crops, and people hunted and fished to survive. Remains of fish bones
from eighth- or ninth-century domestic waste dumps in England,
Flanders, and other parts of northern Europe were dominated by species that lived permanently or seasonally in fresh waters, such as sturgeon, salmon, and whitefish. That all changed in the eleventh century,
when a sea-fishing revival marked the onset of an expansion and
industrialization that continues to this day.

Archaeologists discovered the northern European sea-fishing revolution by sifting through more than a hundred kitchen garbage heaps.[35]
There was a dramatic shift within a few decades around the middle of
the eleventh century from around 80 percent freshwater fish to 80 percent saltwater fish, such as cod, haddock, and herring. Demand was
increasing rapidly from a combination of population growth, urbanization, and the spread of Christianity (Christian practice required periodic or complete abstinence from the meat of quadrupeds). Fresh fish

supplies were in freefall as a result of human-caused habitat change in rivers, lakes, and estuaries. The spread of agriculture meant forests felled and land plowed deeply for crops. Soil erosion soon turned fast-running, cool, clear waters into sluggish, warm, and turbid water that species like salmon did not enjoy. Nor did salmon and other fish, which migrated from sea to rivers to spawn, benefit from the construction of thousands of dams across Europe's rivers to supply power for corn mills and other industry. With the migration routes blocked, freshwater fisheries' production collapsed.

Most medieval fish was caught and eaten locally, but long-distance trade resumed around the thirteenth century, this time of air-dried and salted fish.[36] Those caught in the prolific Arctic cod spawning grounds of Lofoten were dried in the frigid air into yard-long pieces as hard as wood. They could be carried hundreds or thousands of miles to supply towns and cities all the way to the Mediterranean. Stockfish, as it was called, lasted up to two or three years, a perfect convenience food for an age that lacked refrigerators.

An improved method of salting to preserve herring in brine was developed in Amsterdam sometime around the fifteenth century[37] (or, I should say, redeveloped, since the ancients were expert fish salters). Fisheries expanded swiftly after that, first with Dutch then British fleets. By the seventh century more than two thousand Dutch boats were pursuing herring from the Shetland Isles to the Baltic. Interest in cod had also shifted west, to the shores of Canada and New England, where abundant cod of extraordinary size immediately attracted European interest.

Fishing methods in Europe developed gradually, as new ways to capture fish were invented. Beam trawling appeared in the fourteenth century, if not before,[38] and involved dragging a net held open by a beam of wood across the seabed. The idea probably came from smaller dredges that were towed over the seabed to catch oysters.

A French nobleman, Henri Louis Duhamel du Monceau, compiled a meticulous catalog of sea-fishing methods in the 1750s and illustrated it lavishly with engravings.[39] They included: intertidal and subtidal longlines bristling with hooks; baited lift nets manned from platforms over shallow water; drift nets like those used by herring fishers to snare fish by the gills at night; and many others. Most people still fished from small boats close to coasts, with the exception of large-scale cod and herring fishing vessels. Fish spoilage remained a problem, and most of the catch was consumed in towns and villages close to the sea. Well boats, another reinvention of a classical Greek technology, could operate farther afield. They carried a water-filled tank and brought live halibut, cod, turbot, and other fish from the central North Sea into towns such as Hamburg and London. Markets for fish underwent dramatic enlargement between the end of the eighteenth and the middle of the nineteenth centuries, in both the Old and New worlds. Toll roads and railways sped up transportation between the coast and inland cities so fresh fish could reach them. This in turn spurred a rapid increase in fishing efforts. But these fisheries still used medieval technology of sail and oar, hook, net, and trap.

In other parts of the world, developments in fishing technology matched those of industrializing countries in their ingenuity. Perhaps the most remarkable fishing method I know of comes from Melanesia and islands of the West Pacific.[40] A kite made from a dried breadfruit leaf that has been stiffened with the midribs of coconut leaflets is flown on string made of coconut husk fibers. The kite trails a mat of spiders' webs in the water, and a skilled fisherman will make it skip from wave to wave, so exciting the interest of garfish. When they grab the lure, their teeth become hopelessly entangled. I have found myself wondering whether the first kite was invented by a fisherman after a floating leaf tangled in a line and lifted off.[41] But I digress.

The addition of steam power to boats in the 1880s heralded the beginning of the modern era in commercial fishing. Once unleashed from the bonds of wind and tide, fishing fleets multiplied and spread forth across the world's continental shelves. Engines freed people to pursue fish farther offshore and deeper, and to work around the clock in worse weather. Engine power and rising demand expanded vessel size and made possible larger trawls and nets. Steam yielded to diesel in the early twentieth century, and fishing intensity climbed, as vessels multiplied and their captains embraced new technology. Then, after World War II, there was a second burst of industrialization that lasted into the 1970s. Boats increased in size and power, and monofilament nets replaced hemp and cotton during this period, enabling people to fish far more effectively with larger gear. The destructive potential of fishing was hugely increased by these developments. Longlines with thousands of hooks were extended to tens of miles long; drift nets became walls of death of similar enormous length; huge engines made it possible to tow midwater-trawl nets large enough to engulf cathedrals; and bottom-trawl nets spread to three hundred feet wide and were fitted with heavy steel balls on the footrope to ease their passage into areas of rough seabed.

Up to the 1950s, fishing technologists had mainly improved upon ancient traditional methods. They substituted materials, refined designs, added engine power, and deployed more or bigger catching gear. The introduction of echo sounders in the 1950s marked the onset of a new fishing revolution. Echo sounders revealed the presence of fish beyond the dreams of the most skilled captains from prior generations. Electronics like echo-sounders, augmented from the 1980s by computers and satellites, ramped up the stakes yet again.

What strikes me, looking at the grand vista of human history, is how technological development has sped up with time. There are gaps of tens of thousands of years between major innovations made by our earliest ancestors. After the end of the ice age ten thousand years ago, things sped up, and new methods of fishing were introduced at intervals of a thousand years or less. The tempo accelerated in the last thousand years, as people invented clever ways to catch and preserve an ever-expanding variety of fish and shellfish. In the last hundred years, the addition of engine power and modern materials vastly increased the reach of fishing, and in the last thirty years, the lethal edge of fisheries has been sharpened by the addition of computer and satellite technologies. In the next two chapters I will explore how this upward-racing curve of progress has affected life in the sea.

CHAPTER 3

Fewer Fish in the Sea

There is a large-sized fish called Hallibut, or Turbut: some are taken so big that two men have much a doe to hall them into the boate; but there is such plenty, that the fisher men onely eate the heads & fines, and throw way the bodies.

—Captain John Smith, first governor of Jamestown,
History of Virginia, 1624[1]

It is early morning and the sea is smooth and dark. The peace is soon broken by an engine echoing off the hills that rise directly from the sea. A fishing boat heads straight for the shore, where it turns and backs up into water scarcely deeper than its draft. Figures on deck maneuver two sets of four dredges over the side and drop them into the water with loud splashes. The dredges consist of heavy steel frames bristling with vertical teeth that dig into the bottom to knock scallops into attached chain-mail bags. The boat roars into gear and the tow begins. Half an hour later, far out to sea, the dredges are swung back on deck, where they spill their loads of stones, seaweed, starfish, and scallops. While one man sorts the catch, the boat turns back and heads shoreward to begin another tow. Only desperation would drive a captain to risk his boat among the shallow rocks of Scotland's Firth of Clyde, but fishing has been bad in recent years, so he must feel there is no alternative but to scratch his living along the shore.

For all their technological brilliance, modern fishing fleets operate at the margins of profitability. Fishing has always been a hard living: wet, backbreaking, messy, and often dangerous work. But in the past fishermen could at least guarantee a good catch. Nineteenth-century photographs of dockside scenes show landings of fish that appear almost miraculous today. Quaysides are stacked with boxes overflowing with outsize fish, while giant halibut, cod, wolffish, ling, turbot, and others too large to fit in crates cover stone floors. Even in the late nineteenth century, though, fishermen had begun to grumble of declining catches. A commission of inquiry was convened in 1883 to investigate complaints. Unable to resolve the issue of falling stocks in the absence of fishery statistics, the commissioners recommended that the government start to collect data, and recording duly began in 1889.

Strangely, most fisheries statistics collected before the European Commission's Common Fisheries policy began in 1983 were neglected until one of my graduate students, Ruth Thurstan, dusted off the old volumes a couple of years ago. Perhaps they were felt to be too antiquated to be of value. Methods of data collection change with time, so comparing old figures with new was difficult. But it was worth the effort, for these old charts paint a stark picture of the present state of fisheries, and one much worse than is suggested by modern statistics.

The graph of landings from bottom trawlers—boats that drag large bag nets across the seabed to scoop up bottom fish—looks like a steep-sided mountain incised with two deep valleys.[2] Catches rose steeply from 1889 and peaked in the mid–twentieth century, when they leveled off briefly before a dramatic collapse to the present day. In 1889, more than twice as many bottom fish (cod, haddock, plaice, and the like) were caught in British waters compared to today. That is an astonishing fact given the technological gulf between then and now. The peak came in 1938, when the fleet landed over five times more fish than now. The two valleys cut into the graph correspond to the world wars, when catches plummeted because it was too dangerous to fish, and boats were put to other uses, such as laying mines.

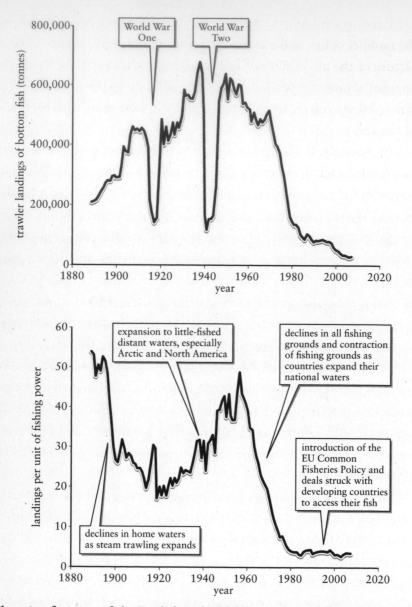

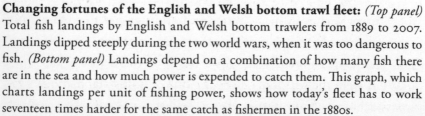

Changing fortunes of the English and Welsh bottom trawl fleet: *(Top panel)* Total fish landings by English and Welsh bottom trawlers from 1889 to 2007. Landings dipped steeply during the two world wars, when it was too dangerous to fish. *(Bottom panel)* Landings depend on a combination of how many fish there are in the sea and how much power is expended to catch them. This graph, which charts landings per unit of fishing power, shows how today's fleet has to work seventeen times harder for the same catch as fishermen in the 1880s.

Landings tell only part of the story, because they depend both on the number of fish in the sea and the time spent fishing. To get a better picture of the availability of fish you have to divide the catch by the amount of power expended. This index of landings per unit of fishing power brings you closer to a real measure of how many fish there are in the sea.

Fortunately, fishing records include the number and size of boats that landed each year's catch. But how can you compare a sailboat to a trawler? Here a long dead nineteenth-century fisheries scientist named Walter Garstang came to our aid. Using the first decade of official statistics he worked out that landings per unit of fishing power fell by 50 percent between 1889 and 1899. He estimated that the new steam trawlers had more than twice the fishing power of sailing vessels to make this calculation. We used similar methods to estimate increases in fishing power by vessels throughout the twentieth century, up to the present day, thereby tracking each advance in technology.[3]

When Ruth showed me the graph of landings divided by the changing power of the fleet, I nearly fell off my chair. I had expected a decline, but this was near annihilation. A fleet that in the 1880s consisted mostly of sail-powered boats open to the elements was far more successful at wresting fish from the sea than we are now. For every hour spent fishing today, in boats bristling with the latest fish-finding electronics, fishers land just 6 percent of what they did 120 years ago. Put another way, fishers today have to work seventeen times harder to get the same catch as people did in the nineteenth century. The simple reason for this stark contrast between past and present is that there are fewer fish in the sea. When we broke the figures down by type of fish, for some the contrast between ninetenth and twenty-first century was even more extreme. Landings per unit of fishing power are down by thirty-six times for plaice, over one hundred times for haddock, and a breathtaking five hundred times for halibut.

What were the seas like before the bite of industrial fishing? Eye-witness accounts from past centuries tell of waters off the coasts of

America teeming with fish. Here is a report from the *Gloucester Telegraph,* a Massachusetts newspaper, from June 4, 1870:

> Accounts from New Jersey say that bluefish came in at Barnegat Inlet last week, sweeping through the bay, over flats as well as through the Channel, driving millions of bushels of bunkers before them and filling the coves, creeks, ditches, and ponds in the meadows full. At Little Egg Harbor Inlet they drove shad on shore so that people gathered them up by wagon-loads. Fish lie in creeks, ponds, etc., along the meadows two feet deep, so that one can take a common fork and pitch them into a boat or throw them on the bank.

The bunkers referred to were menhaden, an oil-rich fish that seemed to have been created to feed all of the oceans' predators, ourselves included. A 1913 report on the United States menhaden fishery found that more than one billion fish were caught that year, producing six and a half million gallons of oil and ninety thousand tons of fertilizer.[4] If all of these fish were placed nose to tail they would have stretched six times around the Earth at the equator.

Meanwhile, nineteenth-century photographs show men on the West Coast thigh-deep in salmon pulled from Puget Sound. Some catches yielded thirty thousand fish. In 1915, four hundred million pounds of salmon were taken from Alaskan waters alone. If they had been packed in barrels, each containing two hundred pounds of fish and placed one on top of another, they would have formed a column 1,200 miles high.[5] Whole fleets in the Gulf of Mexico would set forth in sailing boats and test the water with baited lines until they hit upon a concentration of red snappers. At some sites as many as two thousand fish could be taken in a day.[6]

In 1819, the Reverend Lewis Anspach, a resident of Newfoundland, described the appearance of Conception Bay during the capelin season. The capelin is a small shoaling fish that spawns close to shore. It is prey to a bewildering variety of animals:

It is impossible to conceive, much more to describe, the splendid appearance of Conception Bay and its harbors on such a night, at the time of what is called the Capelin Skull. Then its vast surface is completely covered with myriads of fishes of various kinds and sizes, all actively engaged either in pursuing or avoiding each other; the whales alternately rising and plunging, throwing into the air spouts of water; the codfish bounding above the waves and reflecting the light of the moon from their silvery surface; the Capelins hurrying away in immense shoals to seek a refuge on the shore, where each retiring wave leaves countless multitudes skipping upon the sand, an easy prey to the women and children who stand there with barrows and baskets ready to seize upon the precious and plentiful booty.[7]

While some fish were deemed to be useful, others were considered pests, and great efforts were sometimes taken to annihilate them. Incredible as it may seem now, as recently as the 1950s and 1960s an extermination campaign was mounted in western Canada against basking sharks, the second largest fish in the sea. These docile plankton feeders often got tangled in gill nets set for salmon, so they were hated by fishermen. The fishery protection vessel was mounted with cutting equipment on its bow to slice through sharks feeding at the surface. Several thousand basking sharks died in nets or were slaughtered this way.[8] In Europe, dogfish and porpoises were similarly regarded as pests, because they stole fish from nets and hooks, and eyewitnesses recorded devastating accounts of porpoise massacres going back to the eighteenth century.[9]

With the sole exception of Alaskan salmon, which have been well managed, the species described in these historical vignettes have plummeted since their historic highs. Puget Sound's salmon runs have dwindled to a trickle. Red snapper, bluefish, and menhaden are all overfished in U.S. waters today, while capelin is far below the abundances witnessed in the nineteenth century. In 2010, a quarter of commercial fish stocks assessed in the United States were considered overfished, meaning that they totaled below target levels that are themselves set far below historic abundances.[10] But this misses the real scale of the overfishing. The sad fact is that the status of 275 U.S. fish stocks—which is over half the 528 examined that year—could not be determined: They were too uncommon today for it to be worthwhile collecting data, or so rare that data were unreliable. Many are rare because of past overexploitation. They would have been familiar fare on the tables of nineteenth-century Americans. The picture is much the same throughout the rest of the world. Things are now looking up in the United States due to a welcome shift in direction of fisheries management made in 2007 that I will come back to later, but elsewhere the trends still seem to be in the wrong direction.

We find it hard to believe these descriptions of extraordinary past abundance because it has been so long since such scenes were commonplace. It is a human trait to give greater weight to personal experience than to others' descriptions. The result, as I described in the opening of this book, is an intergenerational shift in the way we perceive the world. Science is particularly susceptible to these shifting baselines, as scientists work at the forefront of knowledge and are always in hot pursuit of the latest ideas.

Many journalists asked me why these old fisheries statistics had been ignored for so long after we published Ruth's study on the decline of UK trawl fisheries. It perplexed them. The answer is that old figures were thought to be no longer relevant to the question that fisheries scientists are asked: How many fish is it safe to catch next year? The answer to that depends on how many fish there are at the moment, which is usually settled by the latest survey, not by reaching back decades or centuries. But it is only by going back many years that a full perspective can be gained on the fortunes of fish stocks and so a judgment reached about their sustainability.

The balance of power between fishers and their quarry has been lopsided in favor of the hunters for the best part of 150 years. We have established regulations in the last few decades to restrain fishing power, but they have failed to give most species the time and space they need to reproduce. Life is extraordinarily dangerous for fish that are caught or killed alongside the target, or "bycatch" species, as they are called. Fishing intensities are now so high that, once some species reach a level at which they can be caught, their chance of death from fishing in any given year ranges from 30 percent to 60 percent, or more.

Every year, estimates suggest, trawlers sweep an area of seabed equivalent to half of the world's continental shelves.[11] Together with dredgers like the Clyde's scallop boats, they have transformed life on the seabed, converting three-dimensionally complex habitats rich in coral, sponge, sea fan, and seaweed into endless, monotonous expanses of shifting gravel, sand, and mud. While trawls and dredges decimate the ocean floor, the hand lines of old have evolved into longlines. The length of hook-studded longlines set every night in our seas is enough to wrap around the world five hundred times over.

Although the scale of the ocean is hard for the human imagination to grasp, life is concentrated into the surface layers, continental margins, and other hot spots where nutrient-rich currents well up to the surface. We focus our fishing on these places, to devastating effect. Even migratory species such as tuna and swordfish, which undertake oceanic odysseys of tens of thousands of miles each year, can be intercepted in a few areas of particular importance to them, such as feeding and breeding grounds.

Throughout history, fishing has followed a familiar pattern. At first we meet all our needs from local sources. But we eventually reduce the populations, and so to sustain catches, we have to develop better ways to outwit the fish, or they move to places that have been less exploited or switch to less favored species. James Bertram recognized this tendency as early as 1873, when he lamented: "We are continually, day-by-day despoiling the waters of their food treasures. When we exhaust the inshore fisheries we proceed straightaway to the deep waters."[12]

This pattern can be read in the changing fortunes of the UK trawl-fishing fleet. From 1889 to the outbreak of World War I, British trawlers fished mainly on home grounds. Landings per unit of fishing power fell steeply as those stocks were depleted. After the war people undertook long voyages, to Iceland, the Arctic, and West Africa, in search of virgin stocks, and landings per unit of power climbed. Rewards from fishing fell off steeply in the 1960s, as these grounds became depleted. This loss was compounded in the 1970s by the enlargement of sovereign waters and the establishment of exclusive economic zones that reached out two hundred nautical miles, forcing boats back to home grounds again, where the fishing hauls were greatly reduced.

Geographic expansion and substitution of species has served fishers well for most of history. In the nineteenth and early twentieth centuries, the U.S. oyster fishery worked its way around American estuaries in sequence, moving south from New York and north from San Francisco, exhausting stocks along the way. But fishing has spread ever wider and deeper in the last century, until it has today reached the limits of the oceans. After World War II, Japanese long-line boats took to the high seas and fanned out across the world's oceans, and by the 1970s were fishing almost everywhere. Soviet fishing boats targeted deep seamounts one after another from the 1960s and 1970s. They have been joined today by fleets from Spain, Taiwan, Sri Lanka, China, and other countries, and they have diversified from tuna and swordfish, to sharks, mahi-mahi, opah, and high-value deep-sea beasts such as the six-foot-long Chilean sea bass found in the freezing depths of the southern ocean. Boats now pursue fish to the edges of polar ice shelves and ten thousand feet into the abyss. The impact of the fishing industry goes even deeper than the maximum limits of our gears.[13] Animals that migrate up and down from underlying layers are also caught, and the palls of mud thrown up by trawls settle farther down. Estimates suggest that most of the world's major fishery species have been reduced in numbers by 75 percent to 95 percent or more. Figures compiled by the UN Food and Agriculture Organization, the

FAO, show that two thirds of the species we have fished since the 1950s have experienced collapse, and the rate is accelerating.[14] It doesn't take a genius to see that with a track record like this we are squandering life.

Life begins at forty, so the saying goes, but for most of us our reproductive years are behind us by this time (although wealthy old men seem to buck the trend!). For fish and shellfish, however, life really does get better with age. Big, old, fat fish produce far more offspring than their young, lithe, and smaller brethren. Their size and experience give them an edge. Their eggs are better provisioned than those of small fish, so more of them survive the dangers of early life. But fishing has dismantled the dominion of the old in the span of a century or two, and evolution has begun to work in a different direction.

A universal rule of fishing is that when you exploit a population, the average size of the animals gets smaller. Most fishing methods are size-selective, which is to say that they catch animals whose bodies or mouths are larger than the size of a mesh or a hook. Even hand-gathered fish and shellfish are susceptible, as people tend to pick out the largest and juiciest ones first. Over time, therefore, fishing alters the balance between young and old in a population. Modern-day fishers are not the first to have caused their quarry to become smaller. You can read the signals of increasing fishing intensity in the ancient shell middens of California: Mussels fell in size by over 40 percent between ten thousand and two hundred years ago, while the average red abalone eaten decreased from nearly eight inches long to under three inches.[15]

Evolution works to maximize the number of descendants that an animal leaves behind. Where the risk of death from fishing increases as an animal grows, evolution favors those that grow slowly, mature younger and smaller, and reproduce earlier. This is exactly what we now see in the wild. Cod in Canada's Gulf of St. Lawrence begin to reproduce at around four today; forty years ago they had to wait until six or seven to reach maturity. Sole in the North Sea mature at half the body weight they did in 1950.[16] Surely these adaptations are good news for species hard-pressed by the onslaught of fishing? Not exactly. Young fish produce many fewer eggs than large-bodied animals, and many industrial fisheries are now so intensive that few animals survive more than a couple of years beyond the age of maturity. Together this means there are fewer eggs and larvae to perpetuate future generations. In some cases the amount of young produced today is a hundred or even a thousand times less than in the past, putting the survival of species, and the fisheries dependent on them, at grave risk.

Over the course of decades, climate cycles like the North Atlantic oscillation shift the probability that young will survive up and down. Many species have adapted by developing long lives and can reproduce many times. They maximize the chances that some of their young will make it by hedging their bets this way. Excessive fishing ramps up the risk of reproductive failure and population collapse by contracting the number of years over which species can reproduce. Fishing drives up the risk of population crashes today by pushing evolution in a direction counter to that of the environment, and in doing so, unravels adaptations developed over thousands of years. Long-term surveys of marine life off the coast of California show that the species we catch now have more variable replenishment than ones we don't.[17]

Intensive exploitation over long timescales leads to the progressive loss of larger animals. This phenomenon is so pervasive that it has been given a name: "fishing down the food web." It is brother to the process of sequential overexploitation—the removal of species in sequence from large to small, predator to prey, high to low value—as they tend to go hand in hand. We begin with the pursuit of large and valuable animals. Big species are often predatory by nature, and as a consequence have firm, succulent flesh that we prize. As predators they tend to be bold and voracious, and they readily succumb to the lure of a baited trap, hook, or net. When they have been depleted we switch to smaller, less-favored, lower-value species.[18]

Where once vast sawfish churned the waters of estuaries in the southern United States, now much smaller fish raise puffs of mud as they feed in the shallows. When rich new cod-fishing grounds were discovered on the Rockall Bank off Scotland's west coast in the 1860s, one report said, "The crews caught the fish as fast as they could bait and haul and it is stated that when any of the cod broke from the hook, great monstrous sharks as blue as if painted with a brush darted round the ship's side and swallowed them in an instant."[19] Today the blue sharks are almost gone, and the cod has been humbled.

In all but the most extreme environments, the web of life is highly complex. Animals and plants interact in a multitude of ways. They hunt or hide from each other, or use the living space others create. When species are reduced in numbers or removed, the effects cascade on others through the web of life, with unpredictable consequences. We know from farms that simplified ecosystems suffer problems. Reducing diverse plant communities to a handful of species raises the likelihood that there will be outbreaks of pests and diseases. On land we control these unwanted effects with chemical sprays or elaborate husbandry. We have no such recourse when things go wrong in the sea.

Think of life as a game with many players. Some players fulfill similar roles, and circumstances dictate their fortune. One set of conditions will favor one species or group of species. Fortunes shift when circumstances change, and others take their places. But the role itself endures. Ecosystems with a higher diversity of species and greater complexity have the resilience to adapt when the environment shifts. Those that have lost species, due to overfishing or other forces such as pollution, will be less able to cope.

The Firth of Clyde, where this chapter began, offers a bracing illustration of what can go wrong. This narrow arm of the sea penetrates sixty miles into southwest Scotland, ending where mountains shade the head of Loch Fyne. In the eighteenth century its banks and bays abounded with fish, shellfish, whales, and porpoises, none more abundant than the herring, which migrated each year into the loch to breed. One eighteenth-century traveler marveled at how, when the herring shoals came in, the sea seemed to consist of two parts fish, one part water. Huge basking sharks and whales feasted upon plankton that bloomed in the tide-churned waters while packs of hunters savaged the herring. Such abundance and vitality attracted a lively fishery. Henry Beaufoy, a British member of Parliament, described the fishing in 1785:

> On conversing with [one of these fishermen], he gave me accounts of the quantities of fish he catched, that appeared altogether incredible; for one article, he assured me, that when he baited his small line for the smaller flat fish, which line contains 400 hooks, it was not at all an uncommon thing for him to take in, at one haul, 350 fish from the 400 hooks. These consist of turbot, soal, and large fine flounders, about two or three pounds weight each. As to skait, he said he seldom fished for them, as they are not a salable article there; but he could easily fill his boat with them, when he chose it, at one haul of his lines—and from the accounts of other people, I had no reason to doubt the man's veracity.[20]

Nets and trawls were introduced in the late nineteenth century that could touch bottom on the shallow spawning banks at Ballantrae upon which herring laid their eggs in the southern Firth of Clyde. The herring fishery there collapsed within a couple of decades. This catastrophe spurred fishers to establish a ban on bottom trawling throughout the Firth of Clyde that stood until the early 1980s. Landings of many bottom fish held up well while it was in place, although herring and saithe, a relative of cod, were wiped out in the 1960s and 1970s through the invention of electronic fish finders and midwater trawls.

With the herring gone, pressure then mounted from prawn trawlers to open up the closed areas. Following repeal of the ban in 1984, cod, plaice, haddock, whiting—in fact, all of the Firth's productive fisheries—declined to virtually nothing within twenty years. Today the seabed is barren, and the only fisheries left are for prawns and scallops, and even they are overfished. It is a marine wasteland.[21]

We can glimpse the endpoint of overfishing in the Firth of Clyde, a time when nothing worth catching is left. Unfortunately, this story is far from unique. The combination of increased fishing power and dwindling areas of refuge is repeated in all of the world's seas and oceans. The Firth of Clyde gives us a stark vision of a future without fish.

When Ruth Thurstan and I published our research on the Clyde in 2010 many fishermen and others familiar with the area agreed with our conclusion. Indeed, we thought the sorry state of Clyde fisheries was, to paraphrase Jane Austen, a truth universally acknowledged. But industry leaders were not pleased. The heads of two Scottish fishing organizations between them were reported in one newspaper deploying the three favorite tactics of those in denial: attack the science; slander the scientists; and blame something else (pollution, seals, and climate change were the straws they clutched).[22] If the consequences of their denial were not so devastating, I would have found more humor in their last comment: "We don't need to do much about this report. . . . [W]e are letting proper marine scientists rubbish it for us."

Two sea anglers who were interviewed for the same piece were not afraid of saying what everyone who lives there recognized. "There's only mackerel left," one said, "and then only in the summer. You used to see people angling from the beach at the West Bay there, even at night. There's no point now, because there's nothing to catch." The other said: "It's not a one-man-band or a guy with a rod that's causing the problem. And it's not down to the Clyde being dirty—it's cleaner now than it was in the seventies—it's the trawlers."

Over the years I have come across spectacular levels of denial among fishing industry representatives. In parliament and senate buildings and committee rooms across the world I have seen them dig their heels in to resist regulations that could help fish stocks recover. Politicians too willingly believe their claims that greater regulation would cause unnecessary hardship. In reality, failure to acknowledge and deal with the problem represents a far more serious risk to their livelihoods. The relationship between politicians and the fishing industry in the European Union has become like that of a doctor assisting the suicide of a patient. For the last twenty-five years politicians have given the industry a one-third larger catch quota on average than scientists have recommended as safe.[23] The only outcome possible from such a policy is the collapse of fish stocks and the fishing industry. If you don't believe me, imagine a farmer who takes ten more sheep to market each year than his flock produces. You can't cheat nature, however good you may be at spinning a story.

To close this tale of demolition, allow me to return to the Atlantic bluefin tuna, for this majestic fish has recently become emblematic of the worst of humanity's destructive tendency. Their abundance has declined by at least two thirds since 1970, and the total decline over the course of the last century is probably closer to 95 percent. By the 1970s, bluefin tuna had already disappeared from the North Sea, perhaps the result of the decimation of a highly migratory subpopulation of the species.[24] By the 1980s, the Black Sea population was all but gone, and it is now considered extinct. This is a species that, by any metric, we have taken from abundance to the verge of extinction. The scientific evidence is unequivocal. Yet the badly misnamed International Commission for the Conservation of Atlantic Tunas continues to award its members catch quotas far above those that would enable this fish to recover.[25] Corporate greed has triumphed over human decency.

Daniel Pauly, a charismatic fisheries scientist from Canada's University of British Columbia, describes world fisheries as a giant Ponzi scheme. Fraudsters in this type of scam pay investors from the capital in a fund rather than from the returns made on their investments.

Ponzi schemes collapse when the flow of new capital dries up. Since the nineteenth century, when fisheries were first industrialized, landings have been sustained by fishing ever farther afield, and deeper. The fishing industry has been dependent on a constant input of new capital. Whenever fish began to run out, fishers moved on or switched to other species. Over time fisheries have eaten up their capital stocks rather than lived within the limits of annual production. But fisheries are now failing because, like in a Ponzi scheme, they are running out of new capital. We now hunt fish to the farthest limits of the oceans, and to depths where productivity slows to a trickle. There is nowhere else to go and few species worth eating remain untouched by fishing. The solution is not for a handful of people to stop eating fish. We need to set up new regulations and police them well. But I will return to this later on in the book, when it comes time to consider solutions.

Hunting and fishing are the oldest of human influences on the sea. Arguably, fishing also remains the most serious. But climate change from greenhouse gas emissions has been building in the background during the last century. It has burst to prominence in the last twenty years as it has begun to affect our daily lives. In the next four chapters I explore the most important of the many ways in which greenhouse gases are changing the oceans.

CHAPTER 4

Winds and Currents

Benjamin Franklin is celebrated as a Founding Father of the United States, an astute diplomat, the cofounder of the first public lending library, and a prolific inventor whose contraptions ranged from the lightning rod to bifocal spectacles. He is known to have had an exceptional range of interests and abilities, but his pioneering contributions to oceanography are not widely recognized outside scientific circles.

Before the revolution, Franklin was postmaster general for the British colonial mail, and as such he applied his formidable intelligence to the problem of how to speed up deliveries between the Old and New Worlds. He wondered, more specifically, why it took ships two weeks longer to sail from London to New York than it took them to travel in the opposite direction. His cousin Timothy Folger, a whaling captain, told him that British ships battled a three-knot current while those coming from America sailed with it. Together the pair tapped the knowledge of experienced American whalers to produce, in 1770, the first map of what is now known as the Gulf Stream.

The Gulf Stream is a fast-flowing surface current that jets into the Atlantic through the Florida Straits and then runs north along the east coast of America. It turns out to sea just south of Cape Hatteras, at which point it crosses the Atlantic to Europe, where it dissipates into a general northerly flow called the North Atlantic Drift. It is part of the "global ocean conveyor," a system of currents that circles the planet and loops from the surface to the deep sea.

The idea that the seas might overturn, with water pouring from the surface into the deep and then back again, was first proposed by Benjamin Thompson, a contemporary who, unlike his more famous counterpart, was a loyalist and fled revolutionary Massachusetts for London. In Europe he made a name for himself as a scientist and statesman. He is best remembered today for his discovery of how to increase the draw of fireplace chimneys, helping cure the centuries-old curse of smoke-filled rooms. Thompson, who in 1791 became Count Rumford (after the New Hampshire town where he grew up, today's Concord), based his proposal that the shallow waters of the oceans overturned and mixed with the deep on his knowledge of the properties of heat. A single measurement of the deep ocean's temperature from 1751, made thirty-six hundred feet below sea level by a British slave ship in the tropical Atlantic, had set him thinking.[1] It was less than fifty-four degrees Fahrenheit, a big contrast to the eighty-four degrees Fahrenheit of the surface.[2] This measurement led him to suppose that, in polar latitudes, water

> deprived of a great part of its heat by cold winds, descends to the bottom of the sea, [where it] cannot be warmed where it descends, [and] as its specific gravity [i.e., density] is greater than that of water at the same depth in warmer latitudes, it will immediately begin to spread on the bottom of the sea, and to flow toward the equator; and this must necessarily produce a current at the surface in an opposite direction. There are most indubitable proofs of the existence of both these currents.[3]

This down-up current is sometimes called the "thermohaline circulation," because it is driven by differences between water masses in temperature and salt content. The reasons for downwelling differ between Arctic and Antarctic. Frigid conditions in the Antarctic cause sea ice to form. Freezing separates freshwater from salt and leaves behind a more briny sea. This bitterly cold and saltier water is denser than normal seawater and therefore sinks, in a process known, rather unimaginatively, as "deep bottom water formation."

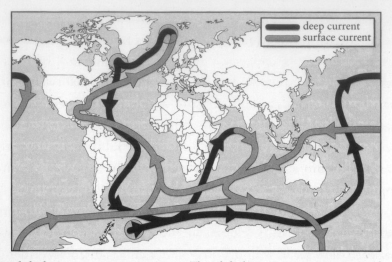

The global ocean conveyor current: The global ocean conveyor current loops around the planet and circulates water between the shallow surface layers and the deep sea. Gray circles show places where water becomes colder and saltier, and therefore denser, and so sinks to create deep bottom water. Water sinking near the poles is counteracted by upwelling from deep to shallow waters at lower latitudes. These upwellings are rich in fish as their nutrients fuel plankton blooms.

In the far north of the Atlantic the Gulf Stream supplies water to polar seas that is more salty than local waters, and it is made even saltier by evaporation from intensely cold, dry winds blowing off Greenland and Europe. At both poles, as cold, salty, and dense water sinks, it pulls in surface water to replace it. This pull is one of the energy sources for the Gulf Stream and North Atlantic Drift (the other being wind). When this water sinks it begins a deep-sea journey that may last fifteen hundred years before it upwells again into the sunlight far away.

You can create a loop current a little like the global ocean conveyor by blowing on a cup of tea to cool it. Your breath pushes the tea across the cup until it hits the edge, at which point it flows down the inside of the cup, crosses the bottom, and upwells again at the side you are blowing from. (If you are ever short of a teaspoon, this is a handy way to mix milk into your tea.)

Like the ideas of many gifted thinkers, Rumford's theory of

ocean circulation was not fully accepted until long after he had died. He never returned to America, although his British sympathies were later forgiven and he endowed a professorship at Harvard University. Oddly enough, despite the fact that they were born just twelve miles apart and had such similar interests—Franklin also invented a stove that Thompson later improved—there is no evidence that they ever met or corresponded. Theodore Roosevelt would later declare that Franklin, Rumford, and Thomas Jefferson were the three greatest minds America had ever produced.

In the 1960s, atmospheric nuclear tests produced radioactive materials that were washed into the Arctic and gave us a way to trace the speed of water flow to the deep sea. Scientists discovered that the global ocean conveyor moved three to four feet per hour.[4] This sinking rate means that 550 million cubic feet of water is displaced per second, which is equivalent to the flow of eighty Amazon rivers, or twelve times the flow of all the world's rivers combined.[5] As powerful as this river in the sea is, it would take nearly twenty-eight hundred years to circulate all of the oceans' waters through what oceanographers call the "North Atlantic pump."

As I intimated, there are two other key areas of deep bottom water formation, in the Ross and Weddell seas of Antarctica. They transfer another 740 million cubic feet of water per second from the surface to the deep sea, reducing turnover time for water in the deep oceans to under twelve hundred years. These pumps are critical to the vertical mixing of water in the world oceans. They carry freshly oxygenated water into the deep sea, helping sustain life there. They also transfer carbon dioxide from the atmosphere to the deep ocean, a point I will return to later on.

One thing that fascinates me is that water bodies in the sea, especially the deep, tend to retain characteristic signatures of temperature and salt content for very long periods. Oceanographers can build up a picture of the three-dimensional structure of the sea from simple vertical profiles of salinity and temperature measured using instruments dangled from boats. It turns out that the oceans are made up of many parcels of water in constant motion, driven by wind and density differences, and their origins and movement around ocean basins can be tracked. At the mouth of the Mediterranean Sea, for instance, water flows in through the Strait of Gibraltar at the surface. Beneath there is a deeper flow back into the Atlantic of dense, salty water concentrated by evaporation as it traveled around the Mediterranean Basin. This pours out across the shallow Gibraltar sill and forms giant turbulent loops that travel west to the Azores and north to Ireland. Similar measurements reveal that the southward flow of deep water from the North Atlantic is not compact, like the surface Gulf Stream, but broad, and it creeps sluggishly along the eastern seaboard of North America.

The North Atlantic and Southern Ocean engines of the global ocean conveyor have recently been identified as possible climate "tipping points" by a multinational team of climatologists,[6] meaning that a certain threshold of water temperature or density must not be passed if the currents are to remain in a stable state. If that critical point is passed we could see a radical and rapid shift to a different state. Since major ocean currents move at the behest of wind and water density gradients, there is a good chance they would shift when the world's climates change. The possibility that this might happen is now a cause of great concern to many scientists.

Unlike past changes, whose causes were geologic or celestial in nature, today's are down to us. People have multiplied and spread to every corner of the globe. We have shifted from being a species governed by nature to one that can harness nature to its own ends. Unwittingly, our ingenuity has unleashed forces over which we have little control and that threaten the way we live.

The concept that some form of global warming might be triggered by the burning of fossil fuels was first proposed in 1896 by a Swedish chemist, Svante Arrhenius.[7] People at that time were fascinated by the ice ages, whose world-shaping influence had only recently become understood. Arrhenius argued that variation in the amount of atmospheric carbon dioxide could have played a key role in glaciations.[8] But he took the idea further. Perhaps his thoughts coalesced while tramping the icy streets of Uppsala one evening as the coal smoke from thousands of hearths wrapped around him. He made the connection that the enormous quantities of fossil fuel being burned could eventually lead to planetary warming. From his rather chilly perspective, he thought warming would be a good thing:

> We often hear lamentations that the coal stored up in the earth is wasted by the present generation without any thought for the future. . . . We may find a consolation in the consideration that here, as in every other case, there is good mixed with the evil. By the influence of the increasing percentage of carbonic acid [carbon dioxide] in the atmosphere, we may hope to enjoy ages with more equable and better climates, especially as regards the colder regions of the earth, ages when the earth will bring forth much more abundant crops than at present, for the benefit of rapidly propagating mankind.[9]

By Arrhenius's calculation, doubling atmospheric carbon dioxide would increase average temperature by 7°F, a figure since revised by the Intergovernmental Panel on Climate Change to the range of 3.6°F to 8.1°F. Not bad for a nineteenth-century chemist!

Today we don't share Arrhenius's optimism about the benign influence of global warming. A little warming will not simply help shrug off the winter cold or lengthen growing seasons. Now we understand that warming alters patterns of wind, cloud cover, and rainfall, changing conditions in ways that are hard to predict and would be difficult and expensive for us to adapt to. Thousands of studies attest to the reality of global warming, now usually called "climate change" because of the far-reaching influences warming has on climate. A record-breaking snowfall or unexpectedly fierce tornado or monsoon flood is as much a product of global warming as scorching summers and prolonged droughts. Measurements from mountains and lakes, glaciers and ice sheets, deserts and rain forests, coasts and oceans all confirm that temperatures are on the rise.

As I explained in the opening chapter, Earth's atmosphere is like a blanket that traps warmth from the sun but also shields us from the ferocity of the sun's heat and harmful ultraviolet rays. If Earth didn't have an atmosphere, temperatures would be like those on the Moon, where there are wild swings between extremes.[10] In sunshine the Moon's surface can top 212°F, while on the dark side it can plunge close to −238°F. (The moon averages −9°F, compared to Earth's clement average of 60°F.) The warmth of the atmospheric blanket depends on the content of its heat-trapping gases. Collectively these greenhouse gases slow the radiation of heat from Earth back into space. The most important are water vapor, carbon dioxide, methane, and ozone, but there are several others in the mix.

Carbon dioxide concentration in the atmosphere has risen by 38 percent since preindustrial times (before 1750), from 280 parts per million (ppm) to 388 ppm.[11] The concentration of methane over the same period has gone up by 150 percent, from 700 parts per billion (ppb) to 1745 ppb. Although there is much less methane than carbon dioxide, it is a far more potent greenhouse gas. You will recall from my description of the early Earth in Chapter 1 that each molecule of methane has twenty-five times the warming potential of a molecule of carbon dioxide.[12] The main source of carbon dioxide is the burning of fossil fuel, while methane comes from livestock (cows and other ruminants fart endlessly), landfills, and rice paddies. It is also a by-product of warming, as trapped methane is released by melting tundra near the poles.[13] Just as Arrhenius predicted, temperatures have risen as we have burned more fossil fuels. Averaged across the globe, they have increased by 1.3°F since preindustrial times. What is alarming is that the rate of rise has now reached 0.4°F per decade. We are on a slippery slope.

So far, much of the heat trapped by greenhouse gases has been taken up by the sea. If it hadn't we would all be sweltering by now. The oceans have sucked heat from the atmosphere because the heat capacity of water is several thousand times greater than air.[14] It is for this reason also that temperature increases in the oceans have been less than on land. Averaged from the surface to seabed, temperatures have risen just one fourteenth of one degree Fahrenheit since 1955.[15] It doesn't sound like much, but most of the warming has been near the surface, where average temperatures have increased by 1.1°F in the last century.[16] This heating has not been even: Some places have warmed quickly, others are little changed. The tropics have warmed least, while the temperatures at the poles have shot up, which is why polar bears and penguins are icons of the changing climate.

On the Antarctic Peninsula, where air temperatures have warmed by 11°F in the last fifty years,[17] Adélie penguins that once nested on frozen ground now huddle in pathetic groups, ankle-deep in mud. The chicks' downy feathers are well adapted to snow, but they lose their insulation in sleet and drizzle. Soaked through and frigid, the chicks die. While it seems obvious that they should move to colder rookeries, it is hard for them to abandon traditional sites used for generation after generation. As a result, Adélie penguins on the peninsula have declined by 90 percent in the last thirty years (those farther south are still doing fine). At the other end of the world, polar bears and seals depend on sea ice, the bears to hunt and the seals to breed. Polar bears are frequently seen these days swimming in open water up to sixty miles from the nearest coast or ice. Many now drown far out at sea as exhaustion overtakes them when ice cannot be found.[18]

A four-degree rise in global average temperature is expected to produce a seven- to eleven-degree warming at the poles. Summer sea ice cover has declined fast in recent years. In the summer of 2000 Jim McCarthy, a Harvard oceanographer who was then a leading member of the Intergovernmental Panel on Climate Change, was left gasping in amazement when the icebreaker he was on found open water at the North Pole. Summer sea ice has shrunk farther since then, and the region is expected to be regularly ice-free by the summer of 2030.[19]

Polar oceans are highly affected by climate change. Less sea ice formation means the surface water at the poles is less salty, and warmer water is also less dense than cold. (The word "warm" is relative—even in a greenhouse world you couldn't enjoy a swim at these latitudes!) Together these changes will slow the rate of sinking and the replenishment of deep bottom water. There's another factor related to climate change that's contributing to slowing the engine of the global ocean conveyor further: More rain and the thawing of frozen tundra have swelled rivers that pour into the Arctic Ocean. This freshwater reduces surface-water density, and so also inhibits deep bottom water formation.

Ever since Benjamin Franklin's time, people have generally understood that the Gulf Stream carries the heat of the Caribbean far into northern latitudes. It is held to be responsible for mild, wet winters in Britain and France and for the warmth of the seas from Long Island to Boston. Without the Gulf Stream, people have come to believe, a deep freeze will descend on London and New York.[20]

The central premise of *The Day After Tomorrow*, a blockbuster thriller released in 2004, was that the Gulf Stream might suddenly shudder to a halt. In the movie, eastern North America and Europe were plunged into ice age conditions by the sudden failure of the Gulf Stream. While the idea may seem far-fetched, and the near instant freeze in the film was definitely science fiction, there is legitimate cause for concern. Oceanographers have already seen deep currents slow in the North Atlantic. Between 1957 and 2004 there was a 30 percent reduction in deep water flow (although this does not yet seem to have slowed the shallow Gulf Stream and the more diffuse North Atlantic Drift). It is too early to link the slowing conclusively to climate change.

Over 120,000 years of climate records held in annual layers of snow cored from within the Greenland ice sheet suggest that this part of the global ocean conveyor has stopped many times in the past. It seems that a breakdown can occur rapidly, in as little as a few decades. The trigger for these stoppages in the past was a sudden drop in the salinity of Arctic seas.[21] The period covered by the Greenland ice cores lies within the last glaciation, from 110,000 years to 10,000 years ago, when much of North America and Europe were ice-bound. The ice sheets grew in thickness as snowfall accumulated over thousands of years. Eventually it grew top-heavy and became unstable. Ice surged through the Hudson Strait into the North Atlantic, where it melted, freshened the sea, and switched off deep bottom water formation at the northern extremity of the global ocean conveyor current. More ice poured into the eastern Atlantic from the Baltic Ice Lake.[22] Each event is marked in bottom sediments far out at sea by the appearance of ice-rafted rubble swept offshore as the ice sheets collapsed.

Ocean circulation switched on again abruptly at the end of each shutdown, of which there were at least seven in the last sixty thousand years. We know this because sea surface temperatures suddenly rose more than five degrees as water flowed in from warmer latitudes. This effect extended south at least to Bermuda, where cores through deep sea sediments show sudden rises in surface temperature, from about 60°F to 70°F (today they are about 72°F).[23] Nobody is exactly sure what caused these currents to shift. There are no abrupt changes in any of the factors that might drive the shift, like wobbles in the tilt of the Earth's axis (which change the heat balance of the planet) or fluctuations in atmospheric carbon dioxide that correlate with the timing of North Atlantic temperature flip-flops. However, scientists have come up with several good explanations.

One theory is that the global ocean conveyor current has three different stable states in the North Atlantic: on, off, and partly on, with deep water formation taking place farther south during times when ice sheets were more extensive.[24] One of these states predominates most of the time, and the current varies over time as conditions fluctuate. A change between states occurs when a critical threshold has been exceeded, and the system flips into a different regime that itself might remain stable for hundreds or thousands of years before it flips into a different state again. According to this way of thinking a small change in some driver, like carbon dioxide or methane concentration, could cause a large change to the entire climate system because of its sensitivity to jumps between different states.

If this sounds confusing, a simple analogy might help. Imagine a drunkard walks home late at night along a riverside path. He lurches from side to side but remains on the path. Halfway home someone brushes past him from the opposite direction and shifts him slightly nearer the river. The bank now lies within the ambit of his more extreme lurches. A hundred yards along the path the drunkard falls into the river, and so passes from one stable state (on ground) to another (in the water).

Ice cores from Greenland, Antarctica, and glaciers the world over spell out the reality of abrupt climate change. These cores contain an extraordinary wealth of information on snowfall, atmospheric gas concentrations, and dust levels. The chemical isotope concentrations in ice tell us what the temperature was, while trapped dust layers reveal the extent of the deserts and the frequency and size of forest fires and volcanic eruptions. They allow us to stroll back through time, at first season by season, then year by year, and, deep down where the ice is compressed, decade by decade. How far back these records go depends on the depth of the core—from a few thousand years in some glaciers to over seven hundred thousand years in the deepest Antarctic cores. If there is one lesson here it is that the climate of the last few thousand years of human history has been remarkably stable when set against a long-term background of steep fluctuations and sharp flip-flops. Abrupt changes revealed by the ice record were not so sudden that you wouldn't have had time to buy a polar jacket and upgrade your home boiler. But they were sharp enough that we would really struggle to adjust were one to happen now. If the mighty river that is the global ocean conveyor falters it will have profound impacts on life above and beneath the sea. Radical regional shifts in marine climate alone would precipitate the wholesale reorganization of ecosystems and human societies. There may be some consolation in the fact that the sage Intergovernmental Panel on Climate Change advises us that the North Atlantic arm of the global ocean conveyor is unlikely to fail in the next hundred years. But other shifts are already upon us, like the rapid warming of the Arctic Ocean.

Water flows from the surface to the deep ocean in polar seas at a rate equivalent to twenty thousand Niagara Falls. Since the oceans do not expand by the same volume, this flow must be matched by an equal and opposite upwelling toward the surface somewhere else. Part of that counterbalance is generated by gentle upward mixing in the tropics. In other places upwelling is assisted by wind-driven surface pull. These upwellings are concentrated on the eastern boundaries of the oceans, where winds blow parallel to the coast, like along the western United States.

This upwelling explains why the seas of northern California are at their most bracing in spring and summer, and why San Francisco then wears a foggy shroud: cool deep water upwells under the influence of winds from the north. In a physical quirk of life on a spinning sphere, the average direction of water movement is at a right angle to the direction of the wind.[25] In the northern hemisphere the twist is to the right, while in the southern it is to the left. This effect is called the Coriolis force, after Gaspard-Gustave de Coriolis, a nineteenth-century French mathematician who worked out the interplay of forces acting on rotating objects like water wheels. As far as we know, Coriolis never thought to apply his ideas beyond the energy production of rotating machinery, but the Coriolis force has a profound influence on how water and air circulate in both oceans and atmosphere.

Indeed, it applies to everything that moves across the surface of the Earth. If they did not correct for the Coriolis force, airline pilots would miss their destinations, and artillery gunners would shoot wide of their targets.

Winds that blow from the south along the coast of South America drive the world's largest upwelling in the sea off Peru. Upwellings have enormous significance. They generate nearly half of the global fish catch, although they cover just 1 percent of the area of the oceans. Peruvian anchovy alone account for a tenth of world fish landings in good years. This sleek silver fish lives in schools of almost incomprehensible size and feeds mainly on phytoplankton, microscopic plants that live in the open sea. Upwelling fuels spectacular blooms of these phytoplankton (which, as I will come to later, are also caused by nutrient enrichment from human sources).

The ocean has two layers, surface water and the deep sea, over most of its surface area. Surface waters are warmed by sunshine. Since warm water has a lower density than cold water, this layer tends to float on top of the much cooler deep waters. There is little mixing across the boundary between the two, which is called the thermocline because of the abrupt temperature gradient. This creates a problem for marine life. Sufficient light for plant growth only reaches to depths of 30 feet to 300 feet, depending on the water's clarity. Light will sometimes penetrate as far as 600 feet in the crystal-clear waters of the mid-Pacific, but even that is just a few percent of the average depth of the sea (12,070 feet). Plants grow in this sunlit zone by taking up dissolved nutrients and carbon dioxide. These plants are either eaten by animals or microbes, in which case the nutrients are recycled, or they sink. When they pass through the thermocline, which is typically around 100 feet to 300 feet deep, their nutrients are lost to the deep sea.[26] The bright surface layer thus leaks nutrients to the deep, with the result that plant growth is usually limited by a lack of fertilizer. The deep, meanwhile, has plenty of nutrients but no light, so plants cannot grow. Animals living there exist on table scraps that sink down from the surface layer. Life is sparse. Where deep waters are pushed or pulled up to the surface, as they are off Peru, they flush nutrients into the sunlit surface layers to fuel explosive plant growth.

The world's most intense upwellings occur where winds that blow parallel to the coast are reliable and strong, like West Africa, both north and south of the equator, and the Pacific coast of South America. These winds are a consequence of differential heating between land and sea. The land heats up more than the sea during spring and summer in subtropical latitudes and year round in the tropics.[27] Hot air rises above the land, carrying water vapor with it. It cools as it ascends, and the water condenses into those spectacular thunderstorms you can sometimes watch boil upward while sipping piña coladas on a tropical beach. Condensed water falls in showers so intense you feel as if you were underwater.

This condensation of water leaves the air at cloud level warmer and lighter, so it rises even higher, eventually stopping when it is thoroughly chilled. That high-altitude cold air is now pushed sideways by more air welling up from below. North of the equator it blows north, while south of the equator it heads south, carrying heat from the tropics to higher latitudes, where it cools and sinks back down around thirty degrees north or south. If you glance at an atlas you will see that many of the world's great deserts are concentrated around these latitudes, because the downdrafts are very dry, having had all the moisture squeezed out as the air rose: Mojave, Sonora, and Chihuahua in North America; Sahara and Arabian deserts in Africa; Takla Makan in China; Atacama in Chile; and so on. Since air cannot rise at the equator without being replaced by other air, surface winds blow air back from these drier latitudes to the tropics to complete the loop and drive upwellings as they do. Surface winds blow poleward at temperate latitudes, where tropical circulation loops intersect like cogs with other winds, that then carry heat poleward and draw cool air back to warmer latitudes.

The tropics will trap more heat as greenhouse gas concentrations

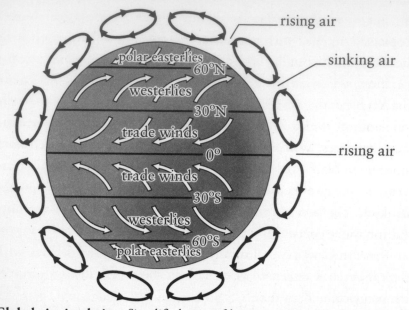

Global air circulation: Simplified view of how air masses circulate on Earth, showing the direction and names of the surface winds prevailing at different latitudes. The loops represent vertical air circulation and show the latitudes at which warm, moist air rises in the atmosphere and cold, dry air sinks.

rise, which will drive a faster redistribution of warmth and strengthen the winds that power upwelling.[28] The productivity boost this will give the oceans could benefit fisheries and help the sea absorb more carbon dioxide (although this is a double-edged sword, as I will come to later). Thumbing his nose at Aristotle, Oscar Wilde once quipped, "Moderation is a fatal thing. Nothing succeeds like excess." He found out in life, to his dismay, that it was actually the other way round, and so it is with upwelling. Excessive upwelling can make unpleasant things happen in the sea. In southwest Africa we may have a taste of the future of upwelling in a greenhouse world.

The coast of Benguela, offshore from the vast red dunes of the Namib Desert (as scorched and thirsty a place as you are likely to find on Earth), sustains one of the planet's most intense upwellings. The lack of plant life on shore contrasts vividly with its abundance in the sea. Copiously supplied with nutrients, phytoplankton flourish, turning the sea green. Normally those plant cells would be consumed by zooplankton, which in turn would be eaten by fish and other predators. The trouble is, the wind is so strong that it pushes water offshore before zooplankton, which take longer to develop, have time to complete their life cycles. While phytoplankton turn over at high speeds and continue to bloom right in the midst of upwelling, zooplankton are swept offshore, so they cannot curtail the explosive growth of phytoplankton. The result is devastating, as decaying phytoplankton sink and rob the sea deep down of its oxygen.

Beneath Benguela's upwelling plumes, the seabed is thick with swirling drifts of dead and decomposing plankton. It can sustain only the ancient microbes that first evolved in the similar toxic broth of the early oceans. This meters-thick sludge is anoxic, or nearly so, and gurgles forth methane and hydrogen sulfide, the waste products of microbial metabolism. Hydrogen sulfide compounds the problem of anoxia, because it reacts with any oxygen present, and strips it from the water as it rises toward the surface. From time to time the sludge effervesces so fast that the rising bubbles rush upward in an eruption of sulfide and methane, turning the sea bright turquoise.[29] The color comes from sulfur particles in suspension that are produced when hydrogen sulfide is oxidized.

Local people on the coast have become inured to the stench of rotten eggs that wafts from the sea. The corrosive pall has been a fact of life for as long as people in the Namibian port of Lüderitz can remember. Sulfur eruptions have been reported here since the nineteenth century. They poison the sea, cause mass death of fish, and trigger "walkouts" of lobsters onto beaches as they try to flee one form of death, only to find another. These walkouts provide a short-lived bonanza for locals, who pile them into every container they can find. Similar walkouts in Mobile Bay, Alabama, when all kinds of creatures are forced to the shoreline by plummeting oxygen, are known as jubilees. But they decimate populations, leaving few animals behind to sustain catches later on. In the early 1990s, one particularly energetic eruption off Namibia killed 80 percent of the population of Cape hake, a bottom-dwelling fish, and led to a collapse of catches.

In recent years sulfide eruptions have increased in frequency and intensity, a phenomenon that is most likely linked to overfishing. A few decades ago sardines (also known as pilchards) were to Namibia what anchovies are to Peru. These tiny fish are filter feeders, and they sweep phytoplankton and some zooplankton from the water as they dart around, mouths agape. In good years, sardines shoaled by the billions. Unlike zooplankton, which are poor swimmers and easily swept offshore and away from the heart of phytoplankton blooms, sardines are powerful swimmers and remained in the thick of it. Voracious sardines ate the phytoplankton as fast as it was produced, so dead plankton did not usually accumulate on the seabed and sulfide eruptions were rare. Sardines in turn were eaten by other predators, like tuna, swordfish, and seabirds, and the Benguela upwelling sustained one of the oceans' most spectacular concentrations of wildlife.

Such productive seas soon drew the attention of the fishing industry, and sardines were targeted heavily beginning in the 1960s. The fishery for sardines collapsed in the 1970s, and it has never recovered. Coastal residents can now blame overfishing for the corrosive sulfide fogs that overwhelm their towns and irritate their eyes and throats. Namibian seas reveal what can happen when we knock holes in the fabric of an ecosystem, whether through fishing or by other means.

Andrew Bakun, from the University of Miami, is the scientist who first made the connection between upwelling intensity and global warming.[30] He predicts that warming will intensify upwelling in other regions of the world that will match or exceed the rates seen in Benguela today. If there are insufficient fish to control phytoplankton blooms, they too could experience the poisonous combination of anoxia and sulfide eruptions. One such place is northern California. There is a real possibility that fifty years from now the homely charm of Mendocino will be tarnished by the stench of rotten eggs and its beaches heaped with putrid fish.

There is a further marine twist to the climate change story. The ocean has two layers, as I mentioned: a warm surface layer floats on top of cooler, denser water below. The additional heat absorbed by the surface layer as the world warms increases the temperature contrast, and hence the density difference between these layers. The steeper the density gradient at the thermocline, the band separating the surface water from the deeper sea, the less water mixes across it. The result is that there are fewer nutrients transferred from deep sea to shallow water, and less oxygen mixed from shallow waters to deep, as warm water holds less oxygen than cool water. As the oceans heat up they will thus release some of their oxygen into the atmosphere.

Dissolved oxygen is plentiful at the surface but drops sharply below the thermocline, and it sags further as you descend, eventually rising again toward the bottom. This pattern has a simple explanation. The surface layer is well ventilated and the deep sea gets its oxygen from the water that sinks down at the poles. In between there are regions where the life-giving gas is scarce.

Low oxygen zones form in places where the water column is strongly separated into layers by density differences (caused by differences in temperature and salinity). Dead plankton and animals sink from the surface—marine snow, as oceanographers like to call it—and are then eaten or broken down by microbes, which use up oxygen. Most of the oxygen in the middle layers of the sea has been there since the water was last in contact with air at the surface (in some cases hundreds to a thousand years or more). The longer ago that was, the more time there has been for animals and microbes to use up its oxygen. There is no sunlight in these inky depths, so no oxygen can be produced to replace what is used up. Deep waters in the Pacific and Indian oceans are "older" than those of the Atlantic, meaning that they have been submerged for longer, and so have less oxygen.

They also have more carbon dioxide, which has been breathed out by the animals and microbes that live there.

The eastern Pacific and Northern Indian oceans have enormous slabs of water in which oxygen concentrations are less than 30 percent of those at surface levels. Fully a quarter of the Pacific is affected by low oxygen at some depth or another.[31] These oxygen-starved zones impose physiological stresses that few organisms can tolerate for long. The warm ventilated surface layer is thin in the eastern Pacific, and the low oxygen zone lies just 650 feet to 2,000 feet below the surface. Fast-swimming tunas are confined to well-aerated surface waters, which is why they are easy to catch there. The upper region of low oxygen water is home to many planktonic animals, which migrate toward the surface to feed under cover of darkness, and then retreat by day to a place where few predators can follow. Sharks and tunas can penetrate low oxygen layers for only minutes or hours before they must return to the surface to catch their breath and warm up.

Every animal is different, of course, and some are better able to tolerate low oxygen than others. But to put the conditions into a human context, to enter water with less than 30 percent oxygen saturation would be like a climber venturing into the death zone at the extremes of altitude. At heights of greater than 26,200 feet, oxygen levels are a bit over a third of those at sea level—too low to sustain human life. Without bottled oxygen, we can only make brief sorties into this zone, because we use oxygen faster than our bodies can take it in.

The challenges that aquatic animals face are different to those faced by land-living animals. Many creatures are neutrally buoyant in water, which means they neither rise nor sink. The major constraint on their movements is not gravity but the frictional drag of water. Other things being equal, big animals can swim faster than smaller ones, and use relatively less energy to do so. But they face a problem extracting oxygen from water. Oxygen demand scales with body volume, which is the cube of fish length (think of the volume of a box—length times breadth times height). On the other hand, the surface area of the gills over which oxygen can be exchanged increases less fast as body size increases, as the square of fish length (the area of a box—length times breadth).[32] While being large-bodied is advantageous hydrodynamically, it is a disadvantage when it comes to breathing. This simple fact explains why predatory fish like tuna don't simply follow schools of their prey, snacking on them when the mood overtakes them until all the prey are gone.[33] In a long-distance race it is the bigger predators that run out of puff first. And when a fish is left gasping for breath it can easily fall prey to something else, like a shark. So marine animals must be very careful not to fall into oxygen deficit.

The huge Humboldt squid is one of the only large predators able to withstand low oxygen for extended periods.[34] These beasts can reach more than ten feet long with their tentacles extended, and they weigh more than a man. Within its crown of muscular tentacles is a parrotlike beak made of two interlocking blades sharp as butchers' knives. They swim by jet propulsion, squirting water through a tube, but most of the time they probably just wait for prey to come to them.

Low oxygen zones will expand in a warmer world as surface warming reduces mixing across the thermocline. The effect will compound as heat absorbed by the oceans penetrates deeper, which means they will hold less oxygen. In combination, the volume of the sea capable of supporting life will shrink. The upper limit of low oxygen waters has already moved up to three hundred feet closer to the surface off North America's west coast.[35]

The West Coast of the United States has seen repeated mass kills of ocean life on the shallow continental shelf since 2002, events unprecedented in the previous fifty years.[36] It turns out that they are due to the strong winds that blow parallel to the coast and drive upwelling, a harbinger of the dark side of strengthening upwelling I just described. As these winds push surface waters offshore, a tongue of deep, oxygen-starved water creeps up over the shelf to replace them and snuff out life. Jane Lubchenco, head of the U.S. National Oceanic and Atmospheric Administration, described the scenes she viewed through video cameras towed through a dead zone off the coast of Oregon in 2002: "We saw a crab graveyard and no fish the entire day. Thousands and thousands of dead crabs were littering the ocean floor, many sea stars were dead, and the fish have either left the area or have died and been washed away. . . . [S]eeing so much carnage was shocking and depressing." One of her colleagues, Francis Chan, said, "One of the areas sampled is a rocky reef not far from Yachats, Oregon. Ordinarily it's . . . swarming with black rockfish, ling cod, kelp greenling, and canary rockfish . . . [but] the fish are gone, and worms that ordinarily burrow into the soft sediments have died and are floating on the bottom."[37] Such mass die-offs of marine life will become a regular feature of future oceans unless climate change can be halted.

The Humboldt squid seems to have benefited from the expansion of its low oxygen habitat (and loss of large predatory sharks due to overfishing).[38] It used to be common only in the seas off Baja California but in recent years it has expanded north to the Gulf of Alaska. Divers increasingly come into contact with them, as the oxygen minimum layer reaches closer to the surface. To meet a squid in the wild is one of life's great pleasures. They follow your movements with intelligent, saucer eyes. Their emotions are written on their skin in quick-fire color changes that pulse and ripple in incandescent waves across their bodies. The Humboldts' uncertain temper adds a little frisson to any encounter. Divers have been roughed up by squid, or had their masks and gear tugged. After the adrenalin rush of initial fear and excitement has passed, most divers feel the squid were more curious than aggressive. If they had really wanted to do harm, their huge strength would have enabled them to do far worse.

The trade-off between body size and oxygen demand explains a pattern that has long been known but never well understood: Fish tend to grow larger in cooler water. This is because they have lower metabolic rates, so they can afford to grow bigger. This suggests that we will have more problems ahead, as the seas warm from climate change. Fish under temperature and oxygen stress will reach smaller sizes, live less long, and have to devote a bigger fraction of their energy to survival, at the cost of growth and reproduction. From our own rather self-centered perspective, their response to the new conditions will be bad because it will reduce production of body mass, which will limit catches. Of course, some exploited species have already seen their body size and growth compromised by overfishing. Thus we see increasing temperature, decreasing oxygen, and high fishing intensities combine to reduce the productivity of fish stocks and undermine the long-term resilience they so desperately need to cope with the changing world around them.

The deep past affords us a bleak warning of what might happen if we fail to control greenhouse gas emissions. The mass extinction at the end of the Permian period 251 million years ago, which I described in Chapter 1, was caused by a planet in the grip of runaway global warming. Life in the sea suffered the one-two punch of anoxia and high carbon dioxide. It took five million years to recover. However, one group of animals that lived within the open waters of the sea, the ceratid ammonoids, which are creatures similar to the beautiful nautilus of our day, rebounded more quickly, returning within a million years of the catastrophe. It is possible that they were adapted to low oxygen conditions. Although the entire group is now extinct, their nearest living relatives are squid, cuttlefish, octopus, and nautilus. Indeed, one of the oldest living forms of squid is *Vampyroteuthis infernalis*, the Latin name of the "vampire squid from hell." It is a blood-red beast that hunts deep within the low oxygen zones of today's oceans. Its success as a predator in these extreme conditions may be a modern legacy of a past catastrophe. Several other mass extinction events in geological history hacked off large branches of the evolutionary tree. The toxic alliance of high carbon dioxide and low oxygen appears to have wielded the ax on more than one occasion.

The medieval Scandinavian and English king Cnut the Great is reputed to have had his throne set down by the sea, where he commanded the tide to halt. The tide, of course, paid no heed, proving that the sea was moved by powers that were beyond the influence of people, even kings. A thousand years later humanity has inadvertently gained the power to move ocean currents. But our powers remain circumscribed, and we must now live with the consequences or find a way to keep the forces we have unleashed in check.

CHAPTER 5

Life on the Move

Jean-Baptiste de Lamarck is famous today, or perhaps infamous, for his wrongheaded theory of evolution. He believed that organisms could pass on to their offspring characteristics that they had acquired during their lifetimes. Giraffes grew long necks, by his account, because generations of ancestors had stretched further and further to reach leaves high on trees. While his evolutionary ideas were discredited by Darwin, Lamarck was in his day a keen observer of life and a distinguished and influential French scientist. One of his other equally mistaken ideas has taken considerably longer to dismantle.

By the end of the eighteenth century it had become clear that people could deplete some species of wildlife to the point of extinction. The aurochs, a huge European wild cow, had been killed off by 1627 and the dodo by 1662. Many animals had been erased from large parts of their ranges, like beavers, wolves, and bears in Europe, and panthers in North America. Aquatic animals seemed more resilient, and Lamarck was convinced that they were in effect impervious to extinction because no natural barriers could contain them. As he wrote in 1809 in his landmark work *Philosophie zoologique*:

> Animals living in the waters, especially the sea waters, are protected from the destruction of their species by Man. Their multiplication is so rapid and their means of evading pursuit or traps are so great that there is no likelihood of his being able to destroy the entire species of any of these animals.[1]

The limit to how far a species will spread from its place of origin depends on its physiology and habitat, and on predators and competitors. A few species do well over wide geographic ranges—tigers, for instance, can survive torrid jungles as well as frozen mountain passes—but most species do well in just one or a few habitats. As they move away from their comfort zone to places where environmental conditions are suboptimal, interactions with better adapted competitors and predators keep them in check.

On land a species' range will often be restricted by physical barriers such as mountains, deserts, or large bodies of water. For some time now we have generally accepted Lamarck's contention that few such barriers exist in the oceans. Most marine species release eggs or larvae into the open sea, where they drift for days or even months before finding a new home, and many of those species swim hundreds of thousands of miles in their lifetimes. This received wisdom had scarcely been challenged until my wife, Julie Hawkins, and I started to plot the ranges of species living on coral reefs. Julie spent years amassing lists of species from all over the world, among them a set of exquisitely illustrated fish lists collected by the naturalist Sir Peter Scott on his travels around the world in the 1970s and 1980s.

It was painstaking work, and after ten years and several thousand species, Julie's mounting grumbles called a halt to our mapping. We found that many coral reef species were not in fact widespread, and some were limited to remarkably small fragments of habitat. The dark indigo Limbaugh's angelfish, to take one example, is known only from Clipperton Reef, a pinnacle of coral in the eastern Pacific that is isolated from neighboring reefs by nearly six hundred miles of blue water. Or there is the pale dottyback, a shy and ghostly fish found only in the Gulf of Aqaba, a narrow finger of water in the northern Red Sea. Or the splendid toadfish, whose only known haunts are the reefs of the resort island Cozumel in Mexico. In recent years a remarkable catalog of coral reef species with small ranges has been built up.[2]

A similar picture is forming for marine species in other habitats and regions. Submerged seamounts, rocky oases amid the vast mud plains of the deep sea, sustain extraordinary numbers of species found nowhere else. The great trenches carved into the ocean floor, where tectonic plates grind back into the Earth's mantle, provide other unique habitats, with their own distinctive plants and animals that are isolated by thousands of kilometers of inhospitable terrain. Brachiopods—they look like bivalve molluscs but are unrelated—have knocked about on this planet since the Cambrian period five hundred million years ago[3] and are mostly scarce and solitary today, but in the fjords of Patagonia they build mounds that are an echo of the reefs they made in their heyday hundreds of millions of years ago, before the mass extinction at the end of the Permian.

What will happen to all these creatures as their environment alters under the influence of global climate change? Unless they can adapt to the new conditions, or move, extinction beckons. Adaptation proceeds via natural selection, as Darwin taught us, since those best suited to prevailing conditions leave more offspring, who carry their genes forward. Genetic mutations take place over generations, and the species imperceptibly shape-shifts into something different. The problem is that these evolutionary changes take time; they require many generations to take hold. If conditions change quickly, the only option may be to move. If you can't move, you die. The history of our planet is punctuated with periods of mass extinction, when geologic or climatic upheavals pushed species beyond their endurance. The comings and goings of life can be read in seaside cliffs, where different fossils are found only in certain layers of rock. While a handful of species seem immortal, most last just one to five million years.[4]

Polar bears and penguins leap off the pages of every circular from an environmental organization that comes through the letter box, and television correspondents reporting on climate change are packed off to icy latitudes to be filmed with them. They are the most telegenic sufferers of global warming, but there are countless other less familiar species whose existence is also in flux. The North Sea has warmed by 2.25°F in the last twenty-five years, and life there is on the move. Fishermen on the coast near where I live in Yorkshire have seen growing numbers of new species pop into their catches over the last decade, like gray gurnard and cuttlefish. Annual surveys made since 1977 using bottom-trawl nets towed across the seabed show how geographic ranges are shifting under the influence of warming.[5] The john dory, for instance, an odd-looking narrow-bodied fish with a long thin jaw, used to be common only in southwest England, but they have moved north, near Scotland, and colonized the North Sea.

Fifteen of thirty-six species surveyed in the North Sea over the last thirty-five years shifted latitudes. The average shift was about 190 miles north. (The temperature difference between the equator and 80°N, roughly the limit of sea ice, is about 54°F; if you divide this difference by the distance, temperature falls by about 1°F for every one hundred miles traveled north.[6]) An interesting difference emerged when shifters were compared to those who had stayed put. Species that responded quickly to climate change tended to be small-bodied animals that reproduce early in life, while those that did not move were larger-bodied and late to mature. This difference echoes one common throughout nature: Disturbed, fluctuating, or rapidly changing environments are dominated by weedy, opportunistic species. The rats and cockroaches of the sea may cope well with climate shifts, while the tigers and elephants are left floundering as their habitats shrink around them.

One reason why animals must move as the climate changes is that their physiology is tuned to temperature. Enzymes that regulate bodily processes work best within a relatively narrow range of temperatures, and species suffer when those limits are breached. But there are other reasons. Species need food and habitat, and these are undergoing their own moves. In the North Sea, for example, many kinds of microscopic plankton that drift in open water have shifted north over the last forty years.[7] Northern zooplankton communities are dominated by a copepod called *Calanus finmarchicus*. This diminutive relative of crab and shrimp is fat and juicy by copepod standards, and reaches a tenth to a sixth of an inch long. Southern zooplankton are dominated by a more puny species, *Calanus helgolandicus*, that measures just a thirteenth to a tenth of an inch. This difference may not seem like much, but to larval cod it is crucial. They thrive on fat *finmarchicus* but starve on scrawny *helgolandicus*. The northward march of zooplankton has been linked to poor replenishment of North Sea cod, a fact that has compounded the decline from over-fishing.

It might seem a simple matter to shift with one's food sources, but most species are driven by another urge—the need to find mates and reproduce. Over thousands of years many have developed intricate life cycles in which they migrate between predictable feeding and spawning areas, tracking patterns in food availability to maximize survival. Scientists playing with equations have discovered that group phenomena like gathering to spawn appear to respond very slowly to shifts in prey availability.[8] Instead of following their food, many fish continue to gather in the same places in order to fulfill their primordial compulsion. A gap thus opens between what they feel compelled to do by their instincts and what they should do to meet their needs. This underscores a point that applies to many of the changes underway in the oceans today: While species can track creeping shifts in conditions, rapid change could overwhelm them.

There is another way for species to stay within their comfort zone: They can go deeper. Water cools with depth, and at the present rate of warming they will soon have to go down as much as eleven feet per year.[9] Some North Sea fish in the survey were found not only to head north but to go deeper. But light falters quickly as you go down. Herbivores will find little to eat beneath the sunlit surface layer, and those that prefer hard bottoms will struggle to find a home in the muddy expanses of the deep.

Warming seas will change marine habitats as the species that build them are killed off or compelled to move. Creatures like oysters and mussels, mangrove trees and salt-marsh grasses, seaweeds and corals are ecological engineers. They develop slowly and take a long time to establish themselves in new terrain, and they are likely to lag far behind the needs of their inhabitants to find more agreeable places to live. The warming seas have already had a profound impact on one such habitat: coral reefs. Shallow-water coral reefs, confined to tropical seas, are the richest of all marine habitats. To drift with a gentle current over a coral reef is one of life's great thrills. The world that unfolds below is crowded with bright dappling colors in constant motion. It feels like gazing at a Monet painting in which every brushstroke has come to life and is swimming around the canvas.

Warm water corals live on a knife edge; wherever they occur, they adapt to local temperatures. If temperatures rise more than a degree or two above the typical maximum, they bleach. Bleaching happens when there is a breakdown in the relationship between corals and the microscopic plants that live within their tissues and provide them with much of their food. At higher temperatures these microscopic plants, or zooxanthellae, as they are called, begin to harm their hosts, at which point the corals kick them out. Since it is the zooxanthellae that give corals their color, the tissue becomes transparent and colonies turn deathly white, as the chalky skeleton beneath is revealed. Bleached corals are starving; they die within weeks unless temperatures revert to normal.

In 1998, sea temperatures soared across the tropics. In some places, like the Seychelles and Maldives, entire reefs turned white, and almost all the corals were killed. If this had been an isolated event it would have been bad enough, as corals can take decades to recover. But such bleaching events have become common. Scientists are now predicting that corals will be exposed to temperature spikes of the same magnitude every year by the middle of this century.[10] If corals cannot adapt, whole reefs will die, taking with them a host of species that call them home. Adaptation is clearly possible. Some of the world's most northerly reefs are found in the seas off Kuwait, where temperatures swing wildly. When winter winds blow hard from the desert, water temperatures plunge as low as 54°F. In the dead calm of stifling summer they top 95°F. Despite these tough conditions, corals have built reefs large enough to support islands. Fast adaptation will be crucial to reef survival, but the prospects are not good. The rate at which species evolve depends heavily on generation time. Species with high turnover rates evolve rapidly (think of flu viruses), while long-lived species will struggle. Like the ancient trees of the High Sierras in the United States, some corals live for hundreds of years, while a few reach a thousand or more.

The Galápagos Islands offer a harsh warning of what could be ahead. In late 1982 a powerful weather phenomenon called El Niño triggered a sharp rise in water temperatures that lasted five months. By the end, Galápagos reefs were all but dead. Thirty years later they still haven't recovered, and structures built over thousands of years have now crumbled to rubble and dust. A few years ago I swam through the remains of what had once been a glorious reef there. It was a scene of devastation. The seabed was strewn with chalky fragments, like a field of bones in the aftermath of massacre. Here and there a mound rose above the wreckage where once some mighty coral had stood. It had survived centuries of lesser temperature spikes, only to die in this one. A few young corals struggled to gain a foothold but were quickly overcome by black-spined sea urchins that swarmed over them like ants on a carcass. What that experience and others like it tell us is that today's bleaching events are unprecedented on millennial timescales.

It is hard to get too animated about a few degrees from our warm-bodied mammalian perspective (although there are ample reasons for us to be anxious as well). But if you live at your thermal limits, as many species do, even this could be intolerable. Warming seas will trigger a global diaspora that will lead to a massive reorganization of life, as long-separated species invade each other's ranges. There will be a march toward the poles, as we have already seen for fish and plankton in the North Atlantic. These invaders will alter polar ecosystems in profound and unpredictable ways. But where do you go if you already live at the poles and prefer your habitat with ice? The answer, some scientists predict, will be a surge of extinctions among polar residents, as their waters heat and ice melts away. Perhaps they will find refuge in deeper layers of the sea, but for some that is not an option. This is why the polar bear and walrus, icons of the realm of floating ice, are now on the World Wildlife Fund's top ten list of species in peril.

This global diaspora is not all about loss. Some species will do well, and some countries may prosper from the changes. Others will suffer. In an elaborate thought experiment, William Cheung, from the University of East Anglia, and his colleagues created a global ocean model to try to predict the consequences of life's reorganization between now and 2055.[11] As they cranked up the heat, species shifted poleward. Those that could not move, for want of suitable habitat or because the way was blocked, declined, in some cases to the point of extinction. The movement of all of these species left some places with fewer inhabitants than before, mostly in the tropics, while the poles suddenly became more populous. There will be a knock-on effect on fish catches. Low-latitude nations like Indonesia, Senegal, and Nigeria could see catches dwindle by 10 percent or 20 percent, compounding the loss of agricultural productivity that terrestrial climate models predict. Fish production in the United States will have to shift, as most of the lower continental states could lose 10 percent of their catches, while Alaska's could expand by a quarter.

Although there are many reasons to worry about the impact of rising temperatures on the sea, the good news is that life can and will move, often quickly. The eggs and larvae of many marine species can travel several tens of miles to more than one hundred and we have erected few barriers to prevent their dispersal. The seas off Greenland experienced a warming spell that lasted between the 1920s and the 1960s. Cod, haddock, and herring expanded their ranges north very rapidly; cod spread more than six hundred miles poleward in less than twenty years and kick-started a fishery that at its peak landed 440,000 tons of fish a year.[12] A host of other species shifted alongside these commercial fish, including bottom dwellers and marine mammals.

Cheung's predictions hold many uncertainties. They assume that the production of food by ocean plants either will not change or might even increase a little. For reasons I will return to in later chapters, how these phytoplankton react as the seas change is largely guesswork right now. There are equally good arguments that productivity will fall. If so, a few countries may break even, but most will see significant reductions in their fisheries.

By now it will be clear where my views lie on climate change. Like thousands of other scientists, I am convinced it is underway. There is abundant evidence of such variety and strength that those who reject it are increasingly isolated. It is important to remain skeptical about every new study and to weigh the solidity of its methods and the rigor of its sampling. Skepticism is the stuff of science. Old theories are overthrown when new data and new ideas provide better explanations for what we see around us. But then there is denial. It is easy to spot those in denial. They use evidence selectively, acknowledging only those findings that support their prejudices and rejecting, or simply ignoring, findings that don't. As evidence mounts against their ideas, their explanations for the discrepancies become ever more tortuous and far-fetched.

Those who claim there is some elaborate conspiracy among scientists to delude politicians and the public should ask themselves a simple question: Why? What could anyone possibly gain from such a hoax? If the answer to "why" is not sufficient, there is the question of how. Science is a highly competitive pursuit. There are no prizes for coming in second. Scientists constantly put each other's ideas to the test. A hypothesis is cast aside as soon as it is found wanting or is reworked on the forge of new evidence. No conspiracy could resist such relentless scrutiny for long.

Much of the controversy around climate change stems from the variability of the Earth's climate over long timescales. We live on a dynamic planet where yearly climatic variations are overlain by multi-year or decadal cycles, known as oscillations. Cores taken through deep sea sediments in California's Santa Barbara Channel show how marine life tracks these changes. In the aftermath of every frenzied predatory attack on a shoal of sardines or anchovies, a rain of glittering scales makes its way down to the seafloor. Scales trapped in sediments allow us to trace the ups and downs of shoaling fish through thousands of years as they respond to climate events like El Niño and the swings of a cycle known to oceanographers as the Pacific Decadal Oscillation.

Modern measurements have been made on timescales of a few decades to a century, so it can be hard to see the signal for the noise. Much of the change to date may fall within the range of natural variation if you extend the timeline out far enough, but the drivers of our current change—greenhouse gas concentrations—have moved far above historic levels. If you are pushing a broken-down car, you have to apply a lot of force before it starts to move. The force of humanity's push on the climate system has been building for over a century, and the consequences are beginning to become obvious.

Life is on the move. This is nothing new. Records locked in ice, seabed sediments, and rocks show that change is normal for life on Earth. As conditions shift, new habitats and ecosystems constructed by better adapted species will re-form in the place of old. It is not change that we must worry about but its speed. If conditions alter too quickly they could exceed the pace at which ecosystems can respond. It could take centuries for life to adjust, and we can't afford to wait that long. In a world of growing human need, that is a big concern.

CHAPTER 6

Rising Tides

As I strolled down Ocean Drive in Miami Beach one evening a few years ago, I could almost imagine myself transported back to the 1930s. The party atmosphere from the nightclubs and bars mingled with the smell of seaweed and the distant boom of breaking waves. Today the Art Deco splendor of South Beach is overshadowed by towering hotels and apartments that line the seafront to the north. It is hard to imagine the place any other way, but in the nineteenth century Miami Beach was little more than a windblown barrier island backed by dense mangrove swamps. Its potential was first appreciated by Henry Lum, a veteran of the California gold rush of 1849 who had a keen nose for a bargain. He bought the land for twenty-five cents an acre and planted coconut palms. But Miami Beach was not really born until 1915, when a bridge was laid across Biscayne Bay to connect it with Miami.

Lum fell in love, as many other Americans have after him, with the wild beauty of the barrier islands. These low-lying islands are made of wave-washed sand and run, on and off, along the East Coast, from Long Island to Miami. The twentieth century saw a coastwide building boom that covered many of them with houses and hotels in places like Kitty Hawk, Surf City, and Wrightsville Beach. But after the glory days of speculation and development, many have begun to fear that the ocean is about to reclaim these lands.

We are reaching the end of an era of around five thousand years of unbroken stability in the volume of water in our oceans. This followed a period of rapid increase that came after the last ice age, when melting ice sheets pushed the height of global seas upward by 420 feet. Some coasts have seen levels rise or fall since then, due to coastal uplift or subsidence (sinking), but the overall amount of water in the oceans has been very stable.

Stable sea levels promote the formation of wetlands where rivers meet the sea. Rivers spread sediment with every flood and regularly discharge mud into coastal waters. Deltas, mudflats, marshes, and barrier islands form where the rate of new sediment deposition exceeds the rate of erosion and subsidence. The land inches out to sea when plants bind and stabilize the sediment. People have given wetland formation periodic spurts over the course of history, when they have cleared land to grow crops and build cities.[1] Denuded lands shed their topsoils into valleys and rivers far more rapidly than those bound by forests or other vegetation. Soil erosion during the early Middle Ages in parts of Europe choked estuaries and built wetlands that stranded once busy ports far inland, such as Luni in northern Italy and Chester in England. In North America, a pulse of wetland creation followed forest clearing for agricultural purposes between the seventeenth and nineteenth centuries.

Today rivers are again running thick with soil, but less of it reaches the coast than in the past. Dams, which have been constructed along almost all of the world's large rivers and countless smaller ones, hold back mud that once would have reached the sea. Crop irrigation in arid regions often diverts water and sediment that would have replenished wetlands. Australia's legendary Murray and Darling rivers, for instance, lose so much water to irrigation that they now reach the coast as mere trickles.

The balance shifts in favor of the sea when sediment supplies are cut off, and wetlands retreat. Deltas naturally subside over time, as sediments settle and the earth sinks below them in a process called isostatic adjustment. In many places today, from Miami Beach and Venice to Tokyo and New Orleans, sea-level rises have combined with coastal subsidence to threaten life and property.

The Earth is ruled by certain immutable physical principles. One such law is that warm substances take up more space than cold ones.[2] To us it hardly matters if soil and rock expands as our planet heats up. But it matters a great deal when that heat expands the oceans and the sea begins to reclaim our most valuable coastal land. Ancient Roman fishponds, built at sea level two thousand years ago, have helped define when the rise in sea level began in earnest.[3] At the time they were constructed the Mediterranean was five inches lower than today. It didn't begin to rise until about one hundred years ago, when global temperatures started to respond to greenhouse gas emissions from the Industrial Revolution. The process has sped up considerably since then. Sea levels rose by approximately eight inches between 1870 and 2000.[4] We know this from thousands of tide gauges across the world. Since 1993 those recordings have been supplemented by satellite observations that measure sea-level fluctuations with great accuracy. Averaged over 130 years, the rise comes to one fifteenth of an inch per year, but in the last twenty years the rate has accelerated, and it now tops an eighth of an inch per year.[5] Three quarters of the rise in sea levels since 1900 has been caused by global warming from carbon dioxide emissions. Water expansion from sea-surface warming contributed a quarter of the rise since 1960, and 30 percent from 1993 to 2009. The rest came from melting glaciers and ice caps, and from groundwater pumping for irrigation.[6]

Sea levels would have risen a fiftieth of an inch per year faster if we had not embarked on wholesale dam building in the mid–twentieth century, which locked up freshwater on land. This beneficial effect of dams has been rather undermined, though, since they have also blocked the passage of vast quantities of mud, an effect that has hastened erosion around river deltas from the Yangtze to the Mississippi. Parts of the Nile Delta have retreated by as much as five hundred feet per year since the Aswan Dam was built in 1964.[7]

So far only the upper layers of the sea have begun to warm up. As heat penetrates deeper, the sea will expand, and sea levels will continue to rise. In 2007 the Intergovernmental Panel on Climate Change (IPCC) predicted that sea levels could rise by another seven to twenty-four inches by 2100, depending on how fast we gain control over the drivers of climate change. Their forecasts combine thermal expansion of seawater with ice loss from mountain glaciers and the Greenland and Antarctic ice sheets. But their figures seem increasingly at odds with the recent acceleration: Sea levels have risen faster than the most rapid rate of change they predicted a decade ago.

New research from the ice caps of Greenland and Antarctica suggest that we are close to, and perhaps past, a tipping point for the rapid melt of land-based ice, which will become the major source of sea-level rise in future. If the Greenland ice sheets were to thaw in their entirety, they would add twenty feet to the height of global seas and trigger mass human exodus from low-lying coasts and cities. The thaw of the West Antarctic Ice Sheet would add another ten feet.[8] A twenty-foot rise would wipe out most of Florida from just north of Miami.[9] It would obliterate the Mississippi delta and drown a third of New York City. It would flood much of London and Hamburg and turn Lagos into a lagoon and Bangladesh into a swamp.

Sea-level records preserved in fossil coral reefs in Central America suggest that we have crossed this tipping point at least once before, 121,000 years ago, when sea levels rose by ten feet within the space of a hundred years (ten times the present rate).[10] Some argue that this period is an analog for what a warmed world might be like. Temperatures then averaged 4°F above today's, and the seas leveled off twenty feet higher than they are now, drowning deltas and plains in many of the low-lying coastal areas that are now densely populated.[11] Of course, such extreme levels would not be reached instantaneously. It will take time for sea levels to respond to a warming planet. A range of estimates made since the 2007 IPCC assessment suggest an upper limit to the rise rate of about six and a half feet in a hundred years.[12]

The melting of Arctic sea ice is perhaps the most striking manifestation of global warming, but even if it were to disappear in its entirety, Arctic ice would not add to sea level, because it floats in the sea. (If you aren't convinced, add an ice cube to a glass of water and you will see how the level remains the same as it melts.) It is the ice sheets sitting on land, mainly over Greenland and Antarctica, that we need to worry about. They have proven less durable than the IPCC had hoped. Ice heated from the surface melts and forms pools and streams. Runnels turn into rivulets that coalesce into blue rivers that course across the frozen surface until they find a crevasse. They then plunge in muted waterfalls that would be magnificent if only we could see them. Beneath the ice sheet, this running water lubricates the stuttering creep of ice across rock and accelerates its slide toward the sea.

Ice caps are often held back at their margins by ridges of undersea rock on which the sheet has grounded. Global warming is gnawing away these barriers, too. Tongues of warm water (at least by polar standards) are creeping under the ice and freeing them from their anchors. The West Antarctic ice sheet has already been loosed from one key holdfast.[13] Measurements from planes and satellites show the seaward edges of both Greenland and Antarctic ice sheets slipping ever faster into the sea. Ice sheet thaw seems to be accelerating through positive feedbacks. Blue pools and rivers of melted water at the surface mean that less of the sun's heat is reflected, and so more ice melts. It is hard to predict the future contribution of ice sheets to sea-level rise because climate change has also increased snowfall over their interior regions, helping counteract losses from the margins. How the balance between ice sheet growth and melt develops over the next century will be critical to countless people who live on low-lying coasts.

Another positive proof of global warming is newly emerging in the Arctic Ocean. The seabed has begun to spew forth bubbles of the greenhouse gas methane on a colossal scale as permafrost melts at the bottom of the ocean where ice sheets are in retreat. Russian scientists announced the findings at a San Francisco conference in December 2011 where the leader of the team, Professor Igor Semiletov, described fountains of methane more than half a mile wide.[14] He said they saw hundreds of such fountains but estimated there could be thousands more. So methane emissions from the seabed are now adding to those from melting permafrost on land to accelerate global warming.

According to recent estimates, even the more modest sea-level rise predicted by the IPCC, a rise of up to twenty-four inches by 2100, could displace hundreds of millions of people and inundate four hundred thousand square miles of the world's agricultural land and coastal cities. Ten percent of the world population lives near coasts on land less than thirty-three feet above the present sea level. Eleven of the world's sixteen megacities, each home to more than fifteen million people, are built on coasts or estuaries: Tokyo, Guangzhou, Shanghai, Mumbai, New York, Manila, Jakarta, Los Angeles, Karachi, Osaka, and Kolkata. That number will grow as the world's population increases, with a further four coastal megacities predicted by 2025: Buenos Aires, Dhaka, Istanbul, and Rio de Janeiro.[15] There are three possible ways we can respond when rising tides threaten: retreat, adapt, or defend. Defense is our usual reaction to threats from the sea.

For centuries Holland has been one of the most densely populated countries in the world. Its flat lands and low elevation mean the Dutch have always been intimate with the North Sea. Storm surges would periodically push the sea far inland, drowning farmland and ruining crops. It is hardly surprising then that the Dutch became masters long ago of the science of coastal defense; they constructed elaborate dikes as early as the fifteenth century to both hold back the sea and increase fertile land. Contemporary illustrations show complex protection schemes of seawalls and groins (walls or fences built at right angles to the shore to intercept the movement of sand) that reveal a detailed understanding of the forces that build and destroy coasts. Dutch engineers were in demand across Europe by the seventeenth century, and they exported the technologies of coastal protection, or "hardening," that remain in use across the world.

In 1953, the Dutch learned through a harsh lesson that sea level was only half of their worries. An intense storm surge from the north combined with an extreme high tide to devastate low-lying areas, killing eighteen hundred people and destroying ten thousand houses. To prevent a repeat they embarked on the world's most ambitious program of coastal defense, raising dikes and levees to close off the North Sea from the Rhine/Meuse delta. Today a quarter of Holland lies below sea level and owes its existence to these defenses. But recent sea-level rises threatens once again, and their coastal defenses have come under renewed scrutiny.[16] The new plans foresee a sea-level rise of up to fifty-one inches by century's end, and work is underway to upgrade their coastal defenses to meet the threat.

The Dutch are better equipped than just about anyone to protect their coast. The problem is, sea-level rises pose urgent threats to many developing countries that don't have half the resources. Nowhere is the battle to protect land more pressing than on low-lying oceanic islands such as the Maldives, Phoenix Islands, and Tuvalu.

Flying into the Maldives's international airport is like landing on the ocean itself. The runway seems such an insubstantial sliver of tarmac amid the endless sea. The capital city of Malé is made up of a clutter of concrete buildings that completely conceal the coral island upon which it perches. A rampart of bleak concrete tetrapods defends the city from the waves of the Indian Ocean. Like Malé, the airport is edged in concrete, a stark reminder that the sea that gave birth to these islands could soon claim them back.

The Maldives is a waterworld. None of its twelve hundred islands rises more than a yard or two above the sea, and each sits atop one of the countless reefs that make up this nation of fish and coral. Seen from the window of a seaplane the islands look like a scatter of emeralds set within white bands of coral sand. Collectively they cover less than one thirtieth of 1 percent of the country. The rest is ocean.

Not surprisingly, Maldivians seem as much at home on their boats as on land. The captain who ferried me from the resort to the dive site on my last visit looked every bit the sea gypsy as he coaxed his boat through the hazard-filled coral labyrinth, foot on the tiller and a cell phone pressed hard to one ear in an effort to hear above the roar of the engine.

The reefs in the Maldives have grown over the course of thousands of years on top of platforms planed flat by the sea during the last ice age. These platforms, now 130 feet to 160 feet below sea level, stretch north to south for five hundred miles in a chain of atolls that follows an ancient wrinkle in the Indian Ocean seabed. Corals grow by absorbing carbonate, the main ingredient of chalk, from seawater. They crystallize carbonate into hard skeletons that form the backbone of reef growth. Swimming across a Maldivian reef crest in 2010 I came upon a coral head freshly smashed open by waves from a passing squall. From the outside, stony corals like this one give an impression of immovable solidity, but inside the skeleton was riddled with tubes and cavities that had been bored and dissolved by a host of creatures that sought refuge from the predatory world outside. This hidden community represents the other side of the reef balance: the erosional yin to the depositional yang. Reefs only grow where deposition outpaces erosion.

The Maldives and thousands of other oceanic islands scattered throughout the tropics exist only because corals have flourished sufficiently to win the race against erosion. Climate change is tipping the balance the other way. As I mentioned earlier, Maldivian reefs bleached badly when temperatures soared in 1998. Corals were on the ascendancy again when I visited twelve years later, but like the regrowth of a forest after a wildfire, there were fewer species than before, and most were shrubby branching forms that grew quickly. Most of the giants of old, corals that might have begun life when Europeans were just setting forth on voyages of global discovery, were dead.

Healthy coral reefs are naturally self-repairing breakwaters, better at protecting coasts than any concrete wall we can throw up to defend cities, resorts, or agricultural land, and far cheaper. The combination of faltering coral growth and rising seas means that nations whose fortunes are founded on coral have begun to contemplate the unthinkable. Tuvalu in the South Pacific has made an arrangement to relocate its people to New Zealand when sea levels rise beyond their capacity to adapt. Already the highest tides wash over islands in Tuvalu that were dry a generation ago.

At least Maldivians can take comfort from the fact that their nation rests on a solid foundation.[17] Rapid relative sea-level rise in places where coasts are subsiding has been a reality for decades, even centuries. For them it is not the eight inches by which sea levels have risen in the last century that matter; it is the amount by which land has sunk. There are several reasons why coasts sink. During the last ice age, vast ice sheets up to ten thousand feet thick lay across much of North America, Europe, and Asia. They pressed the Earth's crust deeper into the mantle, creating depressions. A bit like the hollow on a cushion that is left when you stand up, these indents have gradually disappeared as the Earth has returned to its former shape. Since it is a geological process it is monumentally slow, and it is still underway today, eighteen thousand years after the peak of the last ice age. Northern Scotland, for example, is still rising, while southern England, which was not covered by ice then, is sinking. This helps explain the difference in the amount of hard coastal defenses between these countries. While 8 percent of Scotland's coast is protected by seawalls, rock armor, and the like, half of England's is defended.[18]

Another major contributor to subsidence or sinking is sediment supply. Where supplies are cut off behind dams or levees, coastal marshes and mudflats are starved of the material they need to build upward as sediments settle beneath. In the southern United States, for instance, the Mississippi Delta now loses twenty square miles of land each year as a result of subsidence and sea-level rise. In years of heavy snowfall the Colorado River used to disgorge upward of 220 million tons of mud into its delta at the northern end of the Gulf of California. After the Hoover Dam was built in the 1930s, supplies of mud were choked off, and the river today carries a few million tons at most to replenish delta wetlands. There are now more than forty-five thousand large dams around the world, which together trap a quarter to a third of all the suspended sediment carried by rivers.[19]

Sediment supplies can also be cut off by structures that interfere with coastal currents. Some coasts consist of soft rocks or soil and naturally shed sediments to the sea. That sand and mud is carried by waves and currents to other places, where it settles into barrier islands, lagoons, sandy beaches, and dunes. Such areas are often stabilized by plants such as mangroves, sea grass, and salt-marsh grasses that bind mud together. So coastal development can have far-reaching and un-intended consequences, as it disrupts the natural flow of tides and currents, leading to erosion and loss of sandbanks out at sea.

Coasts also subside where we pump water or oil from beneath the ground. Much of Houston in Texas lies barely three feet above sea level today, having sunk from the removal of both groundwater and oil. Some districts have sunk ten feet since the 1930s, and in parts of the city recent sinking rates exceed two inches a year.[20] It is on the front line of cities threatened by sea-level rise, although it doesn't make the top ten global list of cities with uncertain futures. That list is topped by Miami, while New York/Newark, and New Orleans are second and third. Miami's art deco district survives today only because of regular applications of dredged offshore sand to hold back the sea. The lower floors of high-rise buildings to the north are regu-larly splashed by salt spray from Atlantic storms.

It is inconceivable that we would abandon to the ocean some of the world's most populous cities—New York or Nagoya, Japan, for example. Doubtless we will girdle them with walls to hold back the ocean. London has been defended since 1984 by a barrier across the River Thames whose sail-like gates swing shut when high tides com-bine with strong onshore winds to lift the sea to dangerous levels. The record of how many times the gates have been closed tells its own story. They were shut once a year for their first three years of opera-tion, but now typically close more than twenty times a year.

New Orleans gave us a striking example of the risks faced when we choose defense over retreat. If defenses are breached, as New Orleans' levees were during Hurricane Katrina, the ensuing floods are catastrophic. Bangkok has sunk several meters in the last century due to groundwater pumping in the Chao Phraya River delta. Much of the city had to be evacuated for the first time in late 2011 as floodwaters combined with high tides to overwhelm the ring of dikes thrown up around the city.

Defense becomes increasingly expensive and risky as sea levels rise. If the thaw of ice sheets adds more than three feet to sea levels in the coming century, we will have to abandon parts of many cities and coasts to the sea. Three feet doesn't sound like a big deal, but any seawall would have to fend off the worst the seas could throw at them, and the extremes are expected to rise faster than the average.[21] New York, for example, may soon find itself within the Atlantic hurricane belt as warmer seas sustain hurricanes into more northerly latitudes. New Yorkers got a shock in 2011 when lower Manhattan had to be evacuated due to flooding from Tropical Storm Irene. This storm may be a harbinger of things to come, as there is now broad agreement that hurricanes will intensify over the coming decades.[22]

The great river deltas of the world are humanity's breadbasket. They support production of staple crops, such as wheat, rice, and corn, and are highly vulnerable even to modest rises in sea level. A survey of forty of the planet's largest deltas—areas that collectively receive over 40 percent of the runoff from the world's landmass—showed rates of effective rises in their sea levels of up to half an inch per year, nearly four times the rate at which the sea is rising.[23] The higher rate is due to land subsidence in places where a supply of sustaining sediments has been blocked by dams or there has been groundwater or oil abstraction—or a combination of the two. This has led to a rapid loss of land around the world's great deltas, like those of the Nile, Ganges, Mekong, and Mississippi, and so ends a long period of delta growth that followed the stabilization of the sea level seven thousand years ago.

There is little doubt that sea-level rises will provoke mass migration in the coming century as tens of thousands of square miles of
delta lands are at risk of flooding and submergence.[24] Relative sea-
level rise, accelerated by subsidence, could displace tens of millions of
people from these lands and would cut off agricultural production
from some of the world's most fertile soils just at a time when we will
have the greatest need for it because of population growth.[25]

The human cost of climate change in deltas is becoming more
obvious with each passing year. Take Bangladesh, for example, a land
of floodplains and swamps better known to the world as a land of
natural disasters. Formed at the confluence of two immense rivers,
the Ganges and Brahmaputra, more than a third of the country
barely rises above water. The people who inhabit this aquatic landscape live and often die by water. Monsoon rains and Himalayan
snowmelt nourish their fields and rice paddies, but water the giver
also takes away, as typhoons or deluges periodically bring catastrophe. It is hard to argue with the logic of dikes, levees, and seawalls
when faced with images of desperate people stranded on the roofs of
their houses in the midst of lakes that stretch from horizon to horizon. Yet, ironically, Bangladeshis would become more vulnerable to
inundation from the sea if they build embankments to defend themselves from river-borne floods, as their now sediment-starved lands
would sink faster. We cannot sustain life and society for long on
coastal floodplains where those life-giving floods have been quenched.

When we harden coasts with engineered defenses, rises in sea
level will narrow and steepen shorelines and increase the erosive force
of waves. Some of the most productive marine habitats will therefore
suffer. The shifting matrix of wetlands that separates land and sea on
low-lying coasts will dwindle as it is squeezed between elevated sea
and seawalls. Estuarine mudflats, mangroves, and salt marshes will
find themselves in retreat with nowhere to go, their natural adaptability blocked by our handiwork. The fates of thousands of species
that depend on them, such as those of the wading birds which use
coastal wetlands as migration refueling sites and winter habitat, will

be sealed depending on what happens to them. The Humber Estuary, not far from my home, supports over 150,000 such waders in winter. They enliven the mudflats and lift the spirits as they twist and turn in shape-shifting flocks across the evening sky. A hundred or so species of wetland birds are already at risk of extinction worldwide as a result of wetland loss. Sea-level rises do not bode well for them.

Like coral reefs, wetlands and dunes are living barriers between land and sea. They can be battered and breached by tempests, but the living matrix renews itself with time. I once sat out a terrible hurricane in the U.S. Virgin Islands (in the toilet, whose block-built walls had a comforting solidity!). The wind raged all night with a dreadful roar that left my ears ringing. The ground shook, the house trembled, and through the darkness came muffled crashes as our neighbors' homes disintegrated. When dawn finally came, I emerged to an island whose verdant green had been stripped back to the grays and browns of shattered branches and twisted trunks. Along the shore, mangrove trees stood like skeletons, their crowns smashed by wind and wave. But still their gnarled and knotted trunks, rooted deep in the mud, held the line against the sea.

Mangroves in Indonesia and Sri Lanka likewise seem to have helped dissipate the destructive force of the terrible tsunami of 2004.[26] (The correlation between reduced wave impacts and the presence of mangroves has been challenged, because mangroves colonize sheltered shores that were already likely to be less affected by the tsunami. But where would you prefer to be if a tsunami struck: on a shore fronted by open sea or behind a dense natural stand of mangrove trees?) The incredible debris flows pushed by the Japanese tsunami of 2011 showed that even mangroves will do nothing to stop the worst of the waves. But they could save many lives in places at a greater distance from the epicenter of destruction.

The protective function of wetlands has not been properly appreciated, nor have their many other values, such as filtration and purification of runoff or their role as nurseries for commercially important fishery animals. Instead, they have been, and still often are, perceived as mosquito- and disease-ridden swamp or wastelands. As a result wetlands have been drained, filled, and cleared for agricultural land, marinas, homes, shrimp ponds, airports, malls—you name it. Pretty much any use that can be made of flat coastal land has been.[27] Developers in search of prime seafront see wetlands as easy wins in planning applications. So the wetland expansion that has prevailed since sea levels stabilized has in most places been reversed. Around 30 percent of the world's mangrove forests have been cleared since the middle of the twentieth century, and some countries, such as the Philippines, have destroyed half to three quarters of their mangroves.[28] Nearly a third of the global tally of sea grass beds has gone since the late nineteenth century, and the rate of decline has increased to 7 percent per year since 1990.[29] A similar tale of loss can be told for salt marshes.

There is a physical relationship between sea level and erosion on open sandy coasts that says the shore will retreat two orders of magnitude more than any rise in sea level. A barrier island might lose six hundred feet to the eight-inch rise in sea level we have experienced in the last century, but it could lose three thousand feet to the thirty-nine-inch rise that is predicted in the coming century. This will not bring comfort to those who live on the coastal barrier islands that fringe the eastern and southern coasts of the United States. The American geologists Orrin Pilkey and Rob Young predict that it will become impossible to sustain communities on these barrier islands into the twenty-second century if sea levels rise as predicted.[30] Seafront streets and marinas beloved of vacationers to Nantucket, Long Island, or Martha's Vineyard could be washed away. Seafront towns like Ocean City and Atlantic Beach will have to defend themselves or disappear, while to the south, the sea will probably erase all homes and infrastructure from Cape Hatteras and Sea Island, Georgia.

On less dynamic coasts there is an alternative to expensive construction or abandonment, and it goes under the name of "soft-engineering": nature, in other words. Like coral reefs, wetlands are self-repairing breakwaters. Coastal engineers are just beginning to realize their value even as their areas decline. The complex matrix of roots and stems traps and binds sediment and stabilizes the coast. If the health of these habitats is maintained and the input of sediment is sufficient, they can keep pace with sea-level rises by trapping and consolidating a layer of mud and peat whose thickness will increase with the sea level. Vigorous wetlands are like a wall that builds itself higher as the need arises. Replanting mangroves in Vietnam after they were destroyed by herbicides during the war reduced the cost of dike repair from $7 million per year to just over $1 million.[31] Likewise, where storm surges are a problem, broad and dense salt marshes require little additional coastal protection, while narrow, scrappy ones need to be backed by dikes.[32]

Future wetland growth is expected to get a boost from higher carbon dioxide levels in the atmosphere, since plants use this gas for photosynthesis. Sea grasses and mangroves are among only a handful of marine species that could benefit from higher carbon dioxide concentrations. The new Dutch coastal protection plan calls for much more soft engineering of this kind, even to the point of removing existing dikes to promote the restoration of estuarine habitats and tidal regimes.

Scientists forecast that the sea-level rises of up to twenty-four inches predicted by the IPCC by this century's end will lead to the loss of another third of our coastal wetlands. This loss is comparable to or less than the direct destruction that has already been wrought by conversion of wetlands to other uses. Given their growing value in ameliorating impacts of sea-level rises, we need to urgently rethink our attitudes toward wetlands.

CHAPTER 7

Corrosive Seas

Wrapped in the vapors of early morning, the island of Ischia appears to float atop a sea of perfect aquamarine at the entrance to Italy's Bay of Naples. Its intoxicating beauty and therapeutic thermal springs attract over six million tourists a year. The Victorian war correspondent Sir William Russell captured well the contradiction between the island's tranquil appearance and its volcanic birthright:

> If one could have been aware of the terrible forces which were at work beneath that smiling surface, how delusive would the whole of that bright pageant—the charming little villas nestling in their gardens, the country houses white as snow, with their green jalousies, and the small spires of chapels piercing the mass of foliage—have appeared.[1]

The Castello Aragonese grips the top of a craggy rock at the east end of the island, its ancient ramparts blending into gray cliffs that plunge to the sea. Volcanic gases—mainly carbon dioxide—bubble up from underwater springs under these battlements. Mollusks living close to these springs have paper-thin shells, so weak they can be crushed between thumb and finger. These champagne seas, with their high concentration of carbon dioxide, are a worrying portent of the difficulties that could soon face corals, lobsters, clams, oysters, and other shelled forms of life all across our planet.[2]

Ocean chemistry and that of our atmosphere are inextricably linked. Gases dissolved in seawater are in equilibrium with those in the air, which means that as concentrations of carbon dioxide go up in the atmosphere, so does the carbon dioxide dissolved in the sea. When carbon dioxide dissolves in seawater, it produces carbonic acid (which is where fizzy drinks get their tang). This liberates bicarbonate and hydrogen ions and decreases carbonate ions (a union of carbon and oxygen and a key ingredient of chalk).[3] The hydrogen ions make the sea more acidic, and the decrease in carbonate ions is bad news for anyone or anything that depend for a living on making chalky shells and skeletons.

The oceans have absorbed around 30 percent of the carbon dioxide released by human activity since preindustrial times from, primarily, burning fossil fuels, converting forests and swamps into cities and agriculture, and cement production.[4] Over that period the pH of seawater, a measure of its acidity, has fallen by 0.1 units. Most of this drop has taken place in the last few decades. Since pH is measured on a logarithmic scale in which one unit equals a tenfold change in acidity/alkalinity, this means the acidity has risen by 30 percent. If carbon dioxide emissions are not curtailed, acidity is expected to rise 150 percent by 2050,[5] the fastest rate of increase at any time in at least the last twenty million years, and probably as long as sixty-five million years, which takes us back to the age of dinosaurs.[6] As Carol Turley, an expert on ocean acidification from the Plymouth Marine Laboratory put it to me, "The present increase in ocean acidity is not just unprecedented in our lifetimes, it is a rare event in the history of the planet."

With a pH that currently stands at 8.1, today's oceans are actually slightly alkaline. Dissolving carbon dioxide is bringing their pH closer to neutrality (defined as the pH of pure water, which is 7.0). While our great-grandchildren won't exactly get a chemical peel from a paddle in the sea, animals with shells might. By way of comparison, black coffee is several hundred times more acid than the sea, and a soda drink over a thousand times more acid than the pH of 7.6 that could be reached in some places by 2100.[7] But even this small change could completely transform life in the sea.

Acidification is en route to becoming one of humankind's most serious impacts on the sea, yet it has inexplicably been overlooked until very recently. One of the first marine biologists to realize just how bad more acidic oceans could be for marine life was Joanie Kleypas, an American expert on coral reefs. At a meeting to discuss climate change in 1998 she experienced one of those rare eureka moments. The sudden realization that coral reefs would, by the end of the twenty-first century, be bathed in water corrosive enough to destroy them was so overwhelming she excused herself and ran to the bathroom to be sick.

Despite decades of attention to rising levels of carbon dioxide in the atmosphere, the problem of ocean acidification was first highlighted only in 2003 in a study published in the journal *Nature* by Ken Caldeira and Michael Wickett, American scientists based at Stanford and the Lawrence Livermore Lab. They concluded that if we were reckless enough to burn all of the world's known fossil-fuel store it would result in the rise of the oceans' acidity over the next three hundred years to levels not seen in the last three hundred million, save from rare catastrophic events like the one that caused the mass extinction at the end of the Permian period.[8] This study was an epiphany for the scientific world. Since then scientists across the world have mobilized to study ocean acidification, and hundreds of papers have been published. They make sobering reading. Increased acidity is the last thing marine life needs given all of the other ways in which we are making oceans a tougher place for them to live.

Many marine plants and animals take up dissolved carbonate minerals from seawater and secrete it in the form of a calcium carbonate skeleton, often with small amounts of magnesium mixed in. Pteropods (tiny swimming snails), microscopic coccolithophores, and foraminifera do so among the plankton, while corals, snails, crustaceans (crabs, lobsters, shrimp), sea urchins, and coralline algae do so on the seabed, alongside filter-feeding clams, mussels, and oysters. As the pH of seawater falls, it takes extra energy for all of these forms of marine life to secrete carbonate structures. This is because acidification alters their internal chemical makeup, making it harder for them to crystallize carbonate out of solution. So when atmospheric carbon dioxide rises, corals and other calcifying organisms will produce less robust skeletons.

The industry of countless generations of corals over periods of hundreds of millennia has produced some of the world's most spectacular and diverse marine habitats. Coral reefs form vast geological structures, some of which, like Australia's Great Barrier Reef or the Maldives, can be seen from space. I gained a keen sense of the enormous solidity of coral reefs once when I stood in the calm waters of the lagoon of a Pacific coral atoll just inside the reef crest. Great ocean swells expended themselves impotently in angry foam against the coral buttress while calm water gently lapped my ankles. No castle could enjoy greater security from harm than is provided by these self-renewing breakwaters.

Just as Joanie Kleypas and her colleagues predicted in 1999, recent experimental findings suggest that coral reefs and other habitats built from calcium carbonate could cease to grow within our lifetimes. Scientists are suddenly waking up to the possibility that acidification, in combination with coral bleaching due to global warming, will cause the destruction of reefs worldwide. Indeed, the damage has already begun. The skeletons of corals on Australia's Great Barrier Reef have weakened measurably in the last twenty-five years and now contain 14 percent less carbonate by volume than they did before.[9] Since corals contain annual-growth rings, much like trees, changes in skeletal strength can be measured over time. The results suggest that recent weakening has been unprecedented in the last four hundred years. Experiments with deep sea corals from the Mediterranean indicate that they lay down half as much carbonate today as they did before the onset of the Industrial Revolution.[10] Ocean acidification has been dubbed "osteoporosis for reefs" because of this skeletal weakening.

But corals alone do not make a reef. Without the cement of coral-line algae, a group of calcareous seaweeds, to bind corals together into solid structures, reefs would not develop nearly so well. We see this in the deep sea where, far below the reach of light, coralline algae are absent (so are the algae that live harmoniously within coral tissues and boost their growth rates in shallow water). Deep-sea corals grow at glacial rates; at best, they build mounds of loose rubble that over thousands of years often reach only ten feet thick. Healthy shallow-water corals, by contrast, can grow by as much as eight inches per year. Coralline algae are especially susceptible to a fall in pH because they secrete a form of calcium carbonate that is rich in magnesium and is more soluble than the form of calcium carbonate produced by corals. In addition to the "glue" of coralline algae, reefs are also held together by carbonate cement that precipitates chemically from the water that trickles through the interstices of the reef. This form of cementation also weakens as carbonate saturation—the amount dis-solved in the water—falls.

Under normal conditions large regions of the oceans are "super-saturated" with carbonate, which means that it is relatively easy to crystallize it to form skeletal structures. Unprotected mineral carbon-ate structures cannot form or survive in seawater that is not saturated or supersaturated with carbonate ions. We know this because carbon-ate saturation changes with temperature and pressure. At low tem-peratures and high pressure, water dissolves more carbonate. As you go deeper in the sea, carbonate concentrations fall below saturation once you pass depths of around ten thousand to thirteen thousand feet deep in the tropics, shallowing to depths of six hundred or seven hundred feet in polar regions. Solid unprotected carbonate dissolves below this. These effects are enhanced by the increase in dissolved carbon dioxide with depth and low temperature. Cold water holds more carbon dioxide than warm, and the deep sea has more than the shallows because of respiration by the creatures that live there.[11]

As carbon dioxide levels in the sea rise, carbonate saturation will fall, and the depths at which carbonate dissolves will become shallower.

Recent estimates suggest that this horizon is rising by three feet to six feet per year in some places. So far, most carbon dioxide added by human activity remains near the surface. It has mixed more deeply— to depths of more than three thousand feet—in areas of intense downwelling in the polar North and South Atlantic, where deep bottom waters of the global ocean conveyor current are formed. Elsewhere the sea has been stirred to only a thousand feet deep or less.

All tropical coral reefs inhabit waters that are less than three hundred feet deep, so they will quickly come under the influence of ocean acidification. If carbon dioxide in the atmosphere doubles from its current level, all of the world's coral reefs will shift from a state of construction to erosion. They will literally begin to crumble and dissolve, as erosion and dissolution of carbonates outpaces deposition. What is most worrying is that this level of carbon dioxide will be reached by 2100 under a *low*-emission scenario of the Intergovernmental Panel on Climate Change. The 2009 Copenhagen negotiations sought to limit carbon dioxide emissions so that levels would never exceed 450 parts per million in the atmosphere. That target caused deadlock in negotiations, but even that, according to some prominent scientists, would be too high for coral reefs.[12]

Just as Ischia's carbonated volcanic springs provide a warning of things to come, bubbling carbon dioxide released beneath reefs in Papua New Guinea give us tangible proof of the fate that awaits coral reefs.[13] Reef growth has failed completely in places where gas bubbles froth vigorously, reducing pH there to levels expected everywhere by early in the twenty-second century under a business-as-usual scenario. The few corals that survive today have been heavily eroded by the corrosive water. The collapse of coral reefs in the Galápagos following El Niño in the early 1980s was hastened by the fact that eastern Pacific waters are naturally more acid due to their deep-water upwelling than those in other parts of the oceans.[14] Corals there were only loosely cemented into reef structures and collapsed quickly.

Acidification is an even greater problem at the poles, where carbonate dissolves at only six hundred or seven hundred feet deep. Polar seas are renowned for their enormous productivity, which is why many whales trouble themselves to swim thousands of miles every year to gorge themselves in their waters. Pteropods, tiny snails that swim with a foot that expands into a pair of delicate wings (the name means wing foot), are keystone animals in polar food webs. They have been called the potato chips of the sea because of their critical role as a food source, but the name does them no justice. They live within shell castles sculpted from transparent crystal whose cold beauty seems perfectly fitted to the icy seas. They range from the size of a lentil to a fingernail and reach densities of up to ten thousand per cubic meter of water. To visualize such a density, imagine each snail in a space equivalent to about the volume of a kiwifruit. For a predator, hitting a patch like this would be like being caught in the whiteout of a snail snowstorm. If you are a fish lover and have ever eaten salmon, cod, pollock, or any number of other species from frigid waters, you have tasted pteropods secondhand.

Pteropod chips could be off the menu in less than fifty years, as polar seas become undersaturated with carbonate. Experiments with captive pteropods show that their shells dissolve at acidity levels that may soon be reached. In fact, parts of the sea off northern Canada have already become corrosive to pteropod shells as a consequence of ocean acidification enhanced by sea ice melt.[15] The Southern Ocean around Antarctica is predicted to follow by 2030, and the Bering Sea by 2100.

One of the great unknowns in the future life of acid seas is how fast species will be able to adapt as their environment shifts. The geological record holds clues but does not reassure. Fossils record long stretches of vigorous reef development interrupted by periods with few reefs. Charlie Veron, a veteran Australian coral scientist, believes that periods of elevated carbon dioxide correlate with the periods in the geological record when reef formation ground to a halt. Others are less sure and think that episodes of reef growth are not that closely linked with ocean chemistry. They believe that reefs have developed in the past in seas that were more acidic than the levels we are on target to reach by the end of the century. Even they will concede that periods of high carbon dioxide in the atmosphere and oceans are associated with mass extinctions, followed by bursts of evolutionary diversification. This alone rings alarm bells for me.

In the opening chapter I charted the emergence of life in the sea. A billion years ago atmospheric carbon dioxide levels were much higher and the oceans more acidic than they are today. The evolution of simple photosynthetic life forms in the pre-Cambrian reduced atmospheric carbon dioxide by withdrawing it from the air and releasing oxygen. Around 570 million years ago, 30 million years before the Cambrian explosion of life, carbon dioxide concentrations seem to have come down enough for a signal event in the evolution of marine life. The oceans became less acidic and dissolved carbonate reached saturation levels in shallow seas with sufficient consistency to make it possible for them to construct shells. Fossils reefs in Namibia hundreds of yards across date from this time and are made from giant heaps of tiny calcified tubes produced by *Cloudina*, possibly an early antecedent of polychaete worms, some of which make shelly reefs today.

The reduction in acidity of the world's oceans heralded an evolutionary explosion of animals that produced carbonate skeletons and shells. To begin with, most secreted calcite or calcium phosphate shells. Aragonite, the form of calcium carbonate produced by pteropods and modern reef corals, is more soluble than these compounds. Corals that secreted aragonite began to build reefs about 230 million years ago. They yielded to calcite producers in a period of global warming and high carbon dioxide that prevailed between 145 million and 65 million years ago. Modern corals evolved about 40 million years ago, as carbon dioxide levels fell once more. There is obviously a link between atmospheric carbon dioxide, ocean acidification, and the fortunes of animals with chalky carbonate shells. Some managed to survive, and even thrive, in more acid seas in the deep past. But that does not mean they can adjust fast enough to survive the more rapid acidification of modern times.

The last time marine life was subjected to seas as corrosive as those anticipated by 2100 was in a period geologists call the Paleocene-Eocene Thermal Maximum, fifty-five million years ago. Sea-surface temperatures spiked around nine degrees Fahrenheit to sixteen degrees Fahrenheit higher than today's due to a massive release of methane, and possibly of carbon dioxide.[16] We aren't sure exactly where these gases came from or why they were released. Theories include volcanic eruptions, offgassing of peat, or a massive belch of methane hydrates from the seafloor. A methane belch seems the most likely explanation, given the chemical signature of sediments laid down at the time. Methane would have been rapidly oxidized to form carbon dioxide, which in turn acidified the seas. The amount of carbon released at that time is about the same as that which would come from burning all of the world's fossil fuel reserves today, but it took place more slowly—over a few thousand years rather than the few hundred years in which we are doing it today.[17] You can see this event written in sediment cores drilled from the deep sea. Chalky sediments give way suddenly to a brown layer of clay, which almost as quickly returns to chalk. The whole event is over in just four inches to eight inches of core, corresponding to the hundred thousand years or so it took the Earth to swallow back the excess carbon it had released.[18]

The Paleocene-Eocene ocean acidification spike does not seem to have led to any great cataclysm for marine life. Looked at against the sweep of geological time, it might be reported as "Minor disaster strikes: not many dead." Yes, the few coral reefs around seem to have suffered badly,[19] but the only outright extinctions were of deep-sea foraminifera. These tiny single-celled organisms—neither animal nor plant—succumbed when corrosive water swept over them as the oceans acidified.[20] All the bottom-living foraminifera went extinct, while those from surface waters survived.[21] Nonetheless, conditions for their existence and that of other calcareous plankton became harsh, and sediments from the time are full of stunted and malformed shells. If any creatures are capable of adapting to higher acidity, we would expect to find them among single-celled plankton, for whom a generation can be measured in hours or days. They do seem to have toughed it out pretty well. Although there were major changes in their composition, there was no apparent shutdown in the construction of chalky body structures.[22] Some species fared well under conditions of higher acidity. We can take some comfort from this but have to remember that the rate at which we are adding carbon dioxide to the sea is much faster today, so the impacts will likely be far more severe. Carbon dioxide will accumulate in the surface layers of the ocean that are home to most marine life long before it is mixed into the deep seas. It takes a thousand years or more to flush carbon dioxide into the deep sea, where it is eventually neutralized by dissolving chalky sediments.

Coral reefs, like polar bears, have come to symbolize the perils of carbon dioxide emissions and climate change. They are sentinel habitats that forewarn of dangers ahead. If they perish, millions of people will be left homeless. But I have another concern that gets to the very heart of the planetary life-support system: What will acidification do to phytoplankton? Phytoplankton, or plant plankton, produce roughly half of the oxygen that we breathe and are the foundation of nearly all ocean food webs. Life itself depends on them. We depend on them.[23]

Many phytoplankton, such as coccolithophores, produce chalky skeletons. These ball-shaped microscopic plants armor-plate their bodies with exquisitely ornamented disks of chalk that overlap like porcelain shields. Spring blooms form when the days lengthen, fed by nutrients stirred from deeper water by winter storms. Countless trillions of coccolithophores then turn the sea a milky white, producing swirls that can be seen from space. A single bloom can produce a million tons of calcium carbonate in just a few weeks.[24] These blooms do their bit for our climate by reflecting heat back into space from the

oceans. The effect of acidification on coccolithophores and other photosynthetic plankton is not straightforward. Carbon dioxide boosts
photosynthesis, just as it does with land plants. In experiments at higher
levels of ocean acidity many coccolithophores seem to produce just as
much carbonate, or sometimes more than under normal conditions.[25]
Another plus is that not all phytoplankton produce calcium carbonate
skeletons. Perhaps diatoms, with their silica shells (the basic ingredient
of glass), will take up any slack caused if carbonate-producing phytoplankton struggle—or cyanobacteria, or dinoflagellates. We had better
hope so.

Coccolithophore blooms are often stopped in their tracks by
follow-on explosions of viruses that eat them. Viruses are critical to
nutrient cycling and phytoplankton production in the sea, liberating
nutrients to be used again by other plants and animals. Probably the
most astonishing fact I have ever heard about the sea concerned
viruses. A two-pint bottle of crystal clear seawater contains some four
billion viruses.[26] Added together there are something like 4,000,000,
000,000,000,000,000,000,000,000 viruses in the sea (4 nonillion, if
you want an easier way to say it). They outnumber all other marine
life-forms combined by 15 to 1. To get an idea of what that enormous
number means, imagine placing all the viruses end to end. They
would form a thread less than one two-hundredth the thickness of
the finest spider gossamer that would stretch for two hundred million
light years. You read that right: The thread would reach so far across
the universe it would pass by sixty galaxies and countless millions of
stars. The microbial world of our ancient seas is still alive and well. In
all the research I have read on ocean acidification I haven't come
across a single study of the effects it might have on viruses. As it happens, viruses have slowed the rate at which the oceans are acidifying.
Because they are so good at recycling nutrients in sunlit surface
waters, less carbon is exported to the deep sea via the sinking bodies
of animals and plants. This means less carbon dioxide is sucked from
the atmosphere by the ocean; good for acidification, bad for climate
change.

Most scientists' attention to date has concentrated on what will happen to the production of chalky carbonate body parts in more acid seas. But pH change in itself can alter the availability of micro-nutrients that phytoplankton need to thrive. Iron is a key nutrient that limited phytoplankton productivity in ancient oceans and still does throughout much of the high seas. Iron is less willing to combine with organic matter under more acid conditions, and will therefore be less available to organisms that need it.[27] Ocean acidification could expand areas of the sea where iron limits productivity, which would ultimately reduce the amount of food the sea could produce for us.

The more we look at the effects of acidification, the more ways we find in which it will disrupt life at sea. Dozens of experiments on captive invertebrates show increased acidity affects almost every aspect of life, from growth and behavior to reproduction. One surprise is that predator avoidance in larval fish can turn into fatal attraction.[28]

The orange clownfish lives among the stinging tentacles of large sea anemones on coral reefs. Clownfish hide their eggs under the anemones' frilled edge to keep them safe. After a few days, newly hatched larvae head for the open sea. Weeks later they return under cover of darkness to find an anemone of their own. The most dangerous moment of their lives comes as they thread their way through a thicket of predatory mouths in search of a new anemone. Larval clownfish can smell their predators, which helps them avoid being eaten. But clownfish larvae raised under levels of ocean acidity that will be common in 2100—if we fail to curtail carbon dioxide emissions—lose this ability, and instead they are attracted to their predators. Other experiments have shown that they also lose the ability to smell their anemone hosts.[29] Perhaps they will be able to spot the anemones when dawn arrives, but by then they will be seen by a bevy of other predators that take over from the nightshift.

Unless orange clownfish can adapt quickly, acidification will likely bring their time on this planet to a close. We can only guess at how common such responses are among the many hundreds of thousands of other species with planktonic larvae. These larvae may be more sensitive to acidification than adults, which have been the subjects of most studies to date. What we do know is that the use of smell to detect habitats and predators is widespread among marine life. Once again Ischia offers a glimpse of what the future may hold. Hardly any larvae of species with chalky skeletons set up home or survived near the underwater carbon dioxide vents.[30] Experiments on sea urchins have shown reduced sperm vitality and fertilization at acidity levels that could be reached by century's end.

Oyster farms on the Oregon coast give one indication of how hard life will get.[31] Oysters there periodically experience a complete failure of reproduction linked to more acid water. I described previously how upwelling recently flooded the continental shelf with deep water so low in oxygen it caused mass mortality of animal life. Upwelled waters are also more acidic than surface water. When farms were flushed by these more acid waters, larval oysters died at the point when they had to form their first shells.

Overfishing has greatly increased our reliance on shellfish from the wild as well as from farms. Prawns, lobsters, clams, and scallops are relatively more resilient to overfishing than fish and have come to dominate intensively fished seas. All lay down carbonate shells. In 2006, half the value of U.S. fisheries came from shellfish. The fishing industry is badly exposed to risk from more acid seas. Not only that, acidification threatens the important role that filter-feeding shellfish play in cleansing ocean waters.

Ocean acidification affects a fundamental aspect of almost all existence: the ability of marine life to process oxygen. When carbon dioxide levels increase, it becomes harder to extract oxygen from water. Dissolved carbon dioxide quickly diffuses across gills and into cells, reducing the amount of oxygen in circulation and suppressing metabolism. Animals stressed in this way expend more energy to maintain basic bodily functions, which means they can't afford to give as much to growth and reproduction. More acidic oceans may increase the severity of problems resulting from low oxygen, or anoxia, something I will return to in the next chapter.

Coral bleaching used to be what kept me awake at night; now it is ocean acidification. Warming seas have devastated reefs worldwide by weakening the sunlit coalition of corals and algae. Acidification is a punch in the gut to reefs that are already on their knees. It is chilling to think that within the space of one hundred years humanity could reverse a process of coral reef formation that has flourished since the end of the last ice age.

More acid seas will not be devoid of life; there will be winners as well as losers. But life will change. The seabed around Ischia's champagne hotsprings is covered by lush stands of sea grass, their vivid green leaves unsullied by the chalky crusts and tubes of coralline algae and worms that cover plants farther away. Brown and red seaweeds elbow sea grasses aside wherever rocks punctuate the sand. These plants benefit from the fertilizing effect of carbon dioxide and the lack of sea urchins to eat them. Good news, at least, for the Mediterranean's hard-pressed green turtles. They love sea grass.

CHAPTER 8

Dead Zones and
the World's Great Rivers

Britⁱsh parliamentarians enjoy a long summer break that lasts from July to October. What could be so important to draw them away from the business of running the country for so long? The Houses of Parliament squat in gothic splendor on the bank of the River Thames, and it is to this river that we must look for the answer. Like many cities, London grew organically over the centuries, with little or no central planning. By 1815 it was the largest city in the world, inhabited by 1.4 million people, but it was a modern city with a decidedly medieval approach to sewage and refuse, much of which was dumped in the street or in cesspits. Ironically, it was an invention intended to improve hygiene, the flush toilet,[1] that helped create the problems in the Thames for which nineteenth-century London became notorious, and which culminated in the Great Stink of 1858.

In the late eighteenth century, flush toilets began to replace chamber pots. They greatly increased the volume of waste discharged into cesspits, which regularly overflowed into the street. A law was passed in 1815 to remedy the problem by allowing the discharge of sewage into the Thames through a sewer network then under construction. By mid-century the Thames had become choked with sewage and refuse that sloshed back and forth with every tide past the houses of Parliament. As bacteria went to work on the sludge they sucked up dissolved oxygen and turned the river into a stinking sewer that belched poisonous gases into the heart of London. In 1858 the stench became so bad that the windows of the Houses of Parliament had to be hung with sheets soaked in bleach to drive off the smell. Eventually summer sittings of Parliament were abandoned.

London's problems (mercifully since solved) have been repeated across the world and continue to plague most cities in developing countries today. Many of the world's most populous cities straddle mighty rivers on which people depend to bring water and carry away waste. By the time these rivers reach the sea they are loaded with sewage, toxins, and trash from cities, as well as industrial and agricultural runoff.[2] The problem increased sharply after World War II, when farms became dependent on chemical fertilizers to boost yields. Fertilizers now fuel coastal plankton blooms that can cover thousands of square miles and are often visible from space. Satellite photographs of the Gulf of California in Mexico show the telltale green of phytoplankton growth around the mouth of the Yaqui River soon after coffee growers spray their crops with fertilizer.[3]

Plankton blooms are a natural seasonal feature in spring and early summer, when the days lengthen, and must have occurred ever since the seas were first populated with microbes more than two billion years ago. They are mostly seen in temperate and polar seas, where nutrients are stirred from the bottom by winter storms, and at the mouths of large rivers. When plankton die their decay takes oxygen from the water. As a result, beneath pollution-fueled plankton blooms the water often becomes anoxic, spreading a shroud of death. Dead zones are now permanent or seasonal features at over four hundred places worldwide, all of them coastal or in enclosed seas, like the Baltic.[4] At the last count, in 2008, collectively they covered a hundred thousand square miles, or about 1 percent of the area of the world's continental shelves. This may not sound like much, but these are some of the most productive and biologically rich areas of the sea. We can ill afford their loss.

One of the world's largest dead zones forms every year off the mouth of the Mississippi delta in the Gulf of Mexico.[5] The Mississippi River catches the runoff from 40 percent of the landmass of the lower forty-eight United States and discharges an average of 140 cubic miles of water into the Gulf of Mexico every year, enough to fill Lake Mead, the largest reservoir in the United States, sixteen times over. Its water is loaded with particles of organic matter and dissolved nutrients that plankton need to thrive. In the past, this annual nutrient bonanza sustained rich fisheries around the delta, but in recent years nutrient overload has tipped the balance from benefit to blight.

The dead zone begins to form in the spring, after winter snow melt and rains swell the mighty Mississippi and flush its nutrient-rich water into the Gulf of Mexico. When stultifying summer heat anaesthetizes human life in Louisiana, Alabama, and Tennessee, the plankton begins to decay and oxygen levels plummet. Those that can flee do so, as a suffocating blanket engulfs the ocean. Many don't make it, and the corpses of fish, crabs, and shrimp litter the seabed. Others, like clams, snails, worms, and starfish that cannot leave, die where they are. As the shroud of death expands, fishermen are forced to travel further offshore to make a catch. They pass through waters that are eerily empty—no leaping fish, nothing showing on their fish finders, nothing to interest the birds that pass overhead. Researchers like Melissa Baustain from Louisiana State University dive in the dead zone. She said, "The deeper we go down in the water, it gets kind of scary, because there's nothing there. There's no fish, there's no organisms alive, so it's just us."[6] At its peak the dead zone off the Mississippi River chokes life from nearly eight thousand square miles of sea.

Some people believe that the dead zone is a natural feature of the Gulf of Mexico, but much of the Mississippi watershed is farmland, and the evidence points firmly to agricultural fertilizers. The history of the dead zone can be read in the bodies of microscopic animals trapped in sediments laid down on the seabed near the mouth of the Mississippi and far beyond.[7] The relative numbers of species that depend on oxygen-rich water compared to those that thrive when oxygen is scarce tell the story of when the dead zone first formed and how it has grown with time. Chemical signatures in the sediments add detail to this history.

The Mississippi watershed has seen drastic changes since the

colonization of the Midwest and the conversion of prairie to farmland. That led to a steep change in the rate of soil loss around the
beginning of the nineteenth century, which can be seen in core samples
through a corresponding rise in the productivity of diatoms, a type
of phytoplankton. Sediment loads fell again through the early twentieth century, as dams were built and trapped mud. The application
of artificial fertilizers to cropland soared after 1950, along with the
number of sinking diatoms, and has not yet peaked. The dead zone
took hold only after this, although it was too small to attract much
notice until the 1970s. Since then it has formed nearly every year like
a spreading bruise. Inexorable and inescapable, it chokes life from a
place that was within living memory the most vigorous fishing ground
in the Gulf.

A Sumerian creation myth inscribed on clay tablets five thousand
years ago appears to describe an early effort at dam building:[8]

> *Ninurta, the son of Enlil, did great things*
> *He built a big mass of stones in the mountains . . .*
> *He constructed a barrier at the horizon . . .*
> *The stones competed with the powerful water*
> *Now the water of the mountains will not flow down to the valley*
> *   forever*

The first experiment in major dam building took place forty-six hundred years ago, when ancient Egyptians tried to control Nile floodwaters at Sadd el-Kafara.[9] Their construction was destroyed by a
flood before it was completed, but this early failure didn't hold us up
for long. We have since raised over thirty-five thousand major dams,
transforming freshwaters. Nearly half of the world's largest rivers are
dammed, and over 40 percent of river flows to the sea are intercepted
by reservoirs.[10] Collectively they exert a growing influence over life in

the sea. The attractions of low-carbon energy and greater water security (at least in the medium term, before reservoirs clog with silt) are unstoppable, even if we now have a far better understanding than we once did of the costs of blocking rivers.

Before the Aswan Dam was built, when sediment-laden Nile floods reached the Mediterranean they triggered plankton blooms that fed huge shoals of anchovies and a lively fishery. The dam now blocks most of these nutrients, but a new source has been found in the nourishing broth of effluents that flow into the sea from almost every river and stream in the country. The anchovy fishery, which floundered after the dam was built, is back in business, since today the Nile carries the waste of a population of eighty million as well as fertilizer runoff from the delta—not that this thought would tempt me to a feast of anchovies the next time I visit Alexandria!

Dams and crop irrigation cut off sediments and divert water, reducing nutrient flows into coastal seas. Their impact may in fact be more far-reaching than one would imagine. The Yangtze River carries vast quantities of soil from the Chinese highlands. But even this great river supplied less than a tenth of the phosphorus and a third of the nitrogen—two key plant nutrients—to the East China Sea.[11] The rest upwelled from deep water beneath the river plume. (Buoyant plumes of river water flowing offshore at the surface can drag deeper water in the opposite direction, sucking nutrients into sunlit waters where phytoplankton thrive.) After the Three Gorges Dam was built across the Yangtze River, peak primary production by phytoplankton in the East China Sea fell by 86 percent, due to loss of nutrients.[12]

Dead zones form most readily where water movement is sluggish and it is stratified into layers, cutting deeper water off from life-giving oxygen. If water slops around for long enough, stagnant pools near the seabed lose their oxygen as the sinking "snow" of decomposing plankton builds into drifts. Despite their massive nutrient flows to the ocean, the Orinoco and Amazon rivers have no dead zones, because water is pushed offshore too fast for oxygen to be lost. Where the flows of once great rivers have been staunched to a trickle after dams, cities, and crops have drunk their fill, estuaries may hold water long enough for dead zones to form. This is the case, sadly, for rivers like the Loire in France and Po in Italy, when their lack of oxygen chokes life from thousands of square miles of seabed every summer.

Enclosed seas like the Baltic, Adriatic, and Black also suffer problems of stagnation. Shaped like a deep bowl that is cut off from the Mediterranean by a shallow sill in the Bosporus Strait, the Black Sea was a freshwater lake when sea levels went down fairly dramatically during the last glaciation. It refilled eventually, with saltwater, in a catastrophic flood seven thousand years ago, when the Mediterranean rose high enough to break through the Bosporus. Massive floods triggered a tremendous human migration, and the event may be recorded in the ancient tale of Gilgamesh, and perhaps Noah's flood.[13] Today only a warm, less saline, surface layer is well oxygenated and able to support abundant life. This low-density layer sits over the cooler, more saline waters of the deep basin like a lid and has suffocated life below. Deeper than about five hundred feet, the Black Sea is devoid of oxygen.

This lack of oxygen means the Black Sea conceals some spectacular shipwrecks, as the *National Geographic* explorer Bob Ballard has revealed. He discovered one wreck off Sinop, the ancient tuna fishing port, that dates back fifteen hundred years. Spared the jaws of shipworms, the hull and deck are intact, and even the mast still stands, a ghostly relic of a vanished age. The ship offers an extraordinary peek into the seafaring life of the ancient world, but at over a thousand feet deep it remains tantalizingly beyond the reach of archaeologists, so its secrets, like its timbers, are untouched.

The Black Sea owes its stagnant depths to nature,[14] but other enclosed seas have us to blame. Northern Europe's Baltic Sea connects to the North Sea only by a shallow, tortuous channel that curls around Denmark. The rivers that feed it drain from highly populous countries and run through croplands and intensive pig farms. They receive effluents from heavy industry and paper-pulp mills and sewage from cities great and small. By the time they reach the coast they are loaded with nutrients and organic waste. Sediment samples from one Swedish bay tell a similar story to those from the Mississippi dead zone.[15] Land was cleared and trees felled for agriculture and industry at increasing rates from 1800 onward, which led to an ever-growing rise in nutrient runoff. Problems from plankton blooms and anoxic bottom water emerged after the 1950s, when artificial fertilizers came into widespread use. The number of people in Baltic watersheds has more than doubled since then, and sewage has compounded the effects of fertilizers.

The Baltic is a huge brackish water lake that occasionally gulps saltwater from the North Sea. When conditions are right, the North Sea pours in over the sill around Denmark hugging the seabed as it creeps along. Brackish water (a mix of salt and freshwater) is less dense and thus flows out at the same time along the surface. This density difference restricts mixing between the salty deep and fresher surface layers and so promotes a loss of oxygen when dead plankton sink to the bottom of the sea. Species that need saltwater to thrive are thus confined to the deep layers with dwindling oxygen. One reason why Baltic cod stocks fell so catastrophically in recent years was that their eggs could not survive in these deep pockets of low-oxygen saltwater, although overfishing was the main reason for their downfall.

Intriguingly, there seems to be a link between the collapse of cod and other predatory fish and the vigor of seaweed growth. Thick, choking mats of filamentous seaweed have overgrown the bottom in shallow regions of the Baltic like tufts of wool, blanketing areas where you could once gaze from the surface through crystal water at the waving fronds of bladder wracks and kelp below. This filamentous weed spells trouble for fishers, whose nets clog with drifting clumps. The decline of big fish and seals from hunting and disease led to an increase in their prey, which in turn were predators of animals that grazed on the mat-forming seaweeds that now smother the seabed. After these smaller predators decimated the grazers, the algae spread.

Denmark and many other nations bordering the Baltic have made strenuous efforts to reduce nitrogen runoff to the sea to tackle excess nutrients and the problems they cause. Nitrogen inputs have halved since their peak in the early 1980s.[16] Unfortunately, it seems the Baltic is stuck in a vicious cycle. Low oxygen at the seabed releases phosphorus that stimulates plankton blooms, which leads in turn to worse oxygen depletion. This happens because when oxygen declines, life on the bottom shifts from larger species that live deep in the sediments, like burrowing worms and clams, to smaller, more transient species, like threadworms, that live on or near its surface. In well-oxygenated conditions the larger sediment dwellers suck particles of food from the water and deposit their wastes deep in the mud as feces. They draw nutrients from the open water and so are no longer available to fuel further plankton blooms. By contrast, the smaller species that feed near the surface when oxygen is scarce release nutrients back into the water, where they further boost plankton growth, increasing the likelihood of more severe oxygen loss as the plankton decay. Once again we have come to see the value of an intact, functioning marine ecosystem only after it has been damaged.

The Chesapeake Bay offers another sad example; we rue the loss of the vast maze of filter-feeding oysters whose reefs so challenged the navigational skills of Virginia's first settlers.[17] The reefs were scraped and nibbled away to feed America's appetite for oysters. Over the last century nutrient inputs to the Chesapeake have soared, from sewage, industrial effluent, and pig farms. Dead zones were first noticed in the 1930s, but their size and frequency has increased with time. By the 1960s, the declining health of the bay could no longer be ignored. But it was not until the late 1980s that a program was begun to reduce the amount of nutrients washing into the bay. It was supposed to restore good water quality by 2010, but it hasn't worked.[18] Forecasts made in early 2011 for that year predicted the fifth-largest volume of anoxic water seen in any of the previous twenty-six summers of monitoring. While excessive nutrients set the scene for anoxia, abundant filter-feeding oysters might still have prevented it from happening had we not removed them.

The nutrients that fertilize coastal and enclosed seas create another unwelcome problem, which is that plankton blooms can be toxic. Some marine plants produce potent poisons. Collectively they are known as "harmful algal blooms," but more commonly they are called red or brown tides for the reddish slicks that appear at the surface when a bloom is underway. Red tides cause filter-feeding shellfish such as oysters, clams, and mussels to accumulate toxins that cause paralytic and amnesic shellfish poisoning, sometimes killing people who eat them. They also poison fish and mammals, causing mass die-offs.

Florida is one of the more afflicted regions in the world. It has been troubled on and off by red tides since at least the middle of the twentieth century—when scientists first began to study the phenomenon—and far longer, according to the accounts of sailors and visitors. For most of the last twenty years there has been at least one red tide a year. Sometimes they have persisted for more than a year. Nutrient-rich runoff from intensive agriculture seems to trigger blooms offshore, especially in the southwest, after a heavy rain flushes fertilizers out to sea.[19] The limited circulation of Florida Bay concentrates the blooms.

The dinoflagellate responsible for Florida's red tides is called *Karenia brevis*, and it produces a cocktail of potent toxins—brevetoxins—that act on the nervous system. When these algae are churned up by the wind and the waves, the toxins can be inhaled. Manatees, docile sea cows that graze Florida's shallow sea-grass meadows, are especially sensitive, and red tides kill more of them than any other cause. A particularly severe red tide killed 149 of them in 1996, while over 100 bottlenose dolphins succumbed in 2004. Unsurprisingly, people on land suffer from sore throats and runny eyes, and some develop more serious breathing problems when red tides strike. Hospital admissions for respiratory and stomach problems surge during red tides, and those who live close to the coast suffer the worst.[20] Inhaled toxins caused DNA damage, a precursor to cancer, in tests on laboratory rats. Bottlenose dolphins that wash up dead on beaches often contain high levels of brevetoxins. Life in red tide hot spots, such as Tampa Bay, Naples, and Fort Myers, has become bad enough that there is concern it could drive away tourists and depress coastal property values.

There are many other toxic phytoplankton out there able to make trouble for us in places that are polluted by excess nutrients. A different kind grows in Australia's Moreton Bay on the Queensland coast, in fibrous tufts attached to the blades of seaweeds and sea grasses. The problem, as with Florida's red tides, is most severe when storms churn the sea into a spray that can be inhaled, causing lung inflammation and itchy skin. The same algae can be found in Hawaii, where it is eaten by a type of mullet that has become known locally as the "nightmare weke"—when eaten, toxic fish cause hallucinations and nightmares. The toxins aren't just responsible for short-term health problems. Some promote tumor growth, or produce birth defects. It may be no coincidence that Hawaiians who eat seaweeds have high rates of stomach and intestinal cancers.

The tumor-promoting chemicals in harmful algae have been blamed for the growth of horrendous tumors on green turtles, although the tumors are triggered by viruses. These animals feed on sea grasses, the blades of which support toxic algae. I once lived in the U.S. Virgin Islands, where I worked as a coral reef researcher, and most of the green turtles gliding around me when I took my daily swim suffered tumors that ranged from apple- to melon-size. Moreton Bay's turtles have similar problems. Could coastal pollution cause cancer? Now there's an unpleasant thought that might actually get our attention.

One group of animals has done better than most from the triple combination of nutrient enrichment, low oxygen, and overfishing:

jellyfish. To most of us jellyfish are a nuisance at the beach—half glimpsed but sorely felt. I remember well my childish fascination with the amorphous lumps that sometimes littered the shore. I would prod and chop them nervously with my spade, half expecting they would slide over to sting my feet. When, as a new diver, I first came upon a tiny comb jelly adrift in open water, it dumbfounded me. Comb jellies are about the size of figs and are little more than blobs of seawater wrapped in a transparent glaze. But rows of tiny brushlike cilia beat over that skin, in rainbow waves. Their beauty is fragile, as I soon discovered, for the slightest touch tears them asunder.

Since the Monterey Bay Aquarium in California figured out how to raise them in tanks, the splendors of jellyfish have been enjoyed by millions of people (the trick is to have a tank without sharp corners and to keep the water moving in gentle loops). It is a testament to their loveliness that jellies can hold the limelight against the more obvious wonders of tuna, turtle, and shark. Their rhythmic pulses have a hypnotic quality, like beating hearts. But the allure of jellyfish conceals a darker side. They are ferocious predators.

Jellyfish feed on zooplankton; some hunt other jellies. When abundant they compete for zooplankton with other animals, such as herring, sardines, or menhaden. Not much that is edible is off-limits. They hunt eggs and larvae of fish and shellfish, and when jellyfish "bloom," they can have major impacts on the young stages of commercially important species. A bloom of comb jellies in the Black Sea was implicated in the collapse of several fisheries.[21] Blooms can happen quickly, because jellyfish grow fast (it helps that 95 percent of their body weight is water—by way of comparison, only 55 percent to 60 percent of our bodies are water). They multiply with incredible rapidity when the conditions are right, blossoming from a handful to an infestation in a matter of weeks.

One trick that the familiar moon jellies and their relatives have is a dual life cycle. We know them for their free-swimming stages, but they also live attached to the bottom as polyps a few millimeters across, much like sea anemones. When the time comes these polyps can release dozens of tiny jellies into the water, which can grow from fingernail size to adulthood in a few months. The largest jellyfish in the world by mass—Nomura's jelly—is found in Asian seas; it grows to a couple of meters across the disk and can weigh 440 pounds or more. Perhaps inevitably, Nomuras are caught and eaten, but even so they have caused great nuisances in Japanese waters. They clog fishing nets, even of the boats that target them. In 2010, one boat sank under the weight of its catch. For boats that target other fish, jellyfish just get in the way and ruin the fishing. Set nets and traps are torn away by walls of jellies pressed against them by waves and currents. They choke the cooling water intakes of power stations, forcing them to reduce output, or sometimes even to shut down.

Mediterranean resorts have been plagued by jellyfish outbreaks in the last twenty years. The main problem species there is the mauve stinger, whose tentacles inflict slashing welts on the tender bodies of bathers. In the summer of 2004, an estimated forty-five thousand swimmers were treated for stings in Monaco alone. Things got so bad in Spain's Balearic Islands in 2009 that the local government issued the fishing fleet with bespoke jellyfish nets and put them on notice: Should a bloom occur, sweep up the jellyfish before they hit the tourist beaches. In 2007, Irish salmon farms were overwhelmed by hordes of mauve stingers that slaughtered tens of thousands of salmon in their deathly embrace. Similar mass killings have been reported in Japan, India, and Maryland.

Despite their fragility, jellyfish are remarkably resilient, perhaps a legacy of their ancient origins in the pre-Cambrian, when they commanded the waters virtually unchallenged. When nutrients overenrich coastal waters and phytoplankton prosper, there is a shift in the zooplankton from large to small species. Fish hunt by sight and suffer from the loss of easily spotted prey, whereas jellyfish, which hunt by touch, do fine. When the bodies of dead phytoplankton pile up and oxygen levels plummet, most jellyfish are unperturbed. They can survive in water with scarcely a gasp of oxygen, conditions that would kill most fish. These two attributes mean that jellyfish thrive in the low-oxygen nutrient soup of seriously polluted coasts. They bloom even amid the toxic belch of the Benguela upwelling. But there is more. Most jellyfish seem to enjoy a little warming, and experiments suggest they can tolerate the ocean acidification levels predicted for one hundred years hence. The concrete and timber footings of seawalls, harbors, and marinas provide excellent habitat for the bottom-living polyp stages of many jellies. It is probably no coincidence that most recent explosions of jellyfish have taken place in relatively enclosed seas surrounded by densely populated coasts. Overfishing and bycatch have taken a toll on animals that eat jellyfish, like chum salmon, spiny dogfish, and leatherback turtles. In short, the changes afoot in the oceans today could not have been designed better by some great jellyfish megalomaniac intent on global domination. Looking ahead to a future that is gelatinous, some scientists predict a jellyfish joyride all the way through the twenty-first century.[22]

CHAPTER 9

Unwholesome Waters

Filthy water cannot be washed.

—West African proverb

As a boy I lived in a small town set within a deep bay on the North Sea coast of Scotland. I could breathe the sea from my bedroom window, at least on the few days it was warm enough to open. Harbor seals basked on the shore near my front door. It was the 1970s, and there was another smell on the sea breeze in those days—oil. Through my teens I remember measuring the spread of the oil industry across the North Sea in the rising count of bird carcasses that I found piled up on the shore. The tide of bodies swelled over the years: guillemots, razorbills, fulmars, and once even a great northern diver. On land, the quayside that a hundred years before had thronged with the bustle of the herring fleet was now cluttered with pipes, steel cables, and other industrial paraphernalia. At fourteen my biggest regret over the burgeoning oil industry was to see wildflowers buried beneath pipes in an abandoned coastal quarry where I used to entertain myself amid the orchids and insects, dreaming of a future life as a naturalist that I am now lucky enough to have.

Oil has fueled industry from the late nineteenth century, and oil spills have been with us since then. But they only really grabbed the headlines when they happened at sea, an inevitable consequence of offshore drilling and shipping. Much of the oil taken from the North

Sea when I was young came from rigs in less than three hundred feet of water. As these wells were drained companies looked farther offshore, and deeper. Drilling has now moved off the shallow continental shelves into the deep sea. If you push the limits of technology, problems are inevitable, as we learned on April 20, 2010, when disaster struck BP's Deepwater Horizon rig in the Gulf of Mexico while boring a well a mile down. Human error and slapdash attitudes to safety led to a blowout that turned the rig into a fireball. Eleven people were killed and many more injured.[1] For BP, this was only the beginning of their troubles.

Capping a well a mile below water, they discovered, is not like capping one at three hundred feet. Early on a camera was placed on the seabed beside the wellhead. Over the following months, as BP executives writhed in the limelight, the camera beamed images into homes across the world of a violent geyser of oil and gas spewing into clear Gulf waters. While BP tried to play down the size of the spill, the camera showed the well gushing sixty-eight thousand barrels per day,[2] exposing their mendacity. The oil poured out relentlessly for months, as BP tried one trick after another, spreading hundreds of thousands of gallons of chemical dispersants, and at one point even stuffing the well full of old golf balls (some wits suggested that oil executives would have been a better choice). Spring turned into summer, and the spreading deep-sea oil plume burped slicks to the surface that washed ashore on beaches and marshes from Texas to Florida. In another BP gaffe, turtles trapped behind containment berms were immolated in the fires that were set to prevent oil from coming ashore. Eventually something worked, and the gush was choked off on July 15, eighty-six days and an estimated ten tanker loads[3] of spilled oil later (roughly 4.4 million barrels, after an estimate of oil retrieved is subtracted).

Deepwater Horizon taught us several things. We now know how much harder it is to contain a spill at depth than in the shallows. And we know that we are extraordinarily unprepared for problems at the deepening frontier of oil and mineral exploitation. BP's risk assessments for the rig detailed how walruses and sea lions would be protected in the event of a spill, an obvious cut-and-paste job from an Arctic plan! In fact, the company was shockingly unprepared. It didn't possess any specialized equipment to control a deep-sea blowout. The plan might as well have said, "If something bad happens, scratch your head for a while and wonder what to do." Or, as a friend once said to me, "When in danger or in doubt, run around and scream and shout."[4]

The disaster also taught us that spilled oil doesn't always bubble to the surface, as everyone expected. Plenty of oil did appear, and an area the size of Cuba was covered by sheen for months; but most of it remained out of sight as it spread in a giant plume far below. For more than a month BP executives and some government scientists denied its existence, but repeated studies showed it was there, and that it remained there long after the well was capped, together with seven hundred thousand gallons of chemical dispersant that BP injected at the leaking wellhead.

I first discovered that not all oil floats when, as a newly minted PhD, I took a job as a reef scientist in the Egyptian town of Sharm-el-Sheikh. Soon after I arrived a cargo ship hit a reef and spilled hundreds of tons of fuel, which washed ashore in stinking globs. Apparently, the technical term for this kind of oil is "chocolate mousse," and it certainly had that look. Beneath the waves, puddles of thick oil darkened the seabed, where they slopped around for the next year or two, until they became so covered in sand and bits of sea grass they vanished. Some is probably still there today.

In 1979, there was another massive well failure in the Gulf of Mexico, Ixtoc 1, which was being drilled by the Mexican oil company PEMEX. It gushed oil near the Mexican coast for over nine months. With an estimated half million tons spilled, Ixtoc held the record for the world's largest oil disaster until Saddam Hussein's deliberate environmental sabotage in the Arabian Gulf during the first Gulf War. Thirty years on you can still dig lumps of oil from mud around the stems of mangrove trees near the site of the Ixtoc spill. The oysters that used to crust the trunks and roots of the trees disappeared soon after the accident and have never recovered.[5] We don't know much about the long-term impact of this spill, because once the oil was no longer visible, money for research dried up. If we had kept up our interest we would know a lot more now about what to expect from the Deepwater Horizon spill in years to come.

There is a limit to what we can learn from past spills about what to expect from Deepwater Horizon. Until now our experience has been of surface spills, in which floating oil either evaporates, breaks down at sea, or ends up on beaches. There is no precedent for oil injected into dark, cold, deep-sea water at high pressure. The Deepwater blowout could have profound effects on deep-sea life. Soon after the spill the seabed nearby was covered by a thick flush of microbes that feasted on the carbon bonanza. Later measurements showed a virtually dead bottom, which suggests either that it was too much of a good thing or that dispersants killed them. As of now, we still don't know. While the two million gallons of dispersants almost certainly helped limit oil impacts onshore, they might also trigger a breakdown of the food web. Dispersant increased the oil exposure of miniscule plankton and thereby the creatures that eat them by breaking it into tiny droplets. Some scientists fear that the poisoning of plankton and microbes will interrupt nutrient cycles, perhaps leading to eventual starvation of top predators like sperm whales and dolphins. We know that the bodies of tiny plankton in the Gulf are already contaminated with dispersants, and it is worrying that there was a spike in bottlenose dolphin miscarriages in the aftermath of the spill.

Oil spills are the starkest of all human impacts on the sea: dark, suffocating, inexorable. Images of seabirds struggling through blackened waves are shocking, but in most cases oil is not the worst of their concerns. The toll of seabirds and marine mammals tangled up and drowned by fishing gear is far greater than from oil spills.[6] The Gulf of Mexico's fishing fleets kill more marine life in a day than Deepwater Horizon did in months. After the spill, fishing was banned in many parts of the Gulf. When the fleets were allowed to return several months later, catches were plentiful after the brief and salutary respite from fishing.[7] The environmentalist Carl Safina wrote of how many commentators called Deepwater Horizon "the worst environmental catastrophe in American history," but to his mind, the spill was far less disastrous than the twenty to forty square miles of marshland eaten from the Mississippi delta every year by sea-level rise and subsidence, and engineered defenses that starve the delta of life-sustaining mud.[8]

Some of the most iconic oil spills of the past have been from ships, like the *Exxon Valdez* in Alaska's Prince William Sound or the *Amoco Cadiz* off Brittany, but today the main risks of large spills are from drilling and pipelines. Pipeline spills have increased more than fourfold since the 1970s, partly because so much more oil is pumped today, but also because many pipelines are aging.[9] And yet as Deepwater Horizon has underscored, the real problem is our expansion of the oil industry out into the inky deep. Companies can now tap oil far beyond their capacity to contain a spill when things go wrong.

Back in Egypt in 1987 I asked an expert what he thought was the best way to clean up the brown gloop that smothered the shore in front of us. "A bucket and spade," he answered. Low-tech methods may still be the best we have, but the fried turtles from the Gulf of Mexico remind us that they aren't very effective. The best thing for the planet, of course, would be to wean ourselves off oil, but that is unlikely to happen any time soon. In the meantime, governments must demand much higher standards of safety and greater investments in prevention from oil companies.

Oil companies are easy to demonize, but the biggest source of oil pollution in the sea is not from tanker spills or careless drilling, but from people like you and me. Two thirds of the oil spilled into the sea around North America is carried with runoff from land (from dumped engine oil, fuel, and industrial leaks) or introduced directly by pleasure boats and jet skis.[10] The great majority of recreational boats have two-stroke engines. They are cheap, light, and powerful, but these assets come at a high cost to the environment. A quarter of the fuel-oil blows straight through the engine into the sea. (Many scooters use two-stroke engines, which is why they leave you choking on a cloud of oil and gas as they pass.) The floating fuel and oil concentrate at the sea surface, and wrap around and poison the floating eggs and hungry larvae of hundreds of species. I find it extraordinary that we continue to tolerate such profligate polluters when alternatives exist. Four-stroke engines have a closed system, so they emit ten times less pollution than the cleanest two-stroke motors. It is time to call time on two-strokes.

Oil has actually had a few benefits for wildlife. The highly visible impacts of spills helped galvanize the creation of some of the first marine parks, much as logging spurred the establishment of national parks on land in the late nineteenth century. The Great Barrier Reef Marine Park in Australia and Monterey Bay National Marine Sanctuary in the United States came into being this way. But away from hellish scenes of burning sea and turtles dripping tar, pollution's most insidious effects in the ocean come from chemicals we can't see.

In the late nineteenth century, chemists managed to synthesize organic compounds—chemicals that contain the element carbon—with some very useful properties. (All organic compounds contain carbon, but not all carbon-containing compounds are organic.) These compounds were extremely durable, would not burn easily, did not conduct electricity, and were resistant to sun and weather. Chemists called them polychlorinated biphenyls, but most people know them as PCBs. Naturally people quickly thought of many things they could do with these new wonders, and industrial-scale production began in the 1920s. They were put in glue and hydraulic fluids, used as plasticizers in paint and as heat and fire retardants in furniture, and formed the plastic coating of electrical wires, among other things. There was just one drawback, which by the 1970s could no longer be denied: They were also highly toxic, causing cancer, liver damage, skin lesions, and a catalog of other horrors.

The problem began when manufacturers discharged PCB-laced effluents into rivers, lakes, and estuaries, and chimneys disgorged them into the skies. Once inhaled or eaten, PCBs are not broken down, and they tend not to be excreted. They are highly lipophilic, or fat-loving, which means that they are stored in the fat of animals that ingest them. When big animals eat smaller ones they take on much of their load of toxins; so PCBs bioaccumulate through the food web from prey to predators. In the U.S. Great Lakes, for example, PCB concentrations in predatory herring-gull eggs reached fifty thousand times higher than in lake plankton from the base of the food web. In the 1970s, the carcasses of emaciated seabirds laden with PCBs began to wash ashore.

PCBs are just one class of chemicals among a wide range that are collectively known as persistent organic pollutants, or POPs for short. They include compounds like DDT, introduced to the world as a wonder pesticide in the 1940s—and which lost its gleam when it was found to cause hatching failure in birds and alligators. DDT had decimated birds of prey and fish lovers like the brown pelican before it was banned in the United States in 1972. Other countries soon followed.

In the 1960s and 1970s chemists were shocked to discover traces of DDT, PCBs, and other persistent organic pollutants in polar ice. The poles seemed so remote and pure, their landscapes carved from crisp, clean snow and ice, their skies sharp and clear. How was it possible? Pretty easily, it turned out. Sea and air know few boundaries and can carry chemicals to the most inaccessible corners of the globe. Smoke from Chinese coal plants crosses the Pacific to North America in less than a week. Currents also move pollution, although at a more sedate pace. They transport chemicals thousands of miles in months or years. Unfortunately, certain properties in the makeup of the oceans mean that persistent chemicals find a much faster route to the poles.

Between the ocean and the air is a thin layer of water whose properties are very different from that of the sea beneath.[11] At around one four-hundredth of an inch, this layer is not much thicker than a piece of plastic wrap. This thin membrane is stabilized by surface tension and rich in fats, fatty acids, and proteins; it is the reason why the surface of the sea sometimes has a glassy smoothness. Its stability attracts and concentrates microorganisms, floating eggs, particles of dust, and other materials. Unfortunately, water repelling, fat-loving compounds like POPs accumulate in this "surface microlayer." There they reach concentrations much higher than in the underlying water, often tens or hundreds of times higher.

When storms tear the sea into driven spray, contaminants race across the oceans and can bluster inland to drench coasts and people. Pollution in one region, such as the South China Sea, can leapfrog in a series of storm-whipped aerosols to far-flung regions. In this way, with the help of the wind, remote regions like the central Pacific can become polluted even if they seem to be beyond the reach of serious contamination.[12] Storm spray can leap barriers like the south polar current that circles the Antarctic and separates it from the rest of the ocean. Once pollutants have reached the poles they are swiftly taken up by animals there and work their way up the food chain.

The poles accumulate POPs through a second kind of hopping, known to chemists as the grasshopper effect. Many POPs are semi-volatile, which means they evaporate and condense within the range of seasonal temperature variation. In summer the chemicals evaporate off sea and land and blow around in the atmosphere, until temperatures drop sufficiently for them to condense, and they get redeposited. This happens either as the season changes or they reach somewhere cold at higher latitudes or on mountaintops. Chemicals are trapped within or under ice at the poles, and the year-round low temperatures mean little is lost by further evaporation, so contaminants accumulate.

The surface microlayer is critical for microscopic life. Fish eggs and larvae concentrate in the top millimeter of the sea, where they benefit from better feeding conditions or escape from predators like jellyfish. Invertebrates, such as crab or sea urchin larvae, can be up to ten times more abundant in the microlayer than in underlying water, while microalgae may be ten to a hundred times more common and bacteria a hundred to ten thousand times.[13] There they come into intimate contact with toxic contaminants. Animals feeding in the microlayer will take up pollutants and pass them up the food chain.

Persistent organic pollutants differ in an important way from oil pollution. Although they are invisible, in the long run they are far more dangerous to marine life. Oil and gas have been around since time immemorial. Natural seeps in places like the Gulf of Mexico have leaked from the seabed for millions of years. What to us is toxic and unpleasant is food to vast numbers of microbes. Deepwater Horizon jetted an enormous plume of methane into the deep sea. Before long, a bloom of methane-degrading microbes (our ancient friends) had formed around the leak and within a few months had degraded much of it to harmless carbon compounds and water.[14] Toxins such as DDT, PCBs, and many pesticides are highly complex, and the pathways of breakdown are much longer, which is why they hang around causing trouble for so long. In the extreme North and South Atlantic these pollutants are sucked into the deep sea on the global ocean conveyor current. At the icy polar extremes of the planet, where the weather is cold and the sun weak, they resist breakdown far longer than in warmer climes.

The daily adventures of a group of bottlenose dolphins in Florida's Sarasota Bay have been followed by scientists since the 1970s. Their world closely intertwines with that of thousands of people who use this seaway every day for business and leisure. The dolphins enjoy a carefree life frolicking in the waves close to people who live or vacation in Sarasota. In reality they experience as close to an urban lifestyle as any animal in the sea, and that is stressful. Florida is one of the most populous states in America. It supports its share of heavy industry and chemical production, as well as agriculture dependent on lavish applications of agrochemicals. The emerald gleam of Sarasota Bay is not as pure as it seems.

Tissue samples taken from the bay's dolphins show that they carry a heavy burden of contaminants.[15] They are at the top of the food web, and toxins passed from below stop with them. There is a troubling difference, however, between males and females. The toxic load carried by males increases over their lives. The most contaminated animal found in the bay was a dead forty-three-year-old male. His flesh would probably have been condemned as hazardous waste had officials known its chemical content when they disposed of the body. Toxins in females peak in adolescence and then plunge to much lower levels that remain stable until later life. The difference between the sexes is explained by pregnancy and breast-feeding.

Pregnant dolphins, and all mammals for that matter, including ourselves, experience high energy demands from their growing babies. If food is short, they must mobilize fat reserves to cope, which means that accumulated toxins are liberated into the bloodstream and passed on to the developing fetus. After giving birth, they continue to transfer toxins by lactation. Around 80 percent of toxic contaminants in female dolphins are passed to their first-born calf. Not surprisingly, these calves fare badly. Only half survive their first year, compared to 70 percent of calves born subsequently. We can't blame all of this on the chemical brew that mothers feed their offspring. By definition, first-born calves have inexperienced mothers who may also be smaller than older mothers. But there is a link. In a group of captive dolphins held by the U.S. Navy, mothers whose calves died within twelve days of birth had about two and a half times more PCBs in their bodies than those whose calves survived.[16]

The thing about marine mammals that disturbs me is that they aren't very different from us. Early on in this book I raised the possibility that modern humans long ago developed adaptations to accommodate a semiaquatic life, such as the fact that we have a body fat content similar to fin whales. (Actually, given the rise in obesity, some of us now rival seals and walrus!) All that fat predisposes us to concentrate a nasty cocktail of pollutants, just like the dolphin, the seal, and the whale. And like Sarasota dolphins we can pass them to our children in utero and through breast milk. I once told a lady I met at a conference the story of how first-born dolphin calves receive most of the chemical load carried by their mothers. She said, "You know, that explains so much about my older brother!" Joking aside, these toxins could lead to all kinds of problems in later life, from learning difficulties to cancer.[17]

PCBs, together with a variety of other industrial chemicals, including heavy metals, are hormone mimics that can interfere with the regulation of development and other bodily functions. Highly contaminated animals can therefore be sterilized by these "endocrine disruptors," as they are called, or their offspring affected in subtle and insidious ways. It takes only tiny traces to alter our hormone systems in ways that are detrimental to the health of us and our babies.

Indigenous peoples in the far north often carry heavy loads of toxins in their bodies. This is not because Arctic regions are more contaminated than lower latitudes; actually, they are less so. It is because of what people eat. Traditional diets often contain a great deal of blubber from marine mammals, exactly the food you would want to avoid to minimize exposure to toxins. The most contaminated flesh comes from animals that feed high in the food chain, like narwhal and beluga.[18] Animals like bowhead whales, which are still hunted traditionally on Alaska's North Slope, feed lower down the food web, and their flesh is safer. The highest levels of PCBs, DDT, and other contaminants are found in the Inuit of Greenland. Tests of the mental abilities of children in some Arctic communities have revealed that those who come from mothers with high toxin loads were more likely to have problems, a pattern also found in children from industrialized countries. PCBs and other hormone mimics depress the production of thyroid hormones critical to fetal brain development.[19] Such relationships in the United States and Europe have led people to ask whether our increasingly contaminated world is the cause of the rising incidence of conditions such as autism and attention deficit disorders in children. The Arctic Council, an intergovernmental body that oversees the interests of Arctic peoples, recommends that women there continue to breast-feed their children, because they reckon the health benefits still outweigh the possible harm done by pollutants.[20] But it is a close call.

The good news is that time has been called on PCBs in much of the world. The United States banned them in 1976, and many other countries have followed. The Stockholm Convention on Persistent Organic Pollutants, which was drafted in 2001, prohibits the manufacture or use of a suite of problem chemicals (but, incredibly, DDT is still permitted for pest control in some developing countries). By 2008, 153 countries had agreed to abide by the terms of the Stockholm Convention (fortunately including big emerging polluters like India, China, and Brazil). The PCBs and similar toxins already at large in the sea will cause problems for decades to come, as their release continues from the breakdown of contaminated waste and fixtures in buildings, but levels are falling gradually as they slowly degrade to less harmful chemicals. A survey of the Arctic between 2001 and 2008 found that surface waters contain in total less than half a ton of PCBs, just one thousandth of 1 percent of the world production of these chemicals. Levels in seabed sediments were not tested, and toxins there could keep coming back as they leach from sediments, are resuspended by bottom trawls, or are released by the crumbling remains of contaminated waste.

Heavy metals like copper, lead, zinc, and mercury also contaminate the oceans and have proven far more difficult to control than PCBs. One of the most toxic is methyl mercury, because it is easily absorbed and can cross the blood-brain barrier. Much of the mercury we are exposed to today comes from emissions by coal-fired power stations that countries seem hell-bent on building more of, despite their contribution to global warming. Asian power stations produce over half of the world's mercury pollution, and much of it blows straight over the Pacific, where it combines with particles of organic matter and is converted to methyl mercury by microbes.[21] Mercury concentrations in the Pacific have increased by 1 percent to 3 percent per year in recent decades. Mercury is highly poisonous and accumulates through food webs, so top predators like dolphins, tuna, and swordfish can be heavily contaminated. Tests on Pacific black-footed albatross feathers from museum specimens collected between 1880 and 2002 show an increasing burden of methyl mercury over time, with a surge after 1990, mirroring the growth of the Chinese economy.[22]

Like POPs, heavy metals can act as hormone mimics that disrupt the endocrine system. A much-publicized study of white ibises in Florida found that methyl mercury causes male birds to pair up with other males.[23] It was a gift to headline writers: POLLUTION TURNS BIRDS GAY! White ibises feed in the marshes of south Florida, stalking through the swamp picking up animals such as frogs and insects with their slender, curved bills. Scientists took birds from wild ibis colonies and fed some uncontaminated food and others a diet laced with methyl mercury at concentrations found in the wild. Male birds given mercury-tainted food were less attractive to females, and up to half of them ended up building nests with other males. Not surprisingly, mercury contamination reduced the number of chicks fledged by a third.

Canned tuna is virtually ubiquitous, and the wealthy world has developed a passion for sushi, which is where many of us pick up our mercury. (One estimate suggests 40 percent of bodily mercury in Americans comes from tuna.)[24] Mercury in tuna sampled from restaurants and fish markets regularly exceed levels considered harmful by the World Health Organization. Large tunas, such as bigeye and bluefin, are worst affected, so perhaps it is a good thing that bluefin tuna is far too expensive to be eaten in anything other than sliver-size portions![25] Samples of swordfish from California supermarkets were on average one and a half times above federal guidelines for mercury consumption. While there are obvious health benefits from eating fish, the U.S. Food and Drug Administration is so concerned about the problem it advises pregnant women and children, who are more at risk of harm, not to eat shark, swordfish, king mackerel, and tilefish (a bottom-living fish common off the Carolinas).[26]

Another notorious hormone mimic is tributyltin, which was widely applied to ships' hulls in the 1970s and 1980s as an antifouling compound. Female dog whelks—a type of snail—grew penises when exposed to this compound; oysters developed shell deformities; and some shellfisheries near ports and marinas collapsed. The International Maritime Organization imposed a global ban in 2008 in recognition of its potent toxicity. But there are many other chemicals that remain unregulated.

Despite efforts to rid the world of persistent chemical toxins, our environment isn't getting cleaner—in fact, just the opposite. Chemical industries are constantly creating new products, and phasing out one leads to demand for substitutes. There are about 8.4 million commercially available substances worldwide. Over thirty thousand organic chemicals are produced by industry in quantities bigger than a ton a year, and most have never been formally tested for toxicity.[27] The majority probably pose little risk, but the trouble is we don't find the ones that do until they have been in circulation for years.

Another group of chemicals—brominated flame retardants, or BFRs—was drafted in from around the 1970s to replace some of the roles filled by PCBs.[28] They have found their way into every sphere of modern life. They are slathered over our furniture and packed into circuit boards and plastics used in consumer electronics. Clothes made of artificial fibers are steeped in them, and they are in plastic food packaging and Styrofoam cups. Many governments demand high standards of fire prevention and compel manufacturers to use the best chemicals for the job. BFRs can make up a quarter of the weight of plastic in electronic goods and foam fillers of furniture. Although less toxic than PCBs, BFRs share their propensity to build up in our bodies and concentrate their way up food chains. Mothers pass them to their children during pregnancy and breast-feeding. Just like PCBs, they also seem to be endocrine disruptors that can interfere with child brain development.[29] There is now much anxiety about their safety, and some countries, like Canada, have already banned the most pernicious compounds.

Pharmaceuticals are an emerging class of pollutants that is racing up the ranks of chemicals to worry about.[30] Throughout the world human populations are growing, aging, and increasingly dosing themselves with drugs to combat health problems. Lax controls on pharmaceutical manufacturers in developing countries mean there is an increased potential for release into the environment. Pharmaceuticals are designed to have large biological effects at very low concentrations and, like PCBs, some of them accumulate as they work their way up through the food web. Many are excreted and remain stable in the environment for long periods, during which they can cause considerable mischief. Given the many unexplained declines in wildlife around the world, attention is turning to whether trace contaminants like these could be responsible.

There is little in the way of research on marine life so far, but freshwater species have been better studied. Downstream of sewage treatment works, male fish have been feminized by synthetic estrogen from contraceptive pills and hormone replacement therapy. Ibuprofen, a common anti-inflammatory, reduced reproduction of freshwater fleas in a lab experiment.[31] The good news is that lab tests often produce effects only at higher chemical concentrations than are normally seen in the environment. The bad news is that subtle effects may only manifest themselves over the long term. Many drugs, like antidepressants, are designed to affect mood and behavior. Fluoxetine, the active ingredient of Prozac, can cause symptoms in fish such as erratic swimming, unresponsiveness, and decreased aggression and feeding.[32] Happy shrimps live dangerously, according to another study, swimming away from shelter and into the waiting jaws of predators. Any of these effects could potentially disrupt the life cycle, increase mortality, or reduce breeding success.

Another class of emerging pollutants are nanoparticles, extremely small particles billionths of a meter across that are used to make better cooking oils, improve the efficiency of solar panels, and enhance the drug absorption of pharmaceuticals, among almost countless other possibilities in the pipeline. One familiar use is of silver particles incorporated as antibacterials in underwear and socks. By virtue of their size they can easily pass into emissions and become incorporated into body cells. Research is still in its infancy, but there is evidence that such minute particles can do more damage than larger particles of the same substances.[33] There is currently enormous investment in research and development of nanotechnologies but much less in exploring their possible environmental effects. We should be concerned that this revolution in materials science doesn't outpace caution. For example, filter-feeding mussels exposed to nanoparticles in glass wool developed to clean up oil spills accumulated particles in their gills and vital organs, where they caused cell damage and death.[34] Scientists are diligently trying to engineer nanoparticle-sized pesticides to increase their potency. Although this may decrease the quantities that will need to be applied, the potential for new damage is already clear.

I don't want to leave the impression that the oceans are toxic. In much of the sea, especially far from human habitation, pollution is very low. Dangerous chemicals are concentrated where inputs are high, around estuaries and cities, ports, and shipping lanes. Only a few places are contaminated enough to be dangerous, like New Bedford Harbor in Massachusetts. This site received industrial wastes for so many years that sediments there are loaded with a cocktail of chemical nasties. Fishing is banned and a cleanup is underway to remove contaminated mud. The wider problem is that chemicals concentrate from trace amounts in the water to harmful levels in animals. Tens of thousands of tons of flame retardants are produced per year, and they are already well embedded within marine food webs. Tissue samples taken from a group of false killer whales around the main Hawaiian islands contained high levels of flame retardants, while southern California sea lions had levels forty-five times higher still.[35] The bodies of both species were also laden with PCBs and DDT.

It is easy to be indignant about the ubiquity of chemical pollution and to rail at the rapacious corporations that peddle these products. But we shouldn't forget the thousands of lives they save. Flame retardants have spared many people from being torched in their beds and DDT has prevented countless deaths from malaria. But a balance must be struck between safety and danger. Once chemicals get into the sea it is very difficult to remove them. And we have only recently begun to realize that toxins such as mercury and PCBs combine with another kind of waste that is on the increase in the world's oceans: plastic.

CHAPTER 10

The Age of Plastic

One tempestuous night in 1992 a container ship from China ran into heavy seas in the western Pacific. Huge swells breaking over the ship washed a containerload of plastic bath toys overboard. This set some twenty-nine thousand plastic ducks, turtles, beavers, and frogs free on a voyage that for some has not yet ended. When bath toys began to appear on North American beaches, Curtis Ebbesmeyer, a Seattle-based oceanographer, latched onto the importance of this event as a way to track the movements of the great ocean currents.[1] The toys hitched rides on several currents circling the Pacific, and over the years have made landfall in Hawaii, Alaska, and Washington State. Some even threaded their way through the Bering Strait into the Arctic Ocean, where they froze into pack ice. That ice then bumped and scraped its way around the pole, pushed by winds and underlying currents, before it disgorged the toys into the North Atlantic. These travelers have since been picked up on beaches in Maine and Scotland.

More than two millennia have passed since the great philosopher and naturalist Aristotle walked the shores of the Mediterranean island of Lesbos, deep in contemplation. It was on Lesbos that he laid the foundations for his masterwork of natural history, whose influence would endure until the Age of Enlightenment. Those beaches would have been strewn with natural flotsam, the strandline drawn in palm fronds, seaweed, and seeds. Here and there a fragment of worn plank or the sole of a leather shoe or decayed rope would betray the work of man. Fast-forward to a hundred years ago, and beaches had by this time become littered with the flotsam and jetsam of human societies, but the main difference from Aristotle's time was the quantity rather than the type of rubbish. Fragments of nets, strands of hemp rope, and storm-shattered wooden boat spars would still be there, but this time with glass fishing floats, barrel staves, and bigger heaps of rotting organic refuse carried to sea by rivers. Today a beachcomber is faced with a very different kind of garbage.

Archaeologists excavating the remains of our world two thousand years from now will call this the Age of Plastic. Victor Yarsley, an English chemist, helped usher it in. Yarsley was born in 1901, and early on became interested in the industrial potential of synthetic polymers. He spent his early career trying to develop nonflammable celluloid film, but he eventually abandoned the search, declaring it impossible. He then struggled for decades to find ways to realize the potential of plastics, working long hours from a lab jerry-built in his garden shed. His daughter later recalled having to tape over holes and cracks in her bedroom to keep out the awful smells.[2] Success eventually came when he discovered ways to make plastics without air bubbles and perfected mixtures for new products such as dentures and prosthetic limbs. By 1941 Yarsley felt confident enough to set down his vision for a future world built around the miracle of plastic:

> This plastic man will come into a world of color and bright shining surfaces where childish hands find nothing to break, no sharp edges, or corners to cut or graze, no crevices to harbor dirt or germs. . . . The walls of his nursery, his bath, . . . all his toys, his cot, the molded light perambulator in which he takes the air, the teething ring he bites, the unbreakable bottle he feeds from. As he grows he cleans his teeth and brushes his hair with plastic brushes, clothes himself within plastic clothes, writes his first lesson with a plastic pen and does his lesson in a book bound with plastic. The windows of his school curtained with plastic cloth entirely grease- and dirt-proof . . . and the frames,

like those of his house are of molded plastic, light and easy
to open never requiring any paint.[3]

Predictions about life in the future are often funny and usually wildly
off. According to the books I read as a ten-year-old, by now we should
all be skimming the skies in airborne cars and relaxing in the garden
as robots cook our dinner and clean the house. What is remarkable
about Yarsley's world is that it has come true, and we are living it. By
2008, the latest year for which I have a figure, 286 million tons of
plastics were produced using 8 percent of global oil production in raw
materials and energy.[4] The curve of production over time bends
upward like a cliff face, increasing by 9 percent per year. The stark
reality of this ever-steepening upward climb is that more plastic was
made in the first ten years of this century than all of the plastic created in history up to the year 2000. The world is awash with plastic—
most of us are rarely out of contact with something made of the stuff.
We are literally and figuratively swimming in it. But Yarsley's powers
of foresight deserted him on one key matter. His future was a brighter
and better world:

> a world free from moth and rust and full of color, a
> world largely made up of synthetic materials . . . a world in
> which man, like a magician, makes what he wants for
> almost every need out of what is around and beneath him.[5]

Actually, what is around and beneath us today is an endless vista of
discarded plastic—a world full of color indeed. The wildflower-filled
roadsides of old have for many of us been replaced by a near continuous succession of plastic food wrappers, bags, and lost or abandoned
junk. By weight, about a tenth of our waste is plastic, and a much
bigger proportion by volume. A third of all plastic is turned into disposable packaging, used once and thrown away. We pack our discarded plastic into landfills and recycle some, but a significant amount
finds its way to the sea. There is a river at the bottom of my garden;

when floods rush by I sometimes watch as a stream of rubbish bobs past. One fairly ordinary day I counted them: twenty-seven pieces of plastic in an hour. It is a story repeated in every one of the world's streams and rivers that drains inhabited land. When it rains hard over the Caribbean island of St. Lucia, for instance, the sea turns brown with sediment washed from exposed soil, and the bays fill with bobbing plastic bottles and drifting shopping bags. I once opened my Sunday newspaper to a double-page photo that showed a solid mass of plastic bottles, bags, buckets, shoes, and other refuse. So choked was the scene that it was hard to tell whether this was land or water. Eventually I spotted the face of a little Filipino boy amid the rubbish, swimming in Manila Bay. I was stunned. Years later the image still haunts me.

Rivers launch their garbage into the sea, where some is picked up by offshore currents to join bath toys, sneakers, cans, and stuff lost from boats on transoceanic voyages. Curtis Ebbesmeyer called his book about this floating world *Flotsametrics*, but its subtitle, to paraphrase Daniel Defoe, might be *"The Life and Strange Surprising Adventures of a Plastic Duck."* Actually, there is another book about these toys with the marvelous title *Moby Duck!*[6] The great currents of the oceans move in counter-rotating circles that interlock with one another. Due to the Earth's rotation, winds blow to the right in the northern hemisphere and the left in the southern, giving them an easterly or westerly heading (Coriolis again). The wind tends to come from the east near the equator in both hemispheres—these are called the trade winds—while at middle to high latitudes they blow from the west—the westerlies. Winds move the sea beneath, but again the Earth's rotation gives them a twist, and continents block their movement. The result is that air movements spin up vast rotating currents known as gyres.

Sailors have recognized these winds and currents for thousands of years, but they weren't mapped until the nineteenth century, when a United States naval captain, Matthew Maury, gathered together all of the oceanographic measurements available in a pioneering book,

The Physical Geography of the Sea.[7] Maury recognized at least three different gyres, a number that has increased to nine as modern ocean-ographers have added detail. Curtis Ebbesmeyer added two more small loop currents in the Arctic to bring the total to eleven.

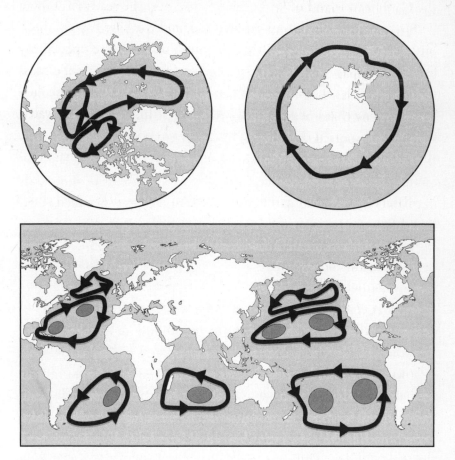

Ocean gyres: Curtis Ebbesmeyer's map of the world's ocean gyres: giant surface loop currents. Gray spots inside the gyres show areas of trash accumulation, the so-called ocean garbage patches. *Redrawn from Ebbesmeyer and Scigliano, 2010.*

In general one finds two main gyres in each ocean, one to the north and another south of the equator. The Pacific Ocean is so vast that it has two in the north. The Indian Ocean has just one to the south, because monsoon winds dominate the north and India blocks water movement there. An unbroken loop of sea circles Antarctica, so the gyre flows all the way around this continent in an endless circle. If you have ever wondered why round-the-world yachtsmen often run into difficulties in the mountainous seas of the Southern Ocean, it's because this is the shortest and fastest way to circumnavigate the world.

Thanks to the Earth's rotation, revolving currents in the gyres push water toward the center, where it piles up into hills. You might have thought the sea was perfectly flat, but there are hills and valleys in the ocean just as on land. Where the crust is thin, such as over ocean trenches, the pull of gravity from the dense mantle beneath is strong and creates hollows in the sea. Admittedly, these are very gentle valleys and hills; the difference between a peak and a valley floor amounts to less than four feet of height over hundreds or thousands of miles. The piled-up water in the middle of each gyre is pulled down by gravity, and there is a downwelling to the deep sea like water flowing down a gigantic bath drain. A nautical almanac published in 1845 described these currents thus:

> It has been observed that the waters of the Atlantic have a greater tendency toward the middle of the ocean than otherwise, and this seems to indicate a reduced level, forming a kind of hollow space or depressed surface . . . in the middle part, although within the region of the Trade-wind, the currents are not regular, but indicate a kind of vortex.[8]

This author mistakenly believed the currents flowed into a hollow in the ocean, when in fact they were pushed uphill by the combined force of the gyre and the Coriolis effect. But he rightly spotted the vortex. The central parts of some gyres—those with high atmospheric pressure air cells above them—gather everything that floats into enormous rafts, which are held there by the unseen hand of downwelling water. In the Atlantic gyre there is a place so distinctive it has a name of its own, the Sargasso Sea. Maury called all of the central regions of the world's gyres Sargasso Seas due to their accumulation of the seaweed *Sargassum*, a floating weed buoyed up by shiny green clusters of floats that to early Portuguese navigators looked like grapes. But the name has stuck only in the Atlantic. Maury wrote:

> Columbus passed through [the Sargasso Sea] on his first voyage of discovery to America. His crew were alarmed at it, for in places it is so thickly matted over with brownish weeds, as to hide the face of the sea, presenting somewhat the appearance of a drowned meadow, and looking as though you might walk on it, without sinking more than shoe deep in water. The sargassos of the southern hemisphere are not so well marked as this, nor are they as rich in drift or floating matter.
>
> You observe that in both the North Atlantic and the North Pacific, the sargasso is in the middle of the pool as it were, or between the warmer current to the west, that flows north, and the cooler one to the east, which returns south. Into this sargasso of the Atlantic are carried the trees and plants and creepers that are brought down from the Andes by the Amazon, and from the Rocky Mountains by the Mississippi and its tributaries. These, with the wrecks of steam-boats that are blown up or snagged, sunk and lost on the Western waters of America, are all drifted out to the sea, and placed within the influence of the Gulf-stream. In the course of time they find their way into this sargasso. On their voyage to it,—their sea of rest,—they become loaded with barnacles and shellfish; arrived there, these die, or, loading their floating domicile too heavily, with increase of inhabitants, it finally sinks, conveying them down with it into the depths below.[9]

When Curtis Ebbesmeyer visited the Sargasso Sea in 1977 he was taken aback not so much by the rafts of weed, but by the number of

Styrofoam cups that blossomed from it like water lilies.[10] His meticulous tracking of ocean garbage has revealed the oceans' remarkable interconnectedness. The great current that circles the planet, the global ocean conveyor, is powered at the surface by many gyres that act together, like the cogs in a clock. Those cogs are of different sizes and spin at different speeds. Slowest of all are the Arctic gyres that rotate about once in thirteen years. The vast gyres of the Pacific Ocean—Ebbesmeyer calls them the Turtle and Heyerdahl gyres, the latter after Thor Heyerdahl, who used it to cross from South America to the Marquesas on a balsa raft—turn once every six and a half years. The loop that circles Antarctica is driven faster by ferocious winds and turns once in three and a third years.

At each cycle some of the floating garbage escapes from the outside edge of the gyre and finds its way to the coast, where it is tossed up onto beaches. Once again, Ebbesmeyer's extraordinary obsession with flotsam gives us a number: About half of the garbage carried is freed at every turn. There are still bath toys out there from the 1992 spill, but not many. At the time of writing, millions of tons of junk that were washed offshore by the devastating Japanese tsunami of 2011 are headed toward Hawaii and North America. The first wave is expected to make landfall by 2012, but as the bath toys show, beaches will continue to receive debris from this disaster for years to come.

For the last several decades, beach cleanups have been a regular activity for marine conservation groups, and an essential task for resort towns. The majority of that junk is plastic. Beach surveys from America to the Southern Ocean, and from the coast of China to remote islands of the central Pacific, show that the ratio of plastics to other trash typically lies in the range of 2:1 to 9:1. Averaged across the European Union there are approximately five hundred to six hundred pieces of litter per 110 yards of beach. This only includes pieces large enough to bother picking up. A survey of Orange County beaches in California that included all visible plastic produced a figure of over 100 million items of trash for the whole county, most of it tiny pieces of plastic.[11] This is equivalent to more than 150,000 items per 110 yards of beach!

Despite the oceans' capacity to carry garbage halfway around the world, plastic is most abundant in populous places, and close to cities. The northern hemisphere is affected much worse than the south, although right now rates of accumulation of beach trash are increasing faster in south polar latitudes than anywhere else.[12] Each year about two thousand items of human-generated trash (most of it plastic) washes ashore in the northern hemisphere, compared to about five hundred in the southern, for every kilometer (0.63 miles) of beach.[13] Enclosed seas like the Mediterranean are affected worse than open ocean coasts. There junk can wash in at eighteen hundred pieces per 110 yards of shore per year.

Most plastic is easily overlooked, too small to be seen by a casual glance, and much too small to be included in these surveys. Next time you are at the seaside draw a square about the length of your forearm on the beach near the strandline. Then pick from inside it every fragment of plastic you can find. You will need to look very closely, since much of it will be less than a quarter-inch across. Many of the pieces will be rounded drops of white or yellowed plastic known as nurdles, or mermaid's tears. They are the raw material of the plastics industry—pellets ready to be molded into finished products. Vast quantities have been lost at sea and in spills on land over the years. Give yourself plenty of time, because you will probably find hundreds, perhaps thousands of them. A survey of some New Zealand beaches found up to three hundred thousand per linear meter (thirty-nine inches) of shore! Of the plastic found in the Orange County beach survey 98.5 percent were nurdles.

Plastic stuff gets thrown away in such abundance partly because it doesn't last very long. I am often annoyed at how something made out of decent materials like wood or metal has to be junked because some critical bit of plastic in it has broken. But that fragility belies an incredible durability in the substance itself. Plastics break down very slowly, but over time they fragment into ever-smaller pieces. When you get bored picking them from your plot of beach, why not relax with a swim? As you enjoy the caress of waves, consider this: The water around you probably contains a multicolored soup of tiny plastic fragments. While there is plenty of the stuff near coasts, it has built up in offshore waters, and spectacularly so in the Kuroshio Current that flows past Japan and East Asia. There densities of plastics were reported up to nine million per square mile of sea.[14]

Not all plastic floats. Roughly half is denser than water and sinks at various rates, while the other half is what we are familiar with from ocean garbage patches, as Curtis Ebbesmeyer dubbed the middle regions of the gyres. Few surveys have been made of seabed litter, so there is little to go on, but it is clear that there is another hidden problem that we will soon have to reckon with. Contamination in some parts of Europe can reach levels of a piece of plastic for every square yard of bottom.[15] Plastics and other rubbish have made it to the deep sea as well. Scientists who deploy instruments on the deep sea bed often return to find them clogged with drifting plastic bags. One submersible pilot reported that plastic bags passed him by like a succession of ghosts. Heavier rubbish thrown from boats reaches the bottom of the sea quickly. The deep sea is not so remote in distance, even if to us it is a profoundly alien environment.

I mentioned that ocean gyres lose about half of their flotsam at every cycle. Some of it comes off the outer edges and escapes toward the coast, but the rest goes inward. Just like the mats of floating seaweed that so amazed the sailors of old, great rafts of plastic, fishing line, nets, ropes, and a thousand and one other bits of junk have accumulated within the gyres. In a story often retold, a yacht captain, Charles Moore, took a different route from usual between Hawaii and Long Beach, California, in 1997 and crossed through the center of the gyre in the northeast Pacific. He was so stunned by the amount of garbage he found there that it changed the course of his life. Since then he has dedicated his life to the study of plastics at sea, and he is now a relentless campaigner for something to be done about them. One of his most famous findings was that there were six times more floating plastic fragments in the surface water by the turn of the millennium by weight than zooplankton.[16] The gyre Captain Moore traversed has gained notoriety as the Great Eastern Garbage Patch, at about the size of Texas one of the world's largest. Surveys across the Sargasso Sea show the scale of contamination. On average, plankton tows there revealed densities of floating plastic fragments in the order of hundreds of thousands of pieces per square mile.[17]

Plastics are completely foreign to life, so animals have been unable to mount any evolutionary defense against them. The Laysan Albatross, which nest on Kure Atoll in the central Pacific Ocean, have the misfortune to live close to the Great Western Garbage Patch, as the center of the gyre on the western side of the Pacific is called. Albatross feed by picking live and dead prey from the sea surface. Unfortunately, they can't distinguish between plastic and flesh. Researchers recently found that adults returning from long-distance foraging trips fed their chicks an average of seventy pieces of plastic per meal![18] After the chicks starved to death and their flesh rotted, they left piles of fishing line, golf balls,[19] pens, bottle caps, and sundry other fragments, each framed by a mute halo of feathers. Some chicks contained more than five hundred plastic items. The plastic content of chicks has increased tenfold in the last forty years, presumably in line with increases in the mass of trash circling the Pacific. As early as 1965, three quarters of Laysan chicks found dead had plastics in them.[20] There is a heartrending possibility that albatross deliberately fly to areas of garbage concentration for the rich "feeding" opportunities they provide.

Autopsies of leatherback turtles in the Atlantic Ocean show that this beast has similar difficulties telling food from plastic. Leatherbacks eat jellyfish and other gelatinous zooplankton. To a dewy-eyed leatherback, plastic bags and balloons look much the same as prey. A study of dead leatherbacks washed ashore since the 1880s found plastics present from 1968 onward.[21] Some contained lumps of tangled plastic bags and Mylar balloons (the helium-filled variety we give to kids) the size of a football. Today over a third of dead animals examined have plastic in their guts. The strain of hauling itself ashore to nest obviously got to one leatherback in French Guiana. It pooped a huge plug of plastic bags and sacks that weighed over five pounds.[22] Normally turtles fast during the breeding season and don't shit on the beach. How many other bags, I wonder, have been recycled through the guts of turtles far out to sea, and how many turtles' lives are threatened by the plastics packed in their bellies? As jellyfish are 95 percent water, they make a pretty meager meal. Adult leatherbacks must consume hundreds of pounds of jellyfish every day to break even, over half their bodyweight. Even if plastics don't kill outright by blocking the gut, they could interfere with digestion and cause animals to starve by tipping their caloric balance into the red.

Hundreds of other species ingest plastics, either deliberately or by eating something else that has. In the North Sea, nineteen out of every twenty fulmars—a smaller relative of albatrosses—that washed up dead on beaches contained plastic. On the Dutch coast, four out of five plastic fragments had peck marks from birds.[23] Plastics have reached some of the remotest places on the planet. Fur seal scats at subantarctic Macquarie Island contain many plastic fragments that were eaten by the deepwater fish they prey on.[24] Even whales have eaten plastics. A pygmy sperm whale stranded on a New Jersey beach in 1993 had its stomach blocked by fragments of plastic bags. The whale recovered after they were removed, and it was released into the Gulf Stream.[25] Others have been less fortunate. A dead pygmy sperm whale stranded in Texas had a plastic garbage-can liner, a bread wrapper, a corn chip bag, and two other pieces of plastic sheeting choking off its stomach.[26]

These whales may have taken in plastics accidentally while feeding, but there are signs that whales target pieces of plastic junk deliberately. Two sperm whales that died on the California coast between them contained over two hundred pounds of net fragments, line, and plastic bags.[27] The pieces packed into their stomachs came from all kinds of nets that may have been up to twenty years old and seem to have been tossed overboard by fishermen mending their gear.

I once came across a photograph taken far out at sea in one of the oceans' garbage patches. A huge ball of fishing net suspended by a couple of hundred floats stretched across the frame for at least ten yards. At one end, a sad looking and exhausted turtle stared at the camera, helplessly entangled. This poor beast had dragged the net with it across how many miles of ocean nobody could tell. Like Sisyphus in Greek mythology, it was condemned to a life of immense hardship undertaking a task that could never be completed. Some of the most disturbing images of the harm done by human detritus are of sea lions that have put their heads through loops of plastic or net as pups. The loops tighten as they grow and cut a deep, bloody slash into their necks. When the sea lion finally dies from this slow-motion throat cut, it decomposes and leaves the loop of plastic to be picked up by another. The number of seals entangled in plastic and fishing line on California's Farallon Islands rose from a handful in the 1970s to more than sixty a year by 2000.[28]

Plastics came into widespread use after World War II, so pretty much all of the plastic at-large on the high seas got there after 1945. Since then the tide of plastic pollution has grown relentlessly, and it is fragmenting into ever-smaller pieces. It is often said that plastics take hundreds or even thousands of years to degrade. Japanese researchers found that plastics floating in the North Pacific for decades were degrading much faster than expected.[29] This apparently good news story has a poisonous twist. These particles are not just harmless roughage. They concentrate toxic compounds on their surfaces, sometimes to levels a million times or more above concentrations in the seawater around them.[30] Recall from the last chapter that the surface microlayer of the sea, in which many plastics can be found, concentrates toxins. They attach to plastics. In one Japanese experiment, polystyrene beads soaked in seawater for several days picked up PCBs. When they break down, plastic particles release toxic compounds like flame retardants, styrene, phthalates, and bisphenol A into the sea. These toxins exert their insidious effects on wildlife, and us, at minute concentrations.

Bisphenol A and phthalates are ingredients of a large variety of plastics. Phthalates are added to increase flexibility and transparency, but they don't bond with the plastic and so are easily leached out. Bisphenol A is added to the resins that coat the inside of food tins, among many other products. Both have become highly controversial in recent years because of health concerns about their endocrine-disrupting properties, and both have been banned or restricted in a number of countries, including Canada, Australia, the United States, Japan, and those in the European Union. Some organizations, like the UK's Food Standards Agency, insist that such substances are safe, but research is ongoing, and many subtle developmental effects have been detected in lab studies. Phthalates are attracted to fats like those they would contact in the surface microlayer. Given the ubiquity of these compounds in plastics, the possibility of harm to marine life (and us) must be taken seriously.

We have only recently been confronted with the plastic confetti that is the legacy of decades of pollution. In the late 1990s Hawaiian beaches were inundated with a multicolored snowstorm of plastic fragments. People there were used to bits of plastic bobbing around but had seen nothing like this before. Periodically loops of water twist off the vast ocean currents that circle our oceans to escape from the pool trapped inside the gyre. In some cases this might happen but once in half a century. It seemed that the ocean gyre near Hawaii had belched a gutful of plastic that had been building within it since plastic came into widespread use.[31]

An alarming new twist is that most cosmetics manufacturers now add submillimeter-sized plastic granules to hand lotions and face creams as exfoliants. They are too small to be filtered out by sewage works, and most of the particles are washed to sea, where they can be ingested by tiny plankton that mistake them for food such as cope-pods or fish eggs. Like other plastic fragments, these granules have a high surface area with which to attract and concentrate poisons like

PCBs and mercury. Since plankton are the foundation of almost all ocean food webs, the problems this could create are clear. It is nearly impossible to buy an exfoliating face or hand cream today that doesn't include plastics. Take a look at the labels on yours. If you find poly-ethylene in the list of ingredients, you are washing yourself with plastic.

Many of the particles at-large in the oceans range from a few hundredths to a few thousandths of an inch in size. They now fall into the size range of planktonic food for a huge variety of life at the bottom of the food web. A third of plankton-feeding fish sampled in the North Pacific Central Gyre had plastics in their guts.[32] Experi-ments with captive animals show that microplastic particles are eaten by almost any species that filters water, picks drifting particles of food, or slurps deposits from the seabed. Few people have looked, but in places where they have, beach and bottom sediments are stuffed full of microplastic particles.

These microplastics have concentrated chemicals that could pass into animals that eat them, providing a fast track for the accumula-tion of toxins all the way up to top predators that we like to eat. Scarcely any research has been done on the transfer of chemicals from plastic fragments to animals that eat them, but this is certain to change in the next few years, because the world's attention has turned to plastics at sea. What we do know at this point is that plastic par-ticles fed to mussels end up in their circulation system and stay there for long periods. Chicks of a scavenging seabird, the great shearwater, that carried higher loads of ingested plastic had higher concentrations of PCBs in their tissues.[33]

Remote regions of the ocean, like the Sargasso Sea and Northeast Pacific, have become slowly rotating graveyards of plastic junk, some of it decades old. The tangled remains of lost and abandoned fishing nets drift past golf balls, toothbrushes, gas lighters, and plastic bags. For how many hundreds or thousands of years will they continue their aimless voyages? In one telling incident, a dead albatross chick from Midway Island in the North Pacific had been fed a piece of plastic engraved with a serial number. It was traced to a U.S. bomber that crashed into the sea in 1944.[34] The amount of plastic at-large is still growing exponentially year by year. You come face-to-face with this truth on virtually any beach in the world. In South Africa, for instance, bottle lids on beaches, used as an indicator of the prevalence of smaller plastic rubbish, increased by over fifty times between 1984 and 2005.[35] That is not 50 percent, but 5,000 percent. The genie is already out of its plastic bottle. The oceans are choking on plastics and will continue to do so for hundreds of years, even if we were to stop dumping plastics today. But it is never too late to start the cleanup.

Problems from plastics, sewage, oil, and toxic chemicals are all too familiar, but there are some things we don't normally think of as pollution, at least in the sea. Noise and "biological" pollution—that is, the spread of species beyond their native haunts—are growing concerns, and I turn to them in the next two chapters.

CHAPTER II

The Not So Silent World

When I was a graduate student I spent three unforgettable summers exploring Saudi Arabia's coral reefs. I was part of a team sent there to map its biological riches for the first time, and we were allowed to visit places long closed to foreigners. I remember well the keen thrill of slipping into the water, knowing that I was the first person ever to scuba dive there and wondering what I might discover. In 1984, we reached Tiran Island by sailboat after days beating up the Red Sea against bone-jarring waves raised by powerful northerlies. It was bliss when at last we found shelter in the lee of Tiran's mountainous flank. We had hardly seen a soul for weeks, and Tiran felt like the end of the world. That night, far across the sea, a few lights from the Egyptian coast suggested we were not alone. But the peace was palpable.

The next day I dived one of the coral reefs that defend the entrance to the Gulf of Aqaba. Below water the bottom was hard to make out at first through a blizzard of plankton-feeding damselfish, basslets, and glassfish that swayed to the silent pulse of unseen currents. Purple sea fans and chocolate sea whips bent and wafted on the liquid breeze amid a tapestry of corals of marvelous shapes and colors. Surrounded by such tranquil beauty, I lost all sense of time.

A little more than a decade later I returned to dive the same spot, this time setting forth by day boat from the Egyptian resort of Sharm-el-Sheikh with fifteen other divers. We charged north with ten other boats, all hoping to be the first to tie up to one of the moorings that marked a string of dive sites. The Egyptian coast behind me had been utterly transformed. Whereas before there had been only a few dwellings, now resorts and hotels studded the clifftops, coves, and beachfronts. At the dive site I had to pick my moment carefully to leap into the sea to avoid the propeller blades of two latecomers jostling for space.

I expected to find immediate relief from the confusion and fumes above water, but even thirty feet below the surface I was overwhelmed by the engine roar from boats overhead and others coming and going from more distant sites. The near-constant bombardment made this dive feel like the underwater equivalent of standing in the middle of a multilane highway. I tried to find a quiet corner to while away the dive, but the noise was everywhere. From a rocky ledge a stonefish looked at me with an unblinking scowl. It seemed an appropriate expression. I wondered whether the fish on this reef were relieved when the last boats left in the evening and a semblance of peace was restored.

Jacques Cousteau published his first book in 1953. He called it *The Silent World* but his title was misleading. As well as the timeless notes of whale song, Cousteau's seas were filled with the scratch and rasp of foraging animals, the whistles and clicks of dolphins, the low groans of whales, and the rumble of breaking waves. By that time the thump and growl of boat engines had also invaded the oceans. Noise levels have grown to a roar since the 1950s as the world's globalized economy has launched tens of thousands of new merchant ships. The sound of a supertanker is ear-splitting, and because sound goes further underwater than in air, it can be heard by a whale a day before the ship arrives. Background noise levels in the sea have not been measured systematically, but in the few places we know about, they have increased by about three decibels per decade since the 1950s.[1] The loudness of sounds underwater doubles for every six decibels. So today's seas could be over eight times noisier than those that Cousteau taught us to love in the early 1950s, and in some hot spots are over one hundred times as loud.

A few years ago I had a haunting experience off the tiny volcanic island of Saba in the eastern Caribbean. I had begun a dive to count fish when I heard a loud drawn-out rumbling growl, or rather I felt it as it thrummed through me. I spun around, expecting to see a leviathan bearing down, but none was there. Then came the unmistakable melodious moan of a humpback whale. For the rest of the dive I listened to a song as beautiful as any by Verdi. I never saw whales on this or many subsequent whale-song dives in Saba. They were probably far away, perhaps many miles, for their songs have evolved to carry enormous distances in the sea.

The much greater density of seawater carries sounds farther and faster than air. Noise travels about five times more quickly in the sea, reaching five thousand feet per second in temperate waters. It drops off less quickly in water than in air, so noises can be heard at far greater distances.[2] High frequencies attenuate more quickly than low frequencies, so deep notes go farther. Some of the great whales communicate over hundreds or even thousands of miles using the kind of low frequency rumbles I heard in Saba. Sounds go farthest of all if animals take advantage of a strange property of the sea. Between the warm lighter layer at the surface and the cool dense water below there is a narrow region where sounds travel in a similar way to light in a fiber-optic cable. Differences in density across this boundary—the thermocline, we have come across it before—keep soundwaves produced there in a narrow channel, where they lose little energy as they travel. Before the days of satellite beacons the U.S. Navy issued pilots with an explosive charge. If they had to ditch in the sea, they were to lower the charge on a string and explode it in this acoustic pipeline. The sound would carry to sensors perhaps thousands of miles away in places like California, Hawaii, and Panama that could then work out where the pilot was by triangulation.

Several species of whales are thought to use this aquatic telephone to broadcast calls over hundreds of thousands of square miles of ocean. Their voices are well adapted to be heard from far away. Blue whales are the largest and loudest animals on the planet, and they can bellow at over 190 decibels. Sperm whales are also very loud. Doug Anderson, a cameraman on the BBC's *Planet Earth* series, once came across a newborn sperm whale calf in the Azores that he began to film while the mother hung around below with an older calf. The baby played with him for a while before the camera housing caught in one of its fetal folds. This obviously hurt, because it gave out a stream of clicks and blew hard. Anderson turned to find the mother's head filling his entire field of vision. He thought she would charge, but instead there was a crack like the splintering of a falling tree, so loud it shook his body and left him stunned.[3]

The peace and quiet of the oceans is today broken by far more energetic sounds than ships. All over the world, seismic bangs from oil and gas exploration race through the sea to probe the rock beneath. Survey ships have air guns that produce pulses of sound above 200 decibels to send shockwaves through the seabed. The boom of military sonars used to track submarines can reach 235 decibels. Things sound about 61.5 decibels quieter in the sea compared to air.[4] Sonar noises and seismic blasts therefore come in at an above-water equivalent of around 150 decibels to 175 decibels. For comparison, a rock concert can reach 110 decibels, and people feel pain from 120 decibels, or about the noise of a big chain saw close up. The sound of a seismic gun is over thirty times louder still, and the sound of the most powerful military sonar more than a thousand times louder.

It isn't quite this simple, and I must admit to shameless anthropomorphism in my description of the disgruntled Red Sea stonefish in the chapter opening. There is a second and equally important dimension to sound, which is frequency. High-pitched noises have a high frequency, whereas low-pitched sounds have a low frequency. Most animals can only hear over a limited frequency range. Bats hear higher frequencies than we can. My stonefish would have been stone deaf to many of the sounds that you and I hear, since its hearing range is set lower than ours (although boats produce loud noise at low frequencies, so the reef would probably have been as noisy for the fish as it seemed to me). What would constitute a gut-churning blast of sound for a human might register as only a whisper or not be heard at all by some other creature.

Beaked whales are especially susceptible to sudden loud noises in the sea. These odd animals look like a whale crossed with a porpoise or dolphin. Some have blunt foreheads and thick fleshy lips, while others have drawn-out snouts, like bottlenose dolphins. They are shy, can dive for more than an hour, and rarely spend more than a few moments at the surface, so they are hardly ever seen. We think there are twenty-one species, but there could be more. These whales are so enigmatic they have been called "the least understood group of large animals on Earth."[5] Some species have never been seen alive and are known only from their dead bodies when they washed ashore. They range in size from thirteen feet long and a ton in weight to forty feet and sixteen tons. It is hard to imagine that we could overlook animals the size of elephants or bigger so easily, but it is their lifestyle that makes them elusive.

Peter Tyack from Woods Hole Oceanographic Institution in the United States has devoted his life to understanding how whales use sound to communicate. I once spent a lively bus journey with him crossing Santa Cruz Island in the Galápagos. He completely turned around my notion of what it means to be a whale—in just an hour. Until that time I had thought of whales as animals that lived at or near the surface but dove into the deep blue to find food. Peter said that we should think of beaked whales as deep-living animals that occasionally pop to the surface for breath. The abyss is their home. It is this deep lifestyle that makes intense sonar noise a real problem for them.

In 1996 there was an unusual mass stranding of a species called Cuvier's beaked whale in Greece's Kyparissiakos Gulf. When whales strand together they usually all come ashore at once in the same place. This time the whales came ashore separately along twenty-five miles of coast. No wounds or other signs of harm were found on the bodies. Most of the whales' stomachs were full of squid, which suggested they had fed just before their deaths. For Alexandros Frantzis, then a young biologist at the University of Athens, the smoking gun was a NATO research vessel that had been in the Gulf to test a naval sonar system at exactly the time the whales died.[6] The system used extremely loud sounds at low frequencies—some well within the range of whale hearing—to detect ultraquiet submarines. He became convinced that the sounds had somehow either injured the whales directly or disoriented them so badly that they beached themselves inadvertently.

Beaked whales of many different species have since stranded in similar ways, and almost always the military are in the vicinity testing some extreme noise-making device.[7] Closer scrutiny reveals these victims to have suffered serious internal injury. Some even come ashore bleeding from the ears. Like human divers who have spent too long underwater, many whales had suffered decompression sickness.[8] When divers breathe at depth, nitrogen gas dissolves in their blood. As any diver knows, spend too long too deep or ascend too quickly and the nitrogen will fizz out into your blood, causing the bends. Severe cases can kill you. As the whales surfaced, bubbles of nitrogen formed in their blood, fat, and organs, literally tearing them apart internally and blocking their circulation. How these whales get decompression sickness is still a bit of a mystery. Initially people thought it was because they had been frightened to the surface too quickly. But Peter Tyack doesn't think this would cause injuries of the kind he saw.[9] Instead, he believes that when sonars scare whales they spend too much time bounce diving between 100 feet and 250 feet deep, snatching short breaths at the surface, and so they accumulate dangerous amounts of nitrogen in their blood.

There are other sources of intense sound in the sea. Physicists working with U.S. defense agencies in the 1990s came up with the idea of using sound transmission to measure the temperature of the sea and thereby track global warming. The idea called for 195 decibel pulses of sound to be produced three thousand feet down between Hawaii and California for twenty minutes every four hours . . . continuously. Test transmissions were made between 1996 and 2006. Whose brilliant idea was this? If noise traveled as far above the sea as below, and with as little attenuation, we would have dismissed the scheme after little more than a moment's contemplation. But physicists are sometimes inclined to forget that there is a living world. In one spectacular example, in 2005 they sank an array of highly sensitive hydrophones—instruments that record underwater sounds—into the Mediterranean off the coast of Sicily.[10] They hoped to listen for the sounds that tiny subatomic particles called neutrinos might make as they smacked into water molecules. Because neutrinos are vanishingly small, the chances of that happening are low, even in a huge volume of sea. More than a hundred physicists collaborated on the project, but the possibility that background noise might interfere was not even mentioned until late in the day. Everyone assumed that the deep sea would be virtually silent. Of course, there was far too much noise, even before boat traffic was factored in, to detect the sound of a colliding neutrino. Instead, the hydrophones proved adept at pinpointing sperm whales using echolocation clicks to hunt squid in the liquid darkness. At least the physicists discovered something: There were more whales around than anyone had suspected.

The tests proved the viability of using sound as an acoustic ocean thermometer, but the method remains controversial. Research showed that humpback whales seemed to avoid the sound sources, but nobody can say for certain what the long-term impacts of these sounds might be.[11] Since physicists still have plans to create a web of sound throughout the world's oceans to measure water temperature, this question needs an answer.

The loudest noises in the sea cause the most dramatic effects and have attracted the most concern. But the rise in background noise from boat engines and now wind farms is probably more significant. Sound is one of the most important means of communication for creatures underwater. In the clear water of a coral reef you can see 150 to 200 feet at best. Usually visibility is far less; in the dark and muddy waters of estuaries it can shrink to under three feet. Because sound travels so far and fast, it is the best way to communicate. Many marine animals therefore have excellent hearing and use sounds to visualize their surroundings, to avoid predators, to attract mates, and to find prey. Background noise can mask important sounds and make life more difficult and dangerous.

Whales and dolphins, as their amazing vocalizations suggest, will likely number among those worst affected by noise. The din of boats is shrinking their world. Before we invaded the sea with our noise, fin whales could hear each other over distances of at least six hundred miles. It seems that only male fin whales produce very loud calls, perhaps to find and attract mates over these vast distances.[12] Given the spectacular decline in whale populations caused by two centuries of commercial whaling, it must be much harder now to find mates. Many kinds of whale have nowhere near recovered from whaling, so long-distance calls could be a vital way to reach out to receptive mates. Today noise from ships and wind farms fills the sea with a low-frequency rumble that masks their calls beyond distances of just six miles. Because area scales as the square of distance, the space over which they can contact one another has shrunk ten thousand times.

Sarasota's dolphins are famous for more than just the chemical toxins in their bodies. They know one another by "name." Each dolphin has a distinctive whistle that others recognize.[13] When boats pass by, the dolphins whistle to each other more often and move closer together, which suggests they worry about traffic. Several have been badly injured by boat strikes, so they have good cause. Since boats pass within three hundred feet of a dolphin roughly once every six minutes during daylight hours, they could interfere with the dolphins' ability to feed, breed, and look after their young.[14] Another explanation for the more frequent calls when boats are about is that the deafening sound of engines makes it harder to stay in touch.

Fish exposed to boat noise or other loud sounds suffer stress. Smallmouth bass live in the lakes and streams of North America. The hearts of captive fish beat faster when they were exposed to the sounds of boats, and engine noise produced a stronger and more prolonged response than the sound of a paddled canoe. Several kinds of freshwater fish exposed to boat noise have been found to react by producing more cortisol.[15] This is the fight-or-flight hormone that increases blood sugar levels. It is found in a remarkable array of species throughout the animal kingdom, from fish and birds to ourselves. Animals release it in response to acute danger to mobilize energy, sharpen thinking, and dull pain. But they also produce it in response to longer-lasting stress. Wired executives and harassed mothers have raised cortisol levels. This is when benefits turn to costs. Cortisol can wreak havoc within our bodies. It raises blood pressure, increases risk of heart attacks and strokes, and suppresses the immune system. Fish and other animals in noisy, stressful places may also suffer health problems.

In an experiment that entertained me mightily, Greek researchers played Mozart to captive carp for hours every day over several months.[16] Fish grew faster in the presence of Mozart than with the usual background noise of pumps and filters. The researchers thought that Mozart might have soothed away their stress. Whether or not you believe this, I am sure the composer himself would have had a laugh.

Wind farms are a relatively new source of sound, but given the push for green energy, one that is set to become ubiquitous throughout large regions of shallow sea. Pile-driving noise during construction is loud enough to damage the hearing of dolphins or other marine mammals that stray within a thousand feet. During operation wind farms produce a low-frequency underwater rumble whose volume depends on wind speed. These noise levels are probably similar to those produced by ships, but the difference is that they will affect the same area continuously. There has been very little research on how marine life might respond. On the one hand, it might render large areas of habitat unsuitable for sensitive animals. On the other hand, the footprints of wind farms often exclude fishing and so might benefit some fish and shellfish. It is too soon to tell for sure.

Fish and many other forms of marine life use sounds to visualize the world around them. Think of the way that sound enriches your own appreciation of where you are. The tapping on the window tells you it is raining outside. The whirr from the kitchen lets you know the dishes will soon be clean. The sound of footsteps upstairs shows the kids are in their bedrooms, while the light patter from the corridor tells you to expect a hungry cat to come and chance its luck. We may not think of ourselves as being particularly aural creatures compared to bats or owls, but our ears give us a good sense of the world around us.

The noisy underwater world sets the scene for marine animals in much the same manner but with an important difference. The far greater density of water means that sounds move particles in ways that can be felt. Fish have evolved the lateral line organ along their flanks to sense this particle displacement. Like us and other verte-brates, fish have ears with internal "bones" made of crystalline cal-cium carbonate to hear sounds, too. They share with us the ability to pick out specific sounds from background noise, just as we can hear the voice of a friend through the buzz of a party or tell which instru-ments are being played in an orchestra. For a time it was a mystery why goldfish, which don't make noises themselves, had such finely tuned hearing.[17] Trapped within their paradigms, scientists some-times fail to notice the obvious. Clearly, there are benefits from listen-ing to sounds other than your own voice!

The soundscape underwater has been called "acoustic daylight" for the way it reveals things about the world that cannot easily be seen. Larvae of fish from coral reefs must escape the reef for life in open water where they grow for a period of days to months before they are ready to return as fully formed juveniles.[18] If they cannot find a reef at this point, they die. Many larvae follow sounds they might have heard before they hatched from the egg to guide them back when the time is right—the whisper of breaking waves, the crackle and rasp of shrimp and crab, the scrape of parrotfish beak on coral. Even some coral larvae possess the ability to hear reef sounds.[19] It is easy for us to picture how animals make sense of these sounds, but particle motions from underwater sound are harder for us to comprehend. Different sources probably produce distinctive particle motions that paint the surrounding scene with details we can only imagine.

Most marine animals use sound to tell when predators are about:

the whoosh of a school of baitfish fleeing the thunderclap of a predator's strike; the crack of a snail being crushed; the pistol shot of a snapping shrimp (actually, their claws move so fast to smash prey they produce a sonic boom![20]). Some also call for mates, or use sounds to weigh the risk of a challenge for territory. Drums, as their name implies, are fish that form noisy spawning aggregations in estuaries, thumping out the rhythm of procreation on their swimbladder instruments. Red hind groupers in Puerto Rico surprised researchers by shouting "whoo hooo" as they rushed to spawn.[21] Background noise might make it harder for larvae like those of coral reef fish to follow sounds to the right habitat. Already the number of offspring produced by many fish and invertebrates has fallen precipitously because of overfishing. Noise is yet one more way in which the viability of their populations may be compromised. Few other species have been tested to see how important noises are to navigation, but it seems unlikely that only coral reef animals have this ability, so the problem could be widespread. There can be more immediately harmful consequences of background noise. Hermit crabs in the Caribbean, for instance, became less attentive to possible predators when they were distracted by the sound of passing boats.

The underwater world is now awash with noises that evolution has not prepared marine animals for. The maigre is a kind of drum, a lithe fish of metallic gray that lives in Mediterranean estuaries. In the days before intense fishing, it grew nearly as big as the average man, but these days few reach more than three feet. Maigres exposed to the noise of a powerboat in experiments were one hundred times less responsive to the sound of other maigres. Like right whales, the seascape over which they could communicate would shrink ten thousand times when boats pass nearby.

Male toadfish "sing" by vibrating their swimbladders with "sonic muscles" to entice choosy females to their nests and warn off rivals. Lusitanian toadfish are flat-headed and have wide rubber lips that make them look like their heads have been run over by something heavy. They live along the coasts and estuaries of the east Atlantic and Mediterranean, where males set up territories among rock piles from which they glower at one another. There they display to female fish with a buzz called a "boatwhistle," though a better description would be an extravagant fart. The most appealing males persuade females to leave behind a hoard of perfectly round, golden eggs for them to fertilize and protect. When captive toadfish were played the sound of ferryboats in a 2005 experiment they found it hard to hear one another.[22] Someone has turned up the stereo at the party and now it is impossible to hold a conversation anymore. Just as revelers can switch to sign language, there may be other ways that animals can communicate, but at the expense of subtlety and with an increased risk of misinterpretation.

The Pacific coast of North America is home to a relative of the toadfish whose vocal dexterity throws light on the evolution of sound communication in backboned animals such as ourselves. Midshipmen are fish that look a bit like overgrown tadpoles, with round, flat heads and narrow tails. Some are drab, but others blaze as if sprinkled with gold dust. Lines of glowing photophores decorate their bodies with points of light, like Christmas lights. These fish live in the intertidal zone, where males set up mating territories in pools beneath rocks. From these muddy grottos they sing love songs to passing ladies and grunt warnings to other males. Their songs are short on melody and make Tuvan throat songs seem operatic by comparison. Mostly they just buzz, like lawn mowers drifting through summer suburbs, sometimes interspersed with *boinnngggs* reminiscent of twanged rubber bands. But inside their heads, the part of the brain that controls their songs turns out to be the same part that frogs, reptiles, birds, and people use, suggesting that the ability to "speak" to one another evolved some four hundred million years ago.[23] These fish can get so boisterous that they keep houseboat owners in San Francisco Bay awake at night.[24] One resident described the sound as akin to the noise of a generator; another said it was like ten electric razors run at once.

The clutter of the low-frequency hum we create in the sea overlaps pretty closely with the hearing ranges of fish and whales. There are signs that some try to compensate. Northern right whales call louder when there is background boat noise.[25] The more noise there is, the louder they yell. They also call about an octave higher than they did in the 1950s, probably in an effort to be heard above the low-frequency drone of engines. But there is a limit to how loud you can shout. As noise levels continue to rise, whales will become increasingly isolated. Since there are fewer than four hundred northern right whales left, isolation could seriously threaten their survival. Alternatively, they might be able to compensate by calling during quiet periods or by making longer calls. Killer whales in three British Columbia pods made longer calls when surrounded by boats of whale watchers than when left in peace, which suggests that they struggle to communicate above the clamor.[26]

Birds try to compensate for noise in similar ways. Great tits in cities sing shorter, faster songs at a higher pitch than those in the countryside.[27] Birds next to busy roads often have lower breeding success and are less fit than birds in quiet places. Oven birds nesting in Canada close to a noisy generator couldn't get their messages across and attracted fewer mates than those farther away.

Noise pollution interferes with feeding. If an animal stops foraging every time a boat bears down on it, it may soon go hungry. It also affects species more directly if they use sound to hunt. Beaked whales hunt by echolocation in the deep sea. They produce a stream of clicks, made by blowing air through a structure in their blowpipe affectionately known as "monkey lips." The noise is beamed through enlarged lobes of acoustic fat in their heads, and they receive echoes back in the flesh beneath their jaws. Beaked whales are particularly fond of squid. Aside from a thin, horny internal sheath, squid have a similar consistency to water, which makes them hard to detect. I imagine that you would need to listen closely to find a squid in the dark. A Cuvier's beaked whale recorded hunting when there was background noise had less than half the success others had with no sound interference.[28]

Some whales and dolphins use echolocation to find their way around. If their hearing is damaged they may be prone to stranding in shallow water, or to entanglement in fishing nets. Experiments with captive bottlenose dolphins exposed to high-intensity sonar showed short-term hearing loss.[29] Longer exposure may cause permanent damage. When stranded animals rescued in North America and the Caribbean were tested, a third of rough-toothed dolphins and over half of bottlenose dolphins had levels of hearing loss that would have been judged severe or profound in humans.[30] Endangered northern right whales thread their way through some of the busiest shipping lanes on the planet as they migrate up America's east coast to summer feeding grounds. If noise impairs their hearing, they may more easily fall victim to a ship strike. This is now one of the two main causes of death for these whales, the other being entanglement in fishing gear. Even whales with normal hearing find it impossible to hear one another when ships are closer than a mile or so, which is the case much of the time.

Human noise in the oceans is a relatively recent phenomenon. For most of history the only sounds of any note would have come from the clatter of oars against wood and the hiss and slap of boat hulls as they scythed through waves. That changed in the middle of the nineteenth century, when the first paddle steamers were launched. But it was in the last two decades of the 1800s that the peace was broken in a major way, as tens of thousands of steamboats were built. The first offshore oil wells were drilled off Santa Barbara in California in the 1890s, starting a raucous rush for oil that shows no sign of letting up. Since then, background noise has grown inexorably, tracking global economic growth. Of goods traded internationally, 80 percent are now shipped by sea. As trade volumes increase, so does noise.[31]

It would be a mistake to suggest that all creatures in the sea are adversely affected by noise. Some like to gather beneath floating objects, even if they are noisy ships. And although sounds travel much farther in the sea than on land, there are still plenty of peaceful places in the oceans away from shipping lanes, ports, wind farms, and oil and gas fields. But it is hard to escape the conclusion that the rising clamor is but one more stress piled onto life in the sea. If nothing changes in the way we build or run ships, the din beneath the sea will only get louder.

The great majority of noise in the sea is from boats, and there seems little prospect of hushing them for the sake of marine life alone. Motives for developing quieter boats are more likely to be fuel economy or the comfort of the crew and passengers. There are simple ways to reduce ship noise. The loudest sounds come from propeller cavitation. As a propeller spins it creates a low pressure zone behind each blade. The faster it rotates the lower the pressure, until a point is reached where a gap opens up between the water and the blade. As these cavities spin off they slam shut with an audible slap. When propellers turn at low speeds there is no cavitation, so boats run more quietly. A simple way to reduce noise is just to slow down. That is exactly what the world's merchant shipping fleet has done in recent years to combat the soaring cost of fuel, because cruising slowly is also cheaper. A 20 percent cut in speed can save 40 percent in fuel costs. In fact, many shipping lines have dropped steaming speeds from twenty-five knots to twelve knots, about the speed reached by sailing clippers, the workhorses of international trade and communication in the late nineteenth century.

Slower ship speeds have also been prompted by a growing awareness of the greenhouse gas emissions of the global fleet. The biggest ships emit more carbon dioxide pollution than the world's smallest countries, and collectively, the world's merchant fleet releases about 4.5 percent of global carbon dioxide pollution. Critics argue that speeds will rise again as soon as the price of oil goes down, but there are ways to reduce both noise and fuel emissions through better design. The military has developed propellers that minimize cavitation at high speeds to make ships harder to detect, and cruise ships have also found ways to minimize noise. There is no reason why the merchant fleet shouldn't adopt similar designs, and many ships could be retrofitted. Ships could also reduce the need for noisy engines and fuel by using kites or skysails to catch the wind, which has the added benefit of being free.

The world's shipping fleet is governed by the International Maritime Organization, which has overseen a huge improvement in the safety of oil shipments over the last forty years. Oil spillage has fallen considerably as a result of the imposition of double hulls, the compartmentalization of oil in the hold, pinpoint satellite navigation, and a raft of better sailing practices. It is high time for them to tackle the twin problems of noise and fuel efficiency.

CHAPTER 12

Aliens, Invaders, and
the Homogenization of Life

harles Darwin and Alfred Russel Wallace discovered the principles of evolution by natural selection around the same time, each working entirely alone, by obsessing over patterns in the distribution of species. As both of these astute naturalists traveled the world they noticed that some species came and went, while the characteristics of others changed from place to place. Eventually it became clear to them that species were not fixed by some godly act of creation. Rather, life's forms were constantly reworked by evolution. It is Darwin we celebrate for this theory, but it was Wallace who spurred him to publish it when he later came upon the same idea. Geographic history has left some places with more species than others, and some with more unique species. Other places, by virtue of environmental stress, isolation, or lack of habitat are less rich.

The geographic ranges of species are circumscribed by many factors, most notably physical barriers such as landmasses, currents, or insurmountable gaps in suitable habitat, like ocean basins. When natural barriers divide regions, as the Isthmus of Panama did when it separated the Atlantic from the Pacific three and a half million years ago, evolution may take life in different directions.[1] Over great stretches of time new species arise and others go extinct. Some natural barriers to the dispersal of species rise or fall slowly, others more quickly in relative terms, like shifts in current patterns or the appearance of a new island from a volcano under the sea.

Humankind has lowered or entirely removed these barriers. For the last several thousand years, but especially during the last several hundred, we have made long-distance voyages with animal and plant passengers in our cargo and attached to our ships' hulls or carried in ballast. These journeys opened up new areas for colonization. The distances species could be carried underwent a steep change in the fifteenth and sixteenth centuries as the voyages of Dias, da Gama, Magellan, Columbus, and others began the age of global exploration by sea. When Columbus sailed to America he unwittingly took a bunch of freeloaders with him. Early Arab and Chinese voyages would also have carried species long distances in the Indian and Pacific Oceans.

Getting there is only the first hurdle for a would-be colonist. Conditions must be right for survival, and a sufficient concentration of colonists have to be introduced to establish a self-sustaining foothold in the new territory. In centuries past, ships often anchored or tied up for many weeks at a time on foreign shores (as opposed to only for hours or days today), allowing ample time for attached fouling organisms to shed eggs and larvae. In addition, vessels had their bottoms scraped every few months to keep them clean, and countless such voyagers would find themselves marooned in alien waters. Plants and animals could also be carried from place to place in water that sloshed around the ships' bilges or attached to rocks used as ballast to increase the draft of lightly laden ships. Sailors carried rocks together with their hitchhikers from the Old World to the New, only to dump them on arrival to load cargo. The genes of introduced periwinkles and serrated wrack seaweeds in Nova Scotia give away origins in Scotland and Ireland, from which most of the ship traffic arrived.[2] Ballast became an even more significant means of emigration from the late nineteenth century onward, as large vessels began to take in water as ballast in one port and discharge it in another. Ballast water now ranks as the foremost means by which marine species are introduced to new environments. Many marine organisms begin their lives as tiny eggs or larvae that can be transported in ballast. Even big animals can get around in ballast—fish a foot long have been found in ballast tanks.

We call displaced species alien or nonindigenous to distinguish

Catch of goliath groupers landed from a recreational fishing charter boat into Key West in Florida, in the 1950s. The swollen bellies of the fish show that they are full of eggs and were caught from a spawning aggregation.

Recreational fish catch from Key West in the 1970s.

Recreational fish catch from Key West in 2007. The number and size of fish have become smaller over time. Most of today's anglers are unaware of the extraordinary decline in Florida's reef fish revealed by these images.

Fishing methods were well developed long ago in the Mediterranean as this second-century mosaic panel from the Catacombs of Hermes in Hadrumetum, Tunisia, shows. It illustrates the use of throw nets, octopus traps made from perforated amphorae, drift nets, and hook and line, as well as the wonderful diversity of marine life.

The Fishmarket by the Dutch artist Frans Snyders (1579–1657). Although the composition is fanciful, the animals and their sizes were representative of those available to customers of the day. A huge wolffish gapes from the center of the painting and lies across enormous sturgeon and halibut. To the left are large cod, ling, and salmon steaks while turbot hang from hooks at the back. Other creatures that might surprise present day shoppers include lamprey, porpoise, seals, common skate (now extremely rare in spite of the name), and pike. Musée des Beaux-Arts, Carpentras, France.

Auction scene at Grimsby fishmarket, UK, from the early days of steam trawling around 1900. Not only is the size of the catch extraordinary by today's standards, but so was the size of the fish. The long, slim fish in the background are mainly ling, while those in the foreground are predominantly cod.

Late nineteenth-century catch of thirty thousand salmon from Puget Sound, Washington State.

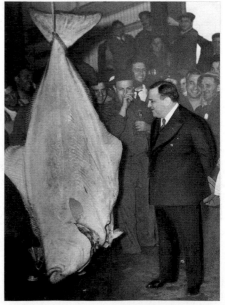

Fiorello LaGuardia, mayor of New York City, posing with a three hundred–pound halibut at the Fulton Fish Market in 1939. By this time, giant halibut had become scarce in the North Atlantic, but they were common at the time the Jamestown colony was founded in Virginia in the early seventeenth century.

Dungeness crabs killed by the encroachment of a deep layer of low oxygen water onto the continental shelf off Oregon. Such events are linked to stronger winds blowing parallel to the shore, a probable result of climate change. As deep water contains more dissolved carbon dioxide, it is more acidic than the shallow water layer that usually bathes coastal seas off the West Coast of the United States.

The "Thames Barrier" is a movable floodgate that protects London from flooding during periods of extreme high tides and storm surge. Rising sea levels will demand similar protective measures for many other cities across the world in the coming century.

Healthy Papua New Guinean coral reefs bathed in seawater of normal pH.

Reefs exposed to carbon dioxide bubbling from submarine vents that are subject to water acidity levels equivalent to those that could be reached by the late twenty-first century under a business-as-usual scenario of carbon dioxide buildup in the atmosphere.

These microscopic coccolithophores are phytoplankton (plant plankton) that produce exquisitely ornamented shells made from calcium carbonate. Each ball is a single cell that measures about 1/6300th of an inch across. Fossilized skeletons of coccolithophores are a major ingredient of chalk. They are expected to be highly susceptible to rising acidity as dissolved carbon dioxide accumulates in the sea.

A kill of menhaden fish in Narragansett Bay, Rhode Island, due to a severe low-oxygen event in August 2003.

Plastic contamination of the Santos estuary, São Paulo, Brazil. Estuaries are especially prone to waste buildup as they often receive garbage washed downstream from highly populous cities.

California sea lion trapped in a piece of monofilament gill net. The fishing net has gradually tightened in a slow-motion noose as the animal has grown in size.

Laysan albatross live in the North Pacific and fly thousands of miles over the open sea in search of food for their young. Unfortunately they cannot distinguish floating plastic trash from food, with heartbreaking results.

Remains of a Laysan albatross chick from Kure Atoll in the Northwest Hawaiian Islands. This chick starved to death on a diet of "junk food."

Humpback whale and calf near Tonga. Whales are highly vocal animals, few of them more so than humpbacks. They use the remarkable sound-transmitting properties of water to call each other over distances of hundreds, sometimes thousands of miles. But their ability to communicate has diminished in the last fifty years due to the rising clamor of human-generated noise in the ocean.

Red lionfish on a shallow water reef in the Bahamas. In the space of less than twenty years, this native of the Pacific and Indian oceans has successfully invaded coral reefs throughout almost the entire tropical western Atlantic, munching its way through countless indigenous fish.

A stand of invasive smooth cordgrass (*Spartina alterniflora*) has taken over mudflats of Seal Beach National Wildlife Refuge near San Diego, California.

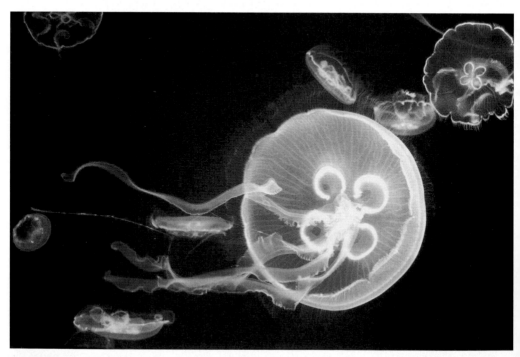

Jellyfish possess an ethereal beauty, which belies their ruthless efficiency as predators of the planktonic realm. They are highly effective opportunists that are beginning to benefit from overfishing, coastal pollution, and climate change. As a result, dense blooms of jellyfish have reached plague proportions in some places, driving people from the water, blocking power station cooling water intakes, and killing stock in fish farms.

A scallop dredge being lifted into a boat in the Irish Sea. Each individual dredge is composed of a steel frame with vertical teeth that dig up the bottom, and a chain-link bag to collect everything that it kicks up (usually a wide assortment of rocks, weeds, fish, and other invertebrates). Scallop dredges are notorious for their collateral damage.

Catch from a prawn trawler in the Torres Strait, Northern Australia, showing a large weight of unwanted bycatch, mainly small fish and invertebrates, that will be thrown away (mostly dead) when the prawns have been removed. Bycatch often outnumbers prawns by 10:1 or 15:1 in trawl catches like this.

Red mangrove in the Galápagos Islands. Many tropical coasts are protected by dense forests of mangrove. Their interlocking roots stabilize sediments and protect coasts from waves and storms. By trapping mud and growing upward over time, healthy mangroves could help shield coasts from the worst effects of sea-level rise. In many countries, however, they have been cleared to make way for shrimp ponds and other coastal developments.

A barge load of retired New York City subway cars destined to be dumped in the sea to create "artificial reefs."

Tires dumped into the sea off Fort Lauderdale, Florida, in the 1970s to create an "artificial reef"—garbage by any other name. Forty years later, there isn't much evidence of flourishing marine life on this "reef." Natural habitats do the job far better than artificial ones.

Aquaculture ponds for shrimp and fish cram the coast in China's southern Bohai Sea as this view from space shows. Much of the Bohai coast looks like this and natural habitats are hard to find in this crowded and highly polluted sea.

Coastal cleanups have become regular events in many parts of the world. This one is from a community near São Paulo, Brazil. Despite long-standing bans on dumping plastics at sea, the amount of trash is still on the increase in most places, some of it dumped illegally, some blown to sea from land, and some washed offshore by rivers.

Cabo Pulmo Marine National Park in Mexico's Baja Peninsula underwent a spectacular recovery of marine life following its protection from fishing in 1999. By 2009 the weight of predatory fish in the park—like many of those in this photograph—had increased by a spectacular eleven times.

It may take many years for protection from fishing to rebuild depleted populations of big, old fish like this black seabass in California's La Jolla Marine Reserve, near San Diego. Such animals are usually the first affected by overfishing and the last to recover.

Shark populations across the globe have fallen into a steep spiral of decline in the last two decades due to demand for their fins, the signature ingredient of a prestige food in Chinese cuisine: shark-fin soup. Economic growth in China has driven a soaring demand for the fins, many of which are stripped from the animals on deck before they are dumped over the side to suffer a prolonged death. This animal was found among many others in a "shark graveyard" on a reef in Sulawesi, Indonesia. Protecting sharks will require measures to curb demand for shark-fin soup alongside restraints on overfishing.

For centuries, green turtles were hunted for their eggs and meat. These days they are protected by law in most countries but nonetheless still often fall victim to other fisheries as unwanted bycatch. This abandoned fishing net with seventeen dead green turtles was discovered drifting off the coast of Bahia, Brazil, days after a storm. Leatherback turtle numbers have plummeted over the last twenty years from such bycatch and they are now on the verge of extinction.

Although protecting the giants of the sea requires decades of dedicated work, there are encouraging success stories. In Brazil, the number of olive ridley turtle nests has increased by fifteen times in northern São Paulo since conservation efforts were launched in 1980 by a national environmental organization, Projeto Tamar.

them from the natives. Some become troublemakers. These "invasive" species run wild, freed from the control of their natural predators, parasites, and competitors in their native haunts. Invasive species are like nature on steroids. They often multiply with extraordinary rapidity and overconsume or outcompete indigenous species for space, food, and other vital resources. The results can devastate ecosystems. Such invaders can eradicate local species and reduce other once dominant life-forms to the role of bit players. One such invasion, underway in the Caribbean right now, will change reef life there forever.

Red lionfish are natives of the tropical western Pacific and are popular aquarium fish. They have broad pectoral and dorsal fins that they hold open like delicately patterned Chinese fans. Their beguiling beauty is deceptive, for they are ruthless predators. They use their fins to usher small fish, ever so gently, into awkward corners of the reef, where they dispatch them with mouths that expand in an instant to engulf their prey. Enough lionfish were released or escaped from aquaria in the late twentieth century in south Florida that a breeding population became established.[3] Consequently lionfish have few Caribbean predators. Their lacy fins conceal sharp spines that deliver a potent shot of venom to the unwary. The Caribbean was poised for invasion.

Early this century, surprised divers began to report sightings of lionfish off the coast of North Carolina. Reports followed from the Bahamas, Turks and Caicos, and Bermuda. The Caribbean has not seen anything like the lionfish for millions of years, if ever. Never having encountered such a belligerent and aggressively defended predator, the local fish don't know what to make of them and show little fear. Lionfish are slurping them up by the millions. Indeed, speared lionfish have been brought up with their bellies packed so tightly they seem scarcely able to move. With food so easy to come by, these lionfish are growing faster and larger than they do in their native range, and with their populations increasing exponentially, they are laying waste to Caribbean reef fishes.[4]

Lionfish have exploded across the Caribbean in less than twenty years. They can now be found from Rhode Island south to Colombia, and from Belize east to the Windward Islands, and may ultimately range along the entire coast of Brazil. Although divers and conservationists soon rallied to challenge the threat, their efforts are futile, except for divers keeping the invader off specific reefs by means of frequent removals. Lionfish are there to stay. Mark Hixon, a biologist at Oregon State University who has studied the lionfish in the Bahamas and the Caymans, thinks "this could well become the most devastating marine invasion in history."[5] Although lionfish are not large—the record-size invader was just seventeen inches—they target small fish and invertebrates that include the young of large species. Such animals are not only valuable as food for people, like groupers and snappers, but also provide important ecological services, such as parrotfish that graze seaweeds that would otherwise overgrow corals. Imagine if some predatory beast invaded America and began to slay millions of birds, to snatch dogs in parks off their leads, and to indiscriminately kill cats from the backyard and possums and deer from the country. To imagine this is to think of what life has become on Caribbean reefs. The lionfish are particularly ruthless, but they are only one of many pernicious invasive species that have begun to cause mayhem in marine ecosystems. One small window of hope may be that lionfish are becoming popular menu items in restaurants in the United States and the Caribbean, but whether harvesting this edible fish will suffice for control remains to be seen.

Ships have long been the most convenient route to a new home. At the last count over fifty thousand vessels grind their way around the world carrying goods.[6] They move about three cubic miles of ballast water from place to place every year, and at any moment some ten thousand species are thought to be in transit.[7] Being slopped around inside the dark belly of a ship, where temperatures can swing widely and food is at a premium, is not the surest way to survive a long journey. But it does select for the most tenacious and resilient, characteristics that serve alien species well in new places. Estuaries, which

support many of the largest ports, have become hot spots for invasions from which animals and plants may spread like plague. Chinese
mitten crabs and Japanese seaweeds are invading European and
North American seas, while Asian kelp is spreading in California. In
Germany's Elbe River, where mitten crabs have been established for
a century, at certain times of the year writhing legions of crabs swarm
downriver like a locust plague on their way to spawn. A West Atlantic comb jelly called *Mnemiopsis* devastated planktonic life and fisheries in the Black Sea after it invaded in the 1980s, and has now reached
the Mediterranean, Baltic, and North seas.[8] European green crabs are
crunching their way through naïve intertidal natives on North America's West Coast and along the shores of Japan, South Africa, and
Argentina. Throughout North America dense stands of phragmites
reeds have begun to block coastal views and choke out native wetland
plants. Intriguingly, the reed is native to North America, so scientists
were puzzled as to why it had turned bad. Genetic analysis gave them
their answer.[9] A variety introduced from Europe is causing the nuisance, and in New England has entirely replaced the native variety.

San Francisco Bay is one of the most invaded ports in the world.[10]
Over the last 140 years, nearly three hundred alien species have settled in the bay and its rivers,[11] and the rate of arrival has increased.
From 1851 to 1960, a new species arrived just over once a year on average, but between 1961 and 1995 there was a new arrival every fourteen
weeks. These new species have transformed the bay, much as San
Francisco itself has become a cultural cornucopia. Hardly any corner
has less than a third of its species count made up of aliens, and in
many places they have complete control, having ousted all the natives.
Atlantic smooth cordgrass has formed dense stands over areas of open
mud that once were rich feeding grounds for birds. Beneath the tranquil surface three kinds of gribbles, tiny isopod crustaceans related to
woodlice, are busy boring their way through wooden docks and pil-
ings. In Seattle they have inflicted $700 million worth of damage.[12]
Clam diggers around San Francisco Bay are now much more likely to
unearth Asian clams than native. So enthusiastic is the feeding of

these clams, they are sucking the water clean of plankton, which could deprive other species of essential food and cause declines of animals whose eggs and larvae drift in bay waters.

The rise in ship traffic is part of the reason why San Francisco has been so comprehensively invaded, but there are other ways to get from place to place. Aquaculture, the farming of marine life for our consumption, is a major source. Rather than farm local species, we like to move the fish and shellfish around that we enjoy most and are best able to raise. In the 1800s, vast numbers of eastern oysters were transplanted from the Atlantic to the Pacific coast of North America, and the same happened in the 1900s with Japanese oysters. Atlantic salmon are grown along cool stretches of the Pacific coast of North and South America. East Coast oysters are cultured on the West Coast. California's red abalone is raised in Chile. But it isn't just the cultured species themselves that get moved: Along with the oysters came scores of unwanted hitchhikers, many of which are now abundant in San Francisco Bay. Fish feeds can also bring unintended guests, and are often responsible for outbreaks of parasites and diseases that I will come to in the next chapter. The expansion of aquaculture across the globe has been extremely rapid and will continue, which means this is a problem that will run and run.

Marsh grasses, like the Atlantic smooth cordgrass, were introduced to San Francisco Bay intentionally by the U.S. Army Corps of Engineers to "restore" wetlands and prevent erosion. It was spectacularly successful, but now state officials want it out and have launched an eradication campaign. Perhaps as much as 6 percent of marine invasions worldwide have come from unintentional introductions from aquaria. After lionfish, the most notorious is a tropical seaweed, *Caulerpa taxifolia*, a dark green weed that scrambles over the seabed in mats of twisted stems and has been dubbed the "killer algae."[13] *Caulerpa* doesn't normally grow in cool places like the Mediterranean. But keepers at the Monaco Aquarium[14] heard of a strain that German aquarists discovered could survive in similar conditions, and they imported it for their own tanks. *Caulerpa* escaped into the

Ligurian Sea almost immediately, through discharge pipes from the circulating seawater system.

In 1984, divers noticed a patch of a strange new weed smaller than a tablecloth growing wild.[15] Curiously, it was not removed. Five years later it covered a couple of acres of seabed. Ten years after its release people knew they had a serious problem. Six square miles of the bottom were blanketed in the weed, which blocked the light and suffocated all beneath. Where meadows of sea grass once waved in sunlight, now there was only *Caulerpa*, like a cancerous growth. And like cancer, this invasion soon metastasized. Fishing boats picked up the weed in their nets and pleasure craft snagged it in their anchors. Soon it flourished in hundreds of new places. Just fifteen years after its release, *Caulerpa* covered 97 percent of available surfaces between Toulon in France and Genoa in Italy, a distance of nearly two hundred miles. It isn't over yet. Much more of the Mediterranean could soon fall victim to the blight. The alga has now secured bridgeheads in Spain, Italy, Croatia, and Tunisia.

Caulerpa racemosa is a related species from southwest Australia that is also highly invasive and has settled in the Mediterranean. It was first reported in 1990, and if anything is an even more vigorous competitor for light and space than its namesake. It even seems to be outcompeting it! Within six months of arrival this plant had gained complete dominance in some places, indiscriminately laying waste to sea grass beds, seaweeds, sponges, and all manner of other creatures that make a living attached to the bottom of the sea. *Caulerpa racemosa* has now spread from Cyprus to Spain, and beyond to the Canary Islands.

Like many invasive species, both these algae have far-reaching effects within the communities they have entered. In the space of decades their arrivals have brushed aside relationships formed among Mediterranean plants and animals over tens of thousands of years. Sea grass beds are rich in fish, crustaceans, starfish, urchins, and hundreds of other species that call them home or use these meadows for a time before moving on. Green turtles graze contentedly on the flexible green blades beneath shafts of light that dance and dapple over the lush pasture. They aren't keen on *Caulerpa*. Most fish also fare badly where *Caulerpa* mats occlude the bottom. The hunters can no longer reach invertebrates living in the sand. Those that once sheltered among blades of grass have found their lives brought to a standstill as the thickets close around them, much as the palace and its dozing inhabitants were overgrown by a dense tangle of briar in the fairy tale "Sleeping Beauty."

So what factors facilitate a successful invasion? The bad news is that many of the problems highlighted elsewhere in this book increase the chances that invaders will establish and make trouble. It turns out that where overexploitation or pollution have depleted local life, competition for space and other resources is less intense and invaders can establish a beachhead from which to spread. Some pollutants may also render conditions more amenable for invaders. Nutrient pollution in the Black Sea, for instance, is thought to have favored the establishment of the *Mnemiopsis* comb jelly. The simplification of marine ecosystems by multiple stresses opens up opportunities for invaders. And the creation of structures such as seawalls, rigs, and wind farms can facilitate the spread of invasive species that require hard bottoms across expanses of otherwise unfavorable habitat.

One of the great paradoxes of invasive species is that they can spring to life from just one or a handful of individuals, like a forest fire caught by a lone spark. Yet we struggle so hard to keep life's spark alive when endangered species have been reduced to the last few dozen individuals. What makes invaders different is freedom, opportunity, and of course luck. We forget that some invasive species may have failed many times to settle successfully before they eventually do. We only see the winners. America drew settlers from all over the world as a land of freedom and opportunity, and still does. When a species is plucked from its homeland and thrust into a new world, it can find itself reborn, freed from the shackles of traditional enemies.

The burden of disease and parasites may be lifted. Likewise, predators and competitors can be thin on the ground for a stranger. Life is good.

The contrast between old haunts and new can sometimes create conservation ironies. Sea lamprey invaded North America's Great Lakes after a canal was opened around the natural barrier of Niagara Falls in 1919. These primitive fish are slippery, slim as eels, and mottled like leopards. They have a ring of conical teeth around a jawless mouth with which they attach to bigger animals. Lampreys rasp holes in the flesh of their hosts, leaving their flanks pocked with bleeding sores. So many now live in Lake Superior that there is hardly a big fish left that has not been attacked. For lamprey the Great Lakes have been one long lunch. Sea lamprey numbers back in their native Europe have plunged in the last half century, a decline usually blamed on poor water quality in the rivers where they spawn. But I suspect the real cause is loss by overfishing of the big animals they use as hosts while at sea. In Europe the sea lamprey is protected; in the Great Lakes it is vermin.

Of course, there are qualities that predispose a species to be a successful invader. Weedy species the world over have been crafted by evolution into skilled opportunists. They often have resistant seeds or spores that can wait as long as it takes for the right moment to germinate. It helps if you grow fast, produce lots of offspring quickly, and don't mind mating with relatives. It is also good not to be too picky about where you live and to enjoy novelty and variety in your diet. I wouldn't have singled out lionfish as likely invaders of the Caribbean from my encounters with them in the Red Sea. They seemed too shy and delicate and were never particularly common. Surrounded by animals that know well to avoid their dangerous beauty, they have to work hard for a meal. There was no hint of the profligate glutton they would become in the all-you-can-eat buffet of the Caribbean.

Few places on this planet have not been reached by some marine invader or other. Not surprisingly, some of the worst affected places have the highest ship traffic. The U.S. West Coast, northern Europe, and the Mediterranean are all hot spots of invasion. But some places are more vulnerable than others. Hawaiian seas have been invaded time and time again.[16] Like the islands above them, Hawaii's seas are rich in species found nowhere else. On land, invasives have devastated plants and animals ill-equipped to deal with aggressive conquerors. In the sea around Oahu many areas are now dominated by invaders: mangroves from Florida line Pearl Harbor; one of the most common shore barnacles is newly arrived from the Caribbean; and blacktail snappers from Moorea and bluestripe snappers from the Marquesas are among the most abundant large fish in shallow waters. Over three hundred alien plants and animals are now known in Hawaiian seas, and these numbers do not take into consideration potentially hundreds more that may be alien but whose histories are less well-known. On the face of it, Asian seas have been less affected by aliens, with reported numbers less than a fifth the number of new arrivals into the U.S. West Coast and Hawaii. But the long-term presence of Southeast Asia at the epicenter of trade in consumer goods tells us to distrust these statistics. Asian waters are probably crawling with invasive species but surprisingly few people have cared to look so far. A Caribbean variety of zebra mussel, the black-striped mussel, has established at high densities all around Singapore and part of Malaysia, for example.[17]

The least invaded seas in the world are at the poles. At the time of this writing there was just one marine invader known in the Southern Ocean and nine in the Arctic. The poles have been spared for several reasons. Not many people live there so boat traffic is sparse. Polar conditions are extreme and few species from anywhere but the opposite pole could cope with them. It is hardly likely that they could find a way between the poles. Times are changing, though, and it is very probable that polar seas will soon find themselves home to many more alien species. Loss of sea ice in the Arctic Ocean due to climate change has led to the mixing of seawater from the North Pacific with the North Atlantic for the first time in eight hundred thousand years. We know this from the arrival in the Atlantic of a species of North

Pacific planktonic diatom that has been absent that long.[18] A far bigger Pacific traveler reached the Mediterranean in 2010, where aston-
ished Israeli scientists spotted a gray whale.[19] The last Atlantic gray
whales were slaughtered by New England whalers in the seventeenth
or eighteenth century. This one must have lost its way and inadvertently navigated the newly ice-free route to the Atlantic.

Global climate change is warming polar seas enough to make life
there possible for species from lower latitudes. On a changing planet,
these range expansions will come to make up a large fraction of new
introductions. They might also gain a helping hand from us. The
prospect of an ice-free Arctic in summer has once more excited great
interest in a northwest passage for ships between the Atlantic and
Pacific. Dozens of explorers lost their lives between the sixteenth and
nineteenth centuries in search of this elusive route, their efforts
doomed by savage ice. As the lost gray whale shows, it might soon be
a reality, and without great care, ships could take many new species
into the Arctic and provide a fast track between the Atlantic and
Pacific for countless others. The increasing quantities of floating plastic and fishing debris to which fouling species can attach themselves
may also carry aliens to polar seas, although there is little firm evidence of that yet.[20] Ocean garbage has multiplied the means of hopping from place to place in the sea and can go where few ships travel.

Not all alien species cause problems. Some may blend into their
new homes with little fuss. They add color and variety, like a new
picture hung on the living room wall. But significant numbers, perhaps the majority, are harmful, as judged by their impacts on native
wildlife or to human interests. An estimate made in the year 2000
suggests that invasive species cost us $1.4 trillion every year in trying
to undo the damage caused, eradicate the invaders, or in economic
losses sustained from degradation of ecosystem services, such as clean
water provision or soil stabilization.[21] Those troublesome Chinese mitten crabs have so filled Germany's levees with burrows that they are
crumbling. The vast majority of this damage can be attributed to
terrestrial invaders. So far little attention has been given to alien

species in the sea, but we can be certain that the costs of aggressive new colonizers, like Seattle's gribbles, are quietly building beneath the waves.

While much of the attention around invasive species has concentrated on costs, some create new opportunities. When the French engineer Ferdinand de Lesseps achieved his life's goal to join the Mediterranean and Red seas by digging a canal between them, he managed what countless others had dreamed of since pharaonic times. The Suez Canal runs at sea level. As soon as it opened in 1869, water began to flow between seas that had been parted for over eleven million years and has carried tide upon tide of emigrants. Since Mediterranean Sea levels are lower than those of the Red Sea by four feet, most of the flow has been northward, and invasions were much facilitated by the lower biodiversity of the Mediterranean. To begin with, the invaders' paths to virgin territory was blocked by a series of briny lakes—the Bitter Lakes—in the middle of the canal. Over the course of a century, flushed by seawater, the lakes became less inhospitable. The rate of colonization of the Mediterranean increased from drip to trickle to steady stream.

Early colonists burst forth into a place with few predators or competitors to challenge them. The eastern Mediterranean had long been thought of as impoverished, with low productivity and few species compared to the richer west. The contrast from north to south through the canal was even greater. The Red Sea hosts an extravagant diversity of life that thrums around vibrant coral reefs. While few corals can tolerate the cool Mediterranean, many reef inhabitants were more cosmopolitan and soon settled in. Several of them now sustain valuable commercial fisheries in Israel, Egypt, and along the North African coast as far as Tunisia.[22] They include rabbitfish, lizardfish, goatfish (only the lizardfish looks anything like its terrestrial counterpart; imagination seems often to have deserted the namers of fish), and a clutch of prawns, three of which were fat and juicy enough to be welcomed.

It is premature to say this without fear of contradiction, but places with few species seem to be easier to invade than richer communities.[23] The idea is that resources are less fully used so would-be invaders can gain a foothold more readily. According to this logic, the eastern Mediterranean was ripe for the taking. The problem is, by extending this logic one has to conclude places where human or other impacts have picked off the natives will be highly vulnerable to alien invaders. Mass mortality of corals and sponges from disease and heat stress has left vast areas of free space, for instance. Given the ubiquity of human sources of stress and their multiplicity of effects, it is a near certainty that most ecosystems in the sea have lost at least some of their resistance to invaders. Their defenses are down just when they need them most. The onslaught has only just begun.

In some ways, the changes to nature that we have unleashed through travel and trade mirror the effects of globalization on human culture. Some cultural influences are like aggressive invaders that, within the space of half a century, have subsumed and erased cultures that flourished for millennia in isolation. Explorers and filmmakers today struggle for days or weeks through sweltering jungle, across torrents and treacherous ravines, only to find on arrival people sporting ragged T-shirts emblazoned with Coca-Cola or Toshiba. So it is that in countryside and town, lake and ocean we find familiar faces in unusual places. What we often miss are the plants and animals they have displaced, that in the past made these environments exotic and strange.

Invasive species are one of the five horsemen of the apocalyptic loss of biodiversity that is sweeping the planet (the others are habitat loss, pollution, climate change, and overkill by hunting or fishing). Yet, strangely, invasives have been neglected relative to the others, particularly at sea. Although they have purged habitats and ecosystems of their rightful owners—just as with habitat loss, pollution, or climate change—they do not yet seem to have caused many outright extinctions in the sea.[24] Native species often hang on in pockets of poor quality habitat, just as peoples quashed by conquering armies find themselves marginalized and browbeaten. Native mullet in the

Mediterranean have been pushed deeper by invasive goatfish. But local population crashes can eventually lead to disappearance, and there are many scientists who believe that invasive species in the sea may yet drive natives to extinction given enough time. The first may be closer than we think.

The Derwent River estuary of Tasmania is home to a little fish with the bizarre habit of walking on its fins. Its pectorals—the fins on either side of the body behind the gills—look like arms, and each ends in a webbed "hand," hence its name, the handfish. Spotted handfish, as this one is called, have flesh-colored bodies speckled red, fins edged in lemon, and dark eyes lined with gold. They display to one another in a semaphore of fin flashes and showy standoffs. Nobody knows why they are found here and nowhere else, but such a limited geographic range means local population loss and global extinction are one and the same. In the 1990s, their universe was upheaved by the arrival of the North Pacific starfish, an undiscriminating predator from Japan and East Asia whose numbers have since swelled to millions. Among their predilections are spotted handfish eggs, laid in nests on the seabed. There could be as few as a couple of hundred to a couple of thousand spotted handfish left in the wild, but it is hard to know, since they are difficult to count. While all species have a right to exist, the loss of one so charming seems especially heartbreaking.

Geologists recognize distinct epochs in Earth's history according to the fossils and rocks present and give them names. The most recent, the Quaternary, covers the last 2.6 million years and is subdivided into the Pleistocene and Holocene. The Pleistocene spans the time of the ice ages, while the Holocene covers the twelve thousand years from the end of the last ice age to the present. Some scientists with a humorous bent have suggested that the present era should be called the Homogecene, because it is a time of unprecedented mixing of animals and plants from one part of the world to another. While efforts are being made to control the spread of invasive species, I wonder about their efficacy in the long term. This is an uncontrolled experiment with life. Who knows what will come of it?

When I was a student in the 1980s my professors made much of the problems of alien species and how they could lay waste to the native fauna and flora in their new homes. Rabbits ate their way through Australia, water hyacinths choked lakes from shore to horizon, killer bees swarmed into the southern United States, and Nile perch chomped their way through an irreplaceable evolutionary showcase of brightly colored African cichlid fish. The solution, in those days, usually involved introducing something else, following the logic of the old nursery rhyme about the woman who swallowed a fly, and then a spider to catch the fly, and a bird to catch the spider, and so on until she had eaten her way up the food chain. Like this hapless lady, who dies in the end as you may recall, the risk is that further introductions will do more harm than good.

There was a dramatic case of such an effect in French Polynesia. Giant African land snails were accidentally introduced and began to damage crops, so a voracious predator was brought from its homeland to control it. This snail, called the rosy wolf snail, developed a taste instead for the colorful tree snails that lived deep in moist valleys. Every valley had its own species of tree snail that had evolved in isolation, separated from the others by craggy mountain ridges they could not cross. The rosy wolf snail soon jumped these barriers to munch its way through an evolutionary cornucopia whose remnants now hang on only in zoos scattered across the world. As far as I know there haven't yet been any tit-for-tat introductions in the sea, so dreadful mistakes like this haven't been made, but the temptation is there.

Introduced species—biological pollution—is one of the hardest kinds to clean up. People have spent lavishly on efforts to rid tiny scraps of land from the scourge of rats so as to spare rare seabirds or unique species like lizards. The biggest islands on which they have been successful to date are little more than a square mile in area, and they are unlikely to get much bigger. If you can catch an invader early and hit it hard, success is possible.[25] The black-striped mussel, a native

of the Caribbean now established in Singapore and Fiji, was detected
in Darwin Harbor in Australia in 1999. Within nine days the harbor
was quarantined and pumped full of bleach and copper sulfate, which
killed pretty much everything, including the mussel. In California, a
parasitic worm that invades commercially valuable abalone was introduced with animals brought from South Africa. It established at
Cayucos and was successfully eradicated by hand removal of infected
abalone and reduction in overall abalone densities below the level
needed to sustain the parasite. This was easy to do for an animal that
lives in shallow water stuck to rocks and is the coveted target of
skilled fishermen. Once an invader has spread beyond a few sites,
however, it is pretty much hopeless to try to eliminate it. Local control is the only option, such as the enthusiastic bands of spearfishers
attempting to rid Caribbean reefs of lionfish, and the chefs who
encourage us to eat them.

Until recently it was almost heresy in conservation circles to
admit that introduced species were here to stay and should be treated
as valid members of their new ecosystems. They are persecuted and
discriminated against and treated as the most reprehensible forms of
life. In the UK, some conservationists still treat species with contempt
that arrived five hundred years ago! When do you draw the line? I
have loved beech trees since I was small, with their muscular gray
trunks, smooth coppery leaves, and funny triangular nuts. It was only
years later that I realized what a fool I had been to be taken in by its
shallow appeal. This tree was an alien! An imposter that had only
made the hop from mainland Europe in the Middle Ages. I am joking, of course; I still love beeches. People mourn loss of the familiar,
but there are good reasons to resist the blending of the world's fauna
and flora as the planet shrinks beneath our globalizing influence. This
mixing represents a real loss of diversity and variety, and sometimes
it takes generations to realize how much intruders have altered their
new environments. When I look at a picture of a coral reef I can tell
from the species present where it was taken, give or take a couple of

thousand miles. Each place has its own distinctive collection of species.

Even if only a small fraction of introduced species cause trouble, their effects on native ecosystems can be hugely disruptive and expensive. Like pollution by toxic chemicals or plastic, the best means of control is prevention. Since the two main vectors for the spread of invasives are ships and aquaculture, the onus is on these industries to deal with the problem. One simple solution for aquaculture would be to stop raising species outside their native ranges. With nearly four hundred species in culture right now, that shouldn't restrict options much and would benefit consumers by giving them a wider range of ingredients to choose from. Ships could solve the problem of stowaways relatively easily by installing purification systems that use spare heat from the engines or UV light to kill unwanted passengers. Another simple measure that can reduce harm is to exchange ballast water offshore rather than in coastal seas. The idea is to avoid exchanging coastal species since these are the ones that cause most harm. Jim Carlton, from Williams College in Connecticut, thinks he has seen fewer invasions of U.S. seas due to ballast water since open-ocean exchange became mandatory in 2004. Similarly, he has heard far fewer reports from Australia or New Zealand, where similar regulations were introduced a decade ago.

On the outside of boats, the usual approach to invasive prevention has been to slather toxic chemicals over the hull, but some of these, like tributyltin, proved terribly toxic to other species. An alternative approach involves coating the hull with a nontoxic layer of vinyl resin reinforced with glass,[26] which reduces drag and limits the ability of fouling species to get a grip. It can be maintained by high-pressure tools that kill any species that do manage to establish on the ship. Like agreements on toxic chemicals, these measures can only succeed with international regulation.

CHAPTER 13

Pestilence and Plague

Soon after North America slammed into South America and the Isthmus of Panama blocked the exchange of water, animals, and plants between the tropical Atlantic and eastern Pacific, a wave of extinctions convulsed the Caribbean and wiped out three quarters of all coral species and many other forms of life. Evolution has since gone its own way in each region and the two oceans now have few species in common. The result is that Caribbean reefs look different from those everywhere else. One feature that distinguishes them is their great abundance of sea fans.

Sea fans are corals with a flexible horny skeleton over which is stretched a thin layer of tissue sprinkled with tiny polyps. These polyps are like miniature sea anemones. By day they are mostly retracted, but at night they expand to catch drifting food with a crown of stinging tentacles. As their name implies, sea fans are flat and fan-shaped, with a mazelike filigree of minor branches stretched between upright ribs. They are almost ubiquitous on Caribbean reefs, their supple forms bending to the movement of wave and current. One day in 1995, on a dive in Puerto Rico, I came across a sea fan that didn't look well. The tissue between its upright ribs was blotched with dark purple stains and several ribs stood bare where parts had died. It was the first of many such sea fans I saw.

Soon after I got home I was contacted by another scientist who had spotted sick sea fans on reefs in Curaçao, some five hundred miles to the south. At the time I was based at the University of the Virgin Islands in St. Thomas, and he was keen to know if I had come across anything like it. By year's end, reports from other countries confirmed that the Caribbean was in the midst of a full-scale epidemic. It would reduce the population of sea fans by as much as 75 percent in the next few years.[1] So what was killing these sea fans and why did it spread so quickly? The pathogen turned out to be a land-living fungus, and so attention swung rapidly to the red soil bleeding onto reefs from deforested island landscapes. But others thought the source could have come from somewhere far more unexpected—Africa.

In their idle moments, shuttle astronauts sometimes contemplated the Earth's great convulsions as they unfold in graceful slow motion far below. The fury of a hurricane is reduced to a twist of bright cloud; the flood of a mighty river a delicate discoloration of blue sea; dust whipped by desert winds a fan of brown streaks above the ocean. It is hard at first to accept that a Saharan storm might affect Caribbean reefs, but jet streams of air far above the sea carry millions of tons of African dust across the Atlantic every year. When I lived in St. Thomas sometimes the rain would fall in dirty red drops that left a thin layer of Africa over everything it touched. Incredible as it may seem, the island of Bermuda owes its red soils to thousands of years of such atmospheric dustfall. Charles Darwin remarked on African dust near the Cape Verde Islands when he stopped there on the *Beagle*:

> The dust falls in such quantities as to dirty everything on board, and to hurt people's eyes; vessels even have run on shore owing to the obscurity of the atmosphere. It has often fallen on ships when several hundred, and even more than a thousand miles from the coast of Africa, and at points sixteen hundred miles distant in a north and south direction.[2]

As Darwin soon discovered, this windblown dust carried the spores of land plants and fungi. Although it is difficult to prove conclusively, some scientists believe that African dust fall was responsible for infecting Caribbean sea fans with the fungus that started the epidemic.[3]

Sea fans are not the only organisms to have been felled by disease on Caribbean reefs. In the 1980s, waves of disease wiped out nearly all the staghorn and elkhorn corals, both signature species for the Caribbean; they were ubiquitous in underwater pictures and resort brochures from the 1950s to the 1970s. Elkhorn coral formed dense arbors of robust branches, as thick as human limbs, whose tips broke the surface at lowest tides. As the bottom fell away to the seaward of these ramparts, they blended into bushy thickets of the more delicate staghorn coral, whose branches sheltered flamboyant troupes of shimmering fish. The cause of the elkhorn epidemic was later discovered to be a human gut bacterium, carried in sewage effluent.[4] This outbreak was one of a collection of pathogens that has contributed to the loss of 80 percent of living coral cover in the Caribbean since 1977.[5] Another disease destroyed almost every long-spined sea urchin in the Caribbean. Waving forests of sea urchin spines blackened the bottom before the epidemic, and often obscured the corals below. Their loss left reefs vulnerable to rampant seaweed overgrowth.

As I mentioned in an earlier chapter, green turtles around the world have in recent decades blossomed grotesque disfiguring viral tumors that bulge from armpits and groin, face and throat. Badly afflicted animals even have tumors that overgrow their eyes and mouth. This is just what you can see. They often have more tumors in their stomachs, throats, lungs, kidneys—pretty much anywhere there is soft tissue. The disease is caused by a papilloma virus and was first reported in the 1930s, but it has reached epidemic proportions since the 1990s in densely populated areas such as Hawaii, Florida, and Barbados.[6] Since then it has been seen in other turtle species, such as the critically endangered leatherback, but green turtles remain worst affected. This disease, as I said, appears to be linked to tumor-promoting chemicals released by harmful algal blooms, which in turn are worsened by coastal pollution.

Other marine habitats have also experienced disease epidemics. A wasting disease has destroyed sea grass meadows from the Gulf of Maine to Florida since the 1930s. Wild and cultivated oysters in Europe succumbed to disease epidemics in the early to mid–twentieth century. In the 1980s, harbor and gray seals in the North Sea were decimated by a distemper virus similar to one carried by dogs, and pilchards in Australia were wiped out by a virus in the 1990s.

Diseases certainly seem more visible today than in the past, but is that just because there are more people looking now? The problem is that we lack a baseline from the distant past against which to measure the prevalence of diseases and parasites today. Although there were reports in the nineteenth century of disease outbreaks in commercial sponges and oysters that destroyed fisheries, mostly people didn't get overexcited about marine diseases, because they couldn't see them. If birds fall from the sky, dying dogs litter the streets, or stands of forest trees wither they are obvious and worthy of attention. But animals and plants in the sea can sicken and die by the millions unnoticed.

Jessica Ward, from Cornell University, and Kevin Lafferty, from the U.S. Geological Survey in California, made an attempt to build a disease baseline of sorts by trawling through all the reports made of marine diseases since 1970.[7] The total number of reports of anything to do with marine life has climbed steadily since then. There are more scientists at work today and more are interested in the sea. So they corrected for this increase by finding the proportion of reports about disease in any given year. This increased over time for turtles, corals, mammals, urchins, and mollusks. There was no upward or downward trend for sharks and rays, sea grasses, or crabs, lobsters, and prawns. Fish appeared less afflicted by disease today than in the past. Of course, this approach has its problems. Science is as amenable to fashion as any human endeavor, and people get excited about different subjects as time passes. Professors train students in the subjects of their own passions, and those students often pursue the same interests in their own careers. But the results have a ring of truth that chimes with experience for those of us who have dived the oceans over this period. We see a lot more pestilence and plague today than we did when we first jumped in the water.

I have said a lot about the downside of parasites and diseases so far, but we must remember that they are a natural and important part of the way the world works. Their presence indicates that an ecosystem is complex and functioning well enough for them to make a living, particularly where parasites or diseases have complex life cycles, perhaps with multiple hosts. One of the richest places Kevin Lafferty has seen for reef fish parasites is the near pristine Palmyra Atoll in the mid-Pacific, where he discovered a particular abundance of parasites that use sharks to complete their life cycles.[8] He thinks that having parasites in the system is akin to having top predators, in the sense that they help maintain checks and balances so that dominant species don't overwhelm their rivals. In other words, while we dismay over coral disease, to a certain degree these diseases probably help maintain the coral diversity that fascinates us.

But something has clearly changed. What is responsible for the rise of illness in the sea? For outbreaks of disease or parasites to occur you need a source of virulent pathogens, a population of susceptible individuals, and a means of transferring infection from one to another. The many ways in which we are changing the oceans offer fertile opportunities for pathogens to become established and spread. People who are stressed tend to be less healthy than their more relaxed colleagues, because stress compromises our immune systems. In an analogous way, multiple stresses leave animals and plants more susceptible to illness. So illnesses endemic to a region and those inadvertently brought in by people will often result in an outbreak. Disease outbreaks illustrate the intersecting effects of many different drivers that increase susceptibility. They show how cumulative stresses sum together to compromise life in the sea. The epidemic that wiped out Caribbean long-spined sea urchins began in Panama, close to the canal that connects the Atlantic and Pacific. Although marine species cannot pass through the canal unaided, because of its freshwater lakes, they can hitch rides on ships. Ballast water not only carries invasive species, it carries disease. When diseases find susceptible animals in abundance, they can cause catastrophic epidemics. As Jared Diamond pointed out convincingly in his book *Guns, Germs, and Steel*, disease was a greater force than weapons in the suppression of indigenous peoples as Europeans colonized the world. Epidemics of influenza, smallpox, and cholera swept through tribes, islands, and entire continents, destroying societies and cultures along the way. Every time contact was made, whether hostile or in friendship, the result was massacre by disease. In some places disease left only one survivor for every ten people originally present.[9] The Caribbean island of Hispaniola (today's Haiti and Dominican Republic) lost 240,000 of its 300,000 inhabitants within twenty years of Columbus's arrival, probably to smallpox.

Diseases had such profound effects on people in the New World and the Pacific Islands because they had never encountered them before and had no immunity. They spread like the wind and felled almost everyone they touched. Much the same is true for life at sea. The ancient disconnect between the Caribbean and the Pacific means that long-spined sea urchins in each ocean developed their own diseases. Pass one to another and the result is an epidemic on a massive scale.

The wave of death from bacterial infection that engulfed long-spined sea urchins in the 1980s sped round the Caribbean so fast that it was over almost before anyone noticed, the disease having run out of hosts. A later effort to trace the timing of this outbreak suggested it swept across eighteen hundred to three thousand miles of reefs in a year.[10] Similarly, a herpes virus that infected Australian pilchards in 1995 raced around Australia and New Zealand at six thousand miles per year. It was probably brought in with unquarantined shipments of assorted frozen bycatch fish destined to feed captive southern bluefin tuna being grown for Japanese markets. Anyone who has been to Australia and been doused with insecticides on arrival will be surprised that such shipments are still let into Australia without any precautions.

Epidemics in the sea seem to spread faster than almost any on land. The only outbreaks that have come close are myxomatosis and calicivirus in rabbits and West Nile virus in birds. All these are carried by flying insects that quickly spread the virus through susceptible populations. There are fewer barriers to dispersal at sea than on land—no mountain ranges, deserts, or inland seas to halt the progress of outbreaks. Ocean currents can move pathogens hundreds or thousands of miles in a matter of days or weeks. But epidemics like the ones that attacked long-spined sea urchins or Australian pilchards also spread against the current, which is puzzling until you remember the sea-surface microlayer. This thin skin over the sea can be churned into an aerosol by storm winds, together with any floating particles, and whipped long distances across the ocean surface regardless of the direction of underlying currents. Entire shoals floated belly up in miles-long rafts during the peak of the pilchard epidemic, so it is easy to see how viruses could have found their way into sea spray.

Across the world, populations of many species of marine mammals have struggled back from the brink over the last fifty years or more. Historic hunting of whales, dolphins, porpoises, and seals pushed some to the edge of extinction long ago—animals like the Guadeloupe fur seal in the southwestern United States and Mexico, or the Mediterranean monk seal. Others did succumb and are no longer with us, like Steller's sea cow from the North Pacific and the Caribbean monk seal. Porpoise meat was greatly admired in places like the Baltic Sea. The name itself derives from the French *porc poisson*, or pig fish. In England it was served at banquets and declared a "royal fish" in the early fourteenth century, presumably to protect stocks from overexploitation by common folk. These animals and others like them were among the first to benefit in the twentieth century from conservation efforts in the sea.

Over the course of decades, mammals like elephant seals, harbor seals, and sea lions have gradually repopulated their former haunts.

Once again they fill breeding beaches, laze on drying rocks at low tide, or frolic in the waves. This change in fortune has cost some populations, because it has made them more susceptible to epidemics. Diseases thrive where there are many susceptible hosts. To trigger an epidemic there must be a sufficient density of vulnerable animals to sustain transmission; otherwise, a disease would fizzle after felling a handful of animals. In 1988, harbor seals began to sicken and die across northern Europe.[11] After much head scratching, the culprit was finally isolated. It was a virus very close in identity and nature to canine distemper virus and was named phocine distemper virus. This disease depresses the immune system of seals. Victims become lethargic, and their eyes runny, but usually it is a secondary infection like pneumonia that kills.

Phocine distemper swept through the harbor seal population and took a few hundred gray seals along the way. By the time it was over a year later eighteen thousand were dead, half the population. Epidemics end when disease transmission rates fall below a critical threshold. The mass death of the seals caused their population size to plummet, and transmission rates fell. Those left behind probably have some resistance so it is harder for the pathogen to find susceptible victims and the disease disappears . . . for a time. After the outbreak, seal numbers began to recover. By 2002, they had come back sufficiently for distemper to take hold once again. Like all pathogens, distemper is able to mutate over time into novel strains to which its host has little or no immunity. The epidemic in 2002 was a near exact replay of the one which preceded it, but this time it cut down over twenty thousand.

The similarity of phocine distemper to canine distemper gives a clue to its possible origin. Other animals provide one of the main sources of emerging diseases. The same is true for pathogens that affect us. Ebola is one of the most feared illnesses, and it leads to near certain death through internal bleeding as blood vessels disintegrate. It is passed to people through contact with primates in Africa, perhaps animals that have been hunted for bushmeat. West Nile virus is passed to people from birds via mosquitoes, while swine flu came from pigs. About three quarters of emerging infectious diseases of people have their origins in other animals. Phocine distemper was probably passed to seals from dogs, which carry the virus without suffering harm. When an epidemic dies down, the dogs continue to harbor the illness until conditions are ripe for another outbreak. In the late 1980s, eighty thousand to one hundred thousand seals were killed in Lake Baikal, this time by the canine distemper virus itself. In Antarctica, crabeater seals were affected by a distemper virus that was probably passed to them by sled dogs.

The much heralded recovery of sea otters in California has recently been stalled by disease.[12] Otters have been dying from a parasite of domestic cats, called *Toxoplasma gondii*. The parasite causes brain inflammation, and while it does not always kill directly, it makes sick otters much more likely to be taken by sharks. Otters living near urban centers with rivers flowing into the sea were three times more likely to be infected than those farther away, which leads some scientists to believe that the rivers carry parasites from cat feces in rainwater runoff. They think that the parasites, once in the sea, could be concentrated from the water by filter-feeding clams and mussels that the otters eat.

Transfer of disease from land to sea is a common theme of recent outbreaks. The African dust suspected to have inoculated Caribbean sea fans with fungus is laden with other spores and microorganisms, many of which remain viable even after their high-altitude journeys. Runoff also carries soil and a mix of other pathogens to the sea. California sea lions have been infected by a soil fungus that causes the occasionally lethal valley fever in people of the southern United States and Mexico. The toxoplasmosis that culled sea otters has also recently been discovered in a spinner dolphin and beluga whales. Spread of disease agents from land to sea is hastened by land clearing, which promotes soil erosion, and by increasing human population densities in coastal areas. It seems likely that these trends will lead to many new instances of disease and parasite transfers in the future.

As I mentioned, diseases thrive where susceptible animals or plants are common. In this time of rapid global change several forces have come together to increase the numbers vulnerable to infection. Earlier I touched on how chemical contaminants like PCBs and methyl mercury can suppress the immune system. Mass mortality of harbor seals in the Baltic raised the possibility that pollution in this highly contaminated sea had weakened their ability to fight off infection. An experiment in which one group of seals was fed herring from the Baltic and another from the less contaminated Atlantic confirmed this hunch.[13] After two and a half years of dining on Baltic or Atlantic herring, the immunity of those from the Baltic group was lower, an effect blamed on PCBs, while the Atlantic group remained healthy.

Those long-suffering dolphins of Sarasota show much the same thing. Animals carrying a heavier burden of toxic PCBs and DDT in their bodies had less responsive immune systems than less contaminated beasts. Concern is growing that immune suppression by pollutants contributed to the mass mortalities of dolphins from *Morbilliviruses* (related to measles) in the Mediterranean and around North America in the 1990s. In the Indian River Lagoon in Florida there have been horrible cases of dolphins whose living bodies are being engulfed by fungus, a result of immune suppression by a cocktail of pollutants. Similar concerns have also been raised about the role of immune suppression by toxic chemicals in people.[14] Perhaps they have contributed to the emergence of several new diseases that have appeared since the 1970s. The jury is still out.

There are other ways in which susceptibility to disease can increase. Physical injury can weaken animals and plants, and wounds provide relatively easy access to the body for infectious agents. In St. Lucia one of my graduate students, Maggy Nugues, noticed that diseased corals were more common in places that were heavily dived than ones that saw divers only occasionally.[15] No matter how careful they are, scuba divers bump into, scratch, and scrape the delicate tissues that cover coral skeletons. These injuries could provide sites for infection. Sure enough, experiments show that injured tissues are much more readily infected than healthy ones. Resorts like Eilat and

Sharm-el-Sheikh in the Red Sea are visited by millions of snorkelers and scuba divers every year, each one of whom knocks, tramples, or crushes some delicate creature or other. In some sites it is hard to find any coral that hasn't been injured. Vast swathes of the northern Red Sea and many other places with coral reefs have been turned over to scuba diving.

There is another form of injury to marine life that in scale and severity dwarfs the impacts from scuba divers: bottom trawling and dredging. When a trawl net is dragged across the seabed it causes immense collateral damage to anything that lives on or near the ocean floor. The footrope of a trawl is designed to run close to the seabed. Sometimes it is weighted and other times it has rollers to enable it to work over rougher areas. In many beam trawls, where the net is held open by a heavy steel beam that drags above the bottom on metal runners, there are several "tickler" chains set ahead of the net mouth to scare up fish that hug the bottom, such as plaice, sole, and flounder. Footropes and chains slice off marine life like corals, sponges, sea fans, and seaweeds. Ahead of the net, they catch boulders that can be yards across, sometimes entire chunks of living reef, and then roll them along the bottom as the net is towed.

Dredgers are even more destructive, although generally smaller than trawls. Those used to catch scallops are made in the form of a steel frame with vertical teeth set beneath to dig up the seabed. Behind is a bag of chain mail and net to hold the catch. Scallop dredges look much like the harrows farmers use to break up clods of earth after a field has been plowed, and their effects below water are not far different. Dredges weave trails of devastation across the bottom. Within these tracks, the bottom is littered with the bodies of dead, dying, and injured animals: three-legged crabs with punctured carapaces struggle over ripped fragments of sponge; chopped pieces of sea fan lie across corals torn by ragged wounds; scallops with chunks of missing shell wait to be eaten by one of the many scavengers that this kind of fishing encourages.

The carnage wrought by bottom trawling and dredging is multiplied from coast to horizon and beyond, from sea to sea and ocean to ocean. Virtually nowhere above three thousand feet deep is spared. Some places get hit once every five or ten years, while places where trawling is unusually intense can be trawled five times in a year. One widely regarded estimate says that an area of nearly six million square miles gets hit by trawls and dredges every year.[16] That is more than fifteen thousand square miles of damaged, dead, and dying bottom life every day. For the sake of comparison, that equals an area about one and a half times the size of Europe or America hit every year. It isn't hard to see how this could increase the number of individuals susceptible to disease.

The hidden hand of climate change also plays its role in the emergence of disease. Raised water temperatures stress host animals, and pathogens thrive where stress prevails. It is often said that a warmer world will be a sicker world, although some people are less sure. However, there are good reasons to think that diseases might flourish in the future. It won't just be fish or plankton on the move as climate changes. Diseases and parasites will also find opportunities as they spread with their hosts, or colonize places where new hosts are found. Such shifts have already been seen. In the early 1990s, an oyster parasite moved three hundred miles north into the waters of New England as the sea warmed enough for it to survive over winter.[17] This single-celled parasite proliferates within oyster tissues and can take several years to kill its host.

Mass coral bleaching is one of the most obvious sentinels of severe warming stress. Disease outbreaks have been reported in some places following sublethal bleaching.[18] Diseases like the fungus that wiped out Caribbean sea fans seem to grow faster in warmer seas.[19] Sponge mass mortality in the Mediterranean has been linked closely with unusually warm water.[20]

Acidification will weaken many life-forms that use calcium carbonate to make shells. Thinner, more brittle shells and skeletons will mean greater chances of injury, and therefore sites for infection. Nutrient pollution from terrestrial fertilizers and sewage can also encourage disease. Diseases on Caribbean sea fans and corals spread faster in an experiment when nutrients were added, which suggests that fertilizer boosts the growth of pathogens as well as the crops.[21] There is a possibility that the general background increase in nutrient pollution of coastal seas has helped to trigger some disease epidemics. But there isn't yet much support for this, and its role probably isn't decisive. Instead, nutrients are just one of a combination of conditions that on their own might be shrugged off easily but together have been lethal.

There is one transformation of the sea that, from the point of view of diseases and parasites, makes it a healthier place. As I said earlier, diseases have not increased in all groups that have been looked at. There was no upward trend for disease in sharks and rays, or crabs, lobsters, and prawns, and fish seem to have less disease now than they did in the 1970s. All of these groups are intensively targeted by fishing, which means that the densities of most have nose-dived in the last hundred years. We can see in ourselves how a change in transmission rates can influence the development of illness. Colds and flu peak in winter in temperate and cold climates because we spend more time indoors and huddled together than in summer. Viruses may be assisted in a few other ways, such as our increased susceptibility in winter, or by greater survival outside the body in weak light. But the basic element is crowding. At low host densities diseases find it harder to get a foothold and spread, so exploited species may be less at risk of epidemics than they were before we fished them so intensely. At last, something positive about overfishing, but not a cause for celebration.

The effect fishing exerts on disease and parasites goes beyond simple reduction of host densities.[22] You will remember that fishing tends to selectively remove the big, old, and toothsome beasts while leaving the young and supple behind. By virtue of age, or diminishing resistance, old animals carry the greatest load of parasites. As we have fished down populations and felled their oldest members, we have inadvertently driven parasites below critical density thresholds that dictate whether they persist or disappear. Put another way, we can fish out parasites well before we fish out their hosts.

When we stop fishing a place we play the tape of decline backward and the area fills up again with life. Corals, sponges, and seaweeds sprout from the seabed and coalesce into glades and thickets. Small fish grow into great stalking predators. Diminutive plankton feeders multiply from twinkling points into constellations of fish. As densities rise, diseases and parasites could make a comeback. I don't know of any research that demonstrates such an effect as yet, but it is a possibility. On the other hand, there are studies that demonstrate the opposite effect in prey or other species not targeted by fisheries. In a California marine reserve sea urchins were less abundant but more healthy, because population explosions were prevented by recovery of their lobster predators.[23] And coral diseases in the Philippines were less common in marine protected areas with healthy fish populations, probably due to prevention of seaweed overgrowth by herbivorous fish.

Disease epidemics are reshaping ecosystems in profound ways. Coral reefs give us a rare opportunity to compare present conditions with what has gone before. Corals constantly rebuild reefs on foundations laid by previous generations. Just as archaeologists piece together the course of history from their excavations, cores drilled through reefs tell us what they looked like tens, hundreds, and thousands of years ago. Cores through Caribbean reefs show that recent branching coral losses from disease are unprecedented on a timescale of several thousand years, as far back in time as the cores were able to show. As coral skeletons crumble in the aftermath of mass mortality, habitable space for fish, invertebrates, and hosts of others dwindles. The reef hemorrhages life, and, as it empties, its ability to support human needs and livelihoods declines. Dense palisades of branching corals once protected islands and coasts from Florida to Panama and South America. Where once divers had to thread their way gingerly through close forests of sharp coral to reach the open water beyond, today they pass over gently undulating hummocks of seaweed and flattened mounds of sponge, sea squirt, and encrusting coral. As coral ramparts tumbled, they left shores open to attack by wave and storm. Beaches in front of Caribbean resorts and condos that reefs once protected have washed away, forcing developers to replace them with ugly concrete defenses.

Diseases can cause other economic losses. The sea grass wasting disease I mentioned earlier destroyed critical nursery habitats for commercial fish all the way up the Eastern Seaboard of the United States. It also caused the extinction of a tiny limpet whose only habitat was stolen from beneath it. As outbreaks of disease in the oceans multiply and spread, we can expect worse to come. They warn us that life there is increasingly stressed.

CHAPTER 14

Mare Incognitum

Like children the world over, my daughters love turtles. At once incongruous and graceful, they connect us to the world of fifteen million years ago, when very similar turtles swam alongside megatooth sharks, or seventy-five million years ago, when they rubbed shoulders with dinosaurs. Only eight species of marine turtle remain from a lineage that stretches back little changed deep into the age of dinosaurs. The largest living reptile is the leatherback turtle, a barnacle-encrusted eminence that can reach ten feet long and weigh two tons. Today we confront the stark possibility that people will drive the leatherback turtle to extinction within the next human generation. Already there is just one leatherback left in the Pacific for every twenty in 1962, the year I was born.

I find this prospect awful to contemplate. What will the world seem like to our grandchildren if the only leatherbacks left are those that inhabit books or computer screens? At the rate we're going the world could by that time have been robbed of the majesty of whale sharks, the frivolity of sea otters, and the frenzy of bluefin tuna, beating the sea to foam as they hunt down prey. These animals are not just icons of children's books and television channels; they are creatures we can pursue with our imaginations far into unknown depths. Without them, life will lose some of its grandeur.

The last thousand years have seen our influences gather like a green wave, scarcely perceptible at first but slowly lifting and steepening over the centuries to burst across the globe in the last sixty years.

Our planetary remodeling did not stop at the shore; it just came a little later to the sea. There is no precedent for the speed and variety of the changes underway today, save perhaps the asteroid impact that ended the reign of dinosaurs sixty-five million years ago. Even the cataclysms of the other great extinction events seem sedate by comparison. There is no sign of any letup. The rate of change continues to accelerate in step with human population and economic growth. Our impacts have intensified with time, and where once habitats and species were subject to just one or two influences, like fishing or siltation, now they are caught amid a morass of stresses whose effects accumulate and reverberate through every layer of the living world.

While life's players differ from place to place, how the play of human impact has unfolded is much the same. The abundance and sheer physical mass of life has declined over time. There has been a shift from dominance by large-bodied animals and plants to small ones. Populations have been truncated so that age and wisdom have given way to youth and inexperience. The web of life has been simplified as predators have yielded to prey and some species have vanished. Habitats built over millennia by the exertions of countless generations of creatures have crumbled from three-dimensional magnificence to two-dimensional monotony. And for our own species, perhaps uniquely gifted with a sense of aesthetics, in too many places there has been a shift from beauty to sordid poverty.

The shifting baseline syndrome has blinded us to these alterations. A trip to the seaside appears to hold the same joy and wonder for my children as it did for me some forty years ago, when there was more to find amid rockpools and waves. But what disturbs me, as my own perspective extends into middle age, is how rapid the alteration has been. The vibrant coral reefs I dived as a youth seem less bright, less imposing, and less wild today, as indeed most of them are. In that time the carpets of stony corals that build reefs and give them much of their vivid color and complexity have shrunk by a third to a half. The fat beady-eyed groupers that skulked beneath ledges have become rare, and the walls of predatory snappers and emperors have thinned.

In many places silt washed off the land has fogged visibility, so the vistas are less impressive and their colors dimmed.

Just fifty years ago there were thirty times more Atlantic bluefin tuna swimming wild than now exist. The shoals that raced into the North Sea each spring to thrash among the herring have disappeared. Vast ice shelves have broken free in the Southern Ocean, the Arctic ice thins, and open water near the North Pole has allowed the surface waters of the North Pacific and the Atlantic to mix for the first time in eight hundred thousand years. Sea grass meadows in the Mediterranean have been swallowed beneath choking blankets of invasive seaweed. Dead zones have grown and multiplied. Mighty currents have slowed. In sum, humanity has achieved dominion over the oceans and marine life is in difficulty.

It is hard to grasp the prospect of seas so compromised that they no longer sustain the ecological processes that we take for granted and upon which our comfort, pleasure, and perhaps even our very existence depends. What will the future look like? Prediction is difficult, given that we have never been here before. In the early days of European seafaring, unexplored areas of ocean were marked on charts as "Mare Incognitum," or Unknown Seas, and the truth is that we are voyaging into such seas again today. But we have a special interest in the outlook, so prediction is worth a shot. After all, as the American engineer Charles Kettering once said, the future is where we will spend the rest of our lives. To see where the ocean world is going is a good place to begin to look at the drivers of change.

Fishing is the most ancient of human influences on the sea, but it is only in the last 150 years that its effects spread beyond the local. The late-nineteenth-century industrialization of fishing came when we added engines to boats. Technological innovation gathered pace through the twentieth century in a race to snag fish and beat others to profit, and has carried on since. The footprint of fishing spread from traditional fishing grounds to distant seas in the early twentieth century, and then in the mid—twentieth century from coasts to the high seas, and shallow water to deep. Over time we have substituted new species as past favorites waned. The price of fish has outstripped inflation for decades, reflecting the increasing difficulty and cost of sustaining supplies. In the last third of the twentieth century, developed countries turned to the developing world to supply their fish after having exhausted their own grounds. The UN Food and Agriculture Organization's scorecard of the world's major fisheries shows an increasing rate of collapses since 1950, when it began to collect statistics.

Growing demand for fish from the rapidly expanding human population has driven these trends. In 1880, on the cusp of the industrial fishing revolution, there were 1.4 billion mouths to feed. By 2011 that number had reached seven billion, five times more. Demand is set to rise by another third in the next forty years with the addition of at least another 2.1 billion people, so pressure for fish protein isn't about to diminish.[1] If we exploit the oceans as we have done for the last century, then overfishing will continue to eat away at the populations of the world's big fish; some will be driven to extinction, while many more will become too rare to play any further meaningful role within their ecosystems. As they disappear, we will continue the switch from large predators like cod and hake to animals low in food webs, like prawns and anchovies. But they too will become overexploited (as some already are), and we will have to seek seafood from other sources, such as Antarctic krill, which will be processed into more palatable looking foods, like fish cakes and fish sticks.

In fact, fisheries for krill and jellyfish are already on the rise. FAO figures show jellyfish catches rose exponentially from almost none in the 1960s to more than half a million tons today. If we fish as we do now, it is doubtful that catches from these animals—which, based on their fleeting lifestyles, could be dubbed the mice and cockroaches of the sea—will be able to substitute for the peak landings reached in the 1980s, so wild seafood will get ever scarcer. A billion people rely on seafood as their main source of animal protein today, most of them in the developing world.[2] The continued decline of wild fisheries threatens malnourishment for many more by the middle of this century.

Just as consumers in the Western world are developing scruples about the sustainability of the fish they eat, economic growth and increased disposable income in Asia, especially China, have released a surge of demand for seafood. Asian buyers have fanned out across the world in search of supplies, especially targeting poor communities in developing countries. Judging by their actions, most have little thought for sustainability. They buy while the fish last and move on. The Asian market has forced prices for some products sky-high, turning once worthless bycatch species such as blue and silky sharks into valuable targets. In 2004, a single large whale shark fin could sell for over US$50,000.[3] The long-term prospect does not look good unless Asian consumers can be persuaded to value sustainability soon.

Things don't have to get this bad. Aquaculture could help sustain supplies of fish and better management could maintain or perhaps increase catches from wild stocks recovered to populations far larger than today's. I will come back to how we could achieve this metamorphosis later on. But let's return to the other drivers of change first.

Human population growth and technological development have also driven the rise in pollution. The Industrial Revolution opened the veins from which toxic metals and other chemicals began to bleed into the sea. While time has staunched some of these flows, others have opened, as miners and heavy industry have shifted their dirtiest operations from developed to developing countries. All the while we have invented new ways to make things. Our enthusiasm for novel materials and processes seems always to outrun prudence. Highly toxic chemicals are set free long before we notice their darker sides. The legacy of industrialization is written in no-go zones too toxic for use. But many chemicals have spread so widely they cannot be avoided. They swirl around all of us. They compromise wildlife health and reproduction and contribute to the rising incidence of disease epidemics in the sea. And as people often sit at the apex of the food web, trumping consummate predators like sleek tuna and grinning shark, we too are vulnerable. Seafood lovers the world over are accumulating toxins in their bodies drawn from waters where we thought they had been laid to rest for good.

Population growth has driven up the demand for food. The world avoided the much anticipated population crash expected in the 1980s and 1990s with the green revolution that industrialized agriculture. We have squeezed ever-greater production from the land by lavish applications of artificial fertilizers and other agrochemicals. They have bought us time, but like fisheries, farmers have won greater production at the cost of the natural capital of their soils.[4] Rains and irrigation water carry soils and their chemical burdens to the sea through estuaries and deltas whose filter feeders have all but disappeared and whose wetlands have been stripped and converted to other uses. Once in the ocean, nitrogen and phosphorus fire plankton and microbial blooms that overwhelm the assimilative capacity of food webs decapitated by overfishing and exceed the physical water mixing needed to keep up with demand for oxygen, as dead plankton accumulate.

Despite the proliferation of dead zones, future seas will not be lifeless. We are creating winners as well as losers. Jellyfish, for example, are great opportunists, and some scientists fear that large parts of our most productive seas will transform into jellyfish empires.[5] Few animals other than leatherback turtles and giant ocean sunfish (both in decline) eat adult jellyfish because they lack substance, and they have to eat huge quantities to break even. There are many more predators that eat snack-size juvenile jellies.[6] With the wholesale decline of marine life from overexploitation, pollution, and climate change, the ranks of these predators have thinned, so more jellies survive.

Jellyfish positively thrive in pollution-enriched seas. Given unlimited food, they can reach adult size fast. With their stinging tentacles, they are formidable predators. Here one of the quirks of ocean food webs comes into play, sealing their dominance. Most animals that might eat jellyfish go through tiny egg, larval, or juvenile stages when the tables turn and they are themselves jellyfish prey. Such role rever-sals of predator and prey are rare on land. In the sea, however, they are prevalent, with surprising effects. The American oceanographer Andrew Bakun invites us to imagine a world in which zebras and antelopes are voracious predators of young of lions or cheetahs. What would the Serengeti look like if this were so?

The jellyfish joy ride begins when high nutrients combine with a fall in abundance of their predators. When plentiful, jellyfish sup-press their predators further by eating more of their young, and so pave the way for a full-blown population explosion. If food runs short, jellyfish don't just die; instead, they shrink and wait until con-ditions improve (although, if nutrient levels fall far enough and for long enough, jellyfish blooms can snuff out). In a future with more acidified seas, jellies won't have troublesome carbonate skeletons to handicap their chances. The altered oceans that haunt our possible future could offer them worlds of opportunity. They have been here before. Enigmatic traces in rocks from the earliest Cambrian period, some five hundred and fifty million years ago, tell of an age of jelly-fish that preceded the great radiation of life that established most of the animal groups alive today. Collectively, the modern reappearance of seas dominated by gelatinous animals, microbes, and algae has been dubbed "the rise of slime."[7] It signals a reversion toward condi-tions that prevailed in the earliest days of multicellular life.

Climate change adds a third set of drivers of change to life in the sea. Heating will thicken the lower density warm layer of water that sits like a lid over much of the ocean to restrict oxygen transfer to underlying waters and constrain upward flow of nutrients. The result will be reduced productivity over great expanses of sea away from coasts and expanded low oxygen zones beneath. Already there are signs that plankton productivity has fallen, probably as a consequence of warmer seas, and the volume of oxygen-starved waters has grown.[8] Nutrient starvation of surface waters may also be due to the collapse in fish stocks from overfishing, causing nutrients to disappear into the depths rather than being recycled in shallow waters. Great whales may once have played a role as a nutrient pump from deep to shallow water.[9] Many feed at depth and poop at the surface. Before whaling, when the oceans thronged with more than a million whales, that could have made a big difference to productivity.

Tropical heating along western coasts will force faster redistribution of heat to the poles, strengthening winds that power upwelling of deep water. Upwelling could benefit fish production by raising nutrients from the deep, but intensified upwelling could push some regions into hyperproduction to create noxious eruptions of hydrogen sulfide like those that plague Namibia's coast today. Upwellings might also pull low-oxygen water of higher acidity onto continental shelves, causing the mass mortality of bottom life like episodes seen recently along the Oregon coast.

The effects of acidification are hard to predict. At the very least life is likely to get much more difficult for species with carbonate shells, which includes some of the most important primary producers in the sea, the phytoplankton that sustain food webs and release life-giving oxygen. We don't yet know whether the photosynthesis boosting effect of extra carbon dioxide might outweigh or compensate its corrosive qualities in the balance between benefit and cost. If not, there are noncalcifying phytoplankton that might step up production. The worst case scenario would be a large-scale depression of primary production, which is the quantity of carbon fixed into plant material, which also means less oxygen produced.

Any fall in the rate of plankton production would reduce the snow of organic debris that sinks from sunlit surface layers to the deep sea. Deep sea communities survive on meager handouts from above, and failure in supply would thin their numbers. As the sea is the ultimate sink for all of the carbon dioxide we are releasing into the atmosphere, it would also reduce the rate at which carbon gets locked away in bottom sediments. This is the "biological pump" that takes carbon from where it can be a nuisance to where it does no harm. If phytoplankton secrete less carbonate because of acidity, their shells will be light and they won't sink as fast. The pump will run more slowly, so the oceans will absorb less of the carbon dioxide we add to the atmosphere, and global warming will accelerate. Limited evidence from coccolithophores in deep sea sediments suggest their mass has actually increased by 40 percent since the Industrial Revolution. For them, the growth-boosting effect of extra carbon dioxide has so far outweighed the acidity cost.[10] Some lab experiments suggest the reverse is true,[11] and in the wild the skeletal weight of foraminifera in the Southern Ocean, another key plankton group, has declined by 30 percent to 35 percent since preindustrial times.[12] The leading edge of science is always fraught with uncertainty. We still have much to learn.

The Roman emperor Marcus Aurelius said, "Look back over the past, with its changing empires that rose and fell, and you can foresee the future, too." There are more than enough portents to be able to see where the oceans are headed. Scotland's Firth of Clyde is a window to a world without fish. Italy's volcanic Ischia lets us dive in an acid sea. China's Bohai Gulf shows how nutrient pollution and overfishing can carpet a sea in weed and spread plagues of jellyfish and toxic plankton blooms. The Gulf of Mexico, Oregon, the Baltic, and hundreds of other places across the world give us a taste of the oblivion spread by anoxic water. Europe's North Sea reveals life on the move as the world warms. The Gulf of Maine and San Francisco Bay are filling with alien species. Sea level rise is eating away at the Mississippi and Nile deltas and threatens densely populated coasts like those of Holland, Germany, and Thailand. And the rotting bodies of Pacific albatross chicks, bloated and starved by a diet of plastics, show us how life struggles as the oceans fill with human refuse.

Each of these places illustrates nature under stress from one or two of the impacts I have described in previous chapters. But these primary stressors are not the only ones they face. The Firth of Clyde, for example, has been polluted by decades of sewage-sludge dumping, the waste from a major city, and heavy industry. The Gulf of Mexico's seabed is raked over year after year by shrimp trawlers. Multiple stresses have more serious effects because they can act in synergistic and often unpredictable ways.

Stresses exert their insidious influences in several ways. Species are sensitive to different stressors to different degrees. If each stress was applied separately, a snail might be noticeably affected by slight acidification, made uncomfortable with a half degree rise in water temperature, and be indifferent to heavy metal pollution. Add two or more together and the effects could be severe. In one dramatic example, brittle stars are able to survive slight acidification, but add trace amounts of the antibacterial triclosan, an ingredient of hand washes and face creams, and it collapses into fragments.[13] Stresses don't usually combine to such spectacular effect, but their collective impact can erode the fitness of individuals and populations. Their effects are best understood in the currencies of life and death. Populations endure when the birthrate equals or betters the death rate. Stresses often drive down births and increase the likelihood of death, so they upset the balance of life's equation.

The extra physiological cost of coping in low-oxygen water, for example, could be met by increasing food intake or reducing calorie demand. Extra food may not be available, so animals could compensate by laying fewer eggs or skipping a breeding season. Or they might grow more slowly, which would also mean fewer offspring, since small animals usually produce fewer eggs. Low oxygen in itself is a powerful constraint on body size, favoring small animals over large because they can breathe more easily, and thereby also cutting egg production as big animals produce more eggs. High body burdens of toxic contaminants, many of them hormone disruptors, might further compromise reproduction through reduced fertility or miscarriage. When the tipping point is passed at which deaths exceed births, a population declines.

The cocktail of human stresses is mixed differently from place to place, but whatever the combination, its collective impact causes worse harm than each one acting alone. In other words, the kill rate from multiple stresses acting in concert is greater than the summed kill rate from each of the same stresses alone. There is a parallel here between the environment and the experience of intensive-care doctors. They must jump in quickly to support patients when they have only one problem to deal with, like blood loss or stroke. When problems begin to cascade through the complex systems of the body, mortality shoots up.

When a species succumbs, the effect multiplies through the ecosystem via the web of interactions that connect species. Some creatures are strongly connected, like salmon sharks and salmon, anemonefish and anemones, or dugongs and sea grass. Others form loose associations. A crab might be happy to live among the branches of half a dozen different types of coral, or a fish might feel equally at home amid the waving fronds of several different seaweeds. Still more are linked indirectly through intermediaries; the food your prey eats is ultimately your food (a truth we too should keep in mind in our relations with the rest of the biosphere).

Stresses act on ecosystems at this higher level when they weaken or knock out links among species. When one species goes down, it takes others with it that, in turn, affect the animals and plants with whom they are connected. The more different stresses we add to the blend, the wider the spectrum of life that is directly affected, so the consequences penetrate further through the tangle of existence. As species dwindle or disappear, ecosystems adapt and re-form, as remaining inhabitants find other ways to make a living. Weak links can strengthen as animals find alternative food or living space.

Knocking out one species, inserting something else, or adjusting the abundance of another lead to effects on species that at first seem completely unconnected with them. The nineteenth-century American national parks pioneer John Muir's famous phrase is apt here: "When one tugs at a single thing in nature, he finds it attached to the rest of the world."[14] The scale of our ignorance of these interconnections is breathtaking. It would take a thousand lifetimes of research to figure out all of the ways in which we are affecting the species in an ecosystem of only moderate complexity. Even within a species, effects differ. While adult lobsters may happily cope with increased acidity, the same change halts development of one of their several juvenile stages.

When species that play key structuring roles disappear, their loss has profound consequences. Longline fisheries in the Pacific Ocean have cut down populations of large ocean-going sharks like the thresher to a tenth of their former abundance.[15] In the wake of their departure, smaller species, such as the pelagic stingray, have proliferated. Off America's East Coast, a similar reorganization has played out. Megapredators, such as hammerhead, great white, and tiger sharks, have collapsed, leaving the field open to their prey.[16] Five species of smaller shark and ray have grown in abundance, some spectacularly so. The cownose ray, blunt-faced stingrays with dark round eyes and long whiptails, travel in groups and raise thick billows of sand as they excavate buried clams. With no big sharks to prey on them, their groups have grown to hundreds strong. Added together there could be as many as forty million up and down the East Coast. Cownose rays are particularly partial to bay scallops, and with this many mouths to feed, there is scarcely a scallop left, leaving U.S. fishing communities empty-handed and bewildered. It isn't an obvious connection, that fishing-out giant sharks would cause the collapse of scallop fisheries, but that is how impacts proliferate. It does illustrate the difficulty of predicting how stresses will reshape ecosystems, even when there is only a single stress involved.

Canadian cod may have fallen victim to another kind of ecological reorganization, since their population crashed in the 1990s and a moratorium on cod fishing was imposed. Cod once reigned supreme in the chilly waters off Newfoundland. Their ferocity and indiscriminate tastes made them almost universally feared and avoided. Lobsters lurked and crabs cowered as throngs of cod advanced over the continental shelf to their spawning grounds. Like most fish, however, cod are vulnerable when young. Eggs and larvae are snapped up by capelin, herring, mackerel, jellyfish, and hosts of others. When cod were mighty they produced so many young that they overwhelmed their predators. It mattered not that tiny cod poured in endless streams into almost limitless mouths because plenty escaped. But when cod were humbled by overfishing, the survivors could only produce a tiny fraction of the former abundance of young, few of which survive the carnivorous gauntlet today. Here is another of those role reversals in which a predator falls victim to its prey.

When stresses take out habitat-defining species like kelp, oysters, or sea grass they can set off chains of events that remodel the habitat so it becomes hostile to the original inhabitants, just as political revolutions sweep aside old regimes and those who thrived in their orbit.[17] When they go, or are pushed out, space frees up for competitors. If strong or predatory interlopers establish, they can change conditions in ways that inhibit recovery of the original species. European coasts were developed at the rate of over half a mile per day between 1960 and 1995, causing widespread wetland loss. Below water, rocky reefs carpeted in lush seaweed forests were subjected to siltation, reduced light, anchoring and dredging, and both the direct effects of destructive fishing and indirect effects of removal of key fish species that kept grazing herbivores in check. When large, canopy-forming seaweeds disappeared from Mediterranean reefs they were replaced by shorter, turf-forming seaweeds. Beneath the old canopies, sediments were swept away by fronds that lashed back and forth under the influence of waves and currents. Turf seaweeds, by contrast, trapped sediment to create an unstable layer over the rock so that it cannot long sustain larger weeds that require firm anchors to hold them in place.

In the Baltic Sea, lush meadows of brown, red, purple, and pink weeds struggled as light penetration faltered under the increasing plankton growth stimulated by excess fertilizing nutrients. The dead bodies from these blooms piled up on the seabed and drove down oxygen levels, so that a thick organic soup began to churn above the bottom, cutting light levels further and driving out animals that once called the weedy forest home. On coral reefs, seaweed again causes problems, when mass coral mortality from high temperatures or disease opens up space. Where grazers have been overfished, sea-weeds rapidly take hold, especially in places where nutrient-rich water also bathes the reef. Just as in the Mediterranean, these seaweeds trap sediment, which makes the coral comeback much harder, and the weeds may themselves harm tiny corals by poison, or by contact abrasion to make them more susceptible to diseases, some of which the seaweeds harbor.[18]

In many cases, loss of habitat-forming species, such as oysters, sponge, or coral, opens the way for invasions by other species. Sometimes it is the invaders themselves that oust the habitat creators, such as the rampant *Caulerpa* weed that engulfed the native sea grasses in the Mediterranean. Invaders complicate the business of projecting the future for marine life still further. In part, the wave of new species advancing into habitats is a product of climate warming. Some arrivals can be predicted quite easily, because they live nearby, but their effects may be hard to judge because of the complex ways in which they ripple through the web of life. Other introductions are more capricious and come about through the chance arrival and unlikely survival of creatures from far afield. The red lionfish that have so rapidly settled into Caribbean coral reefs seem certain to cause a major reorganization of life there whose outcome we can only guess at.

Habitats have been opened up and flattened out and the new regimes usually support fewer species and are less productive than the old. The contrast between kelp forest and turf pavement is like the difference between woodland and pasture. Think of how rich is the life of the forest. Its multiple layers and kaleidoscope of conditions produce communities within communities: the lush and open canopy; the fuzz of lichen and fern on trunk and branch; the dark, mossy wrinkles and cavities; the tangled undergrowth and deep loam beneath. Compare that to open grassland grazed by rabbits or livestock. A close look might reveal a miniature world more varied than is apparent at first glance, but the simple fact is that grasslands offer fewer ways to make a living than woods, and while some species do well (the Serengeti's rivers of wildebeest in full migration come to mind), the abundance of life they support is less.

Complex habitats offer more ways to make a living than the simpler, flatter habitats that remain in the wake of degradation. They also sustain more big fish, who in turn keep productivity high by recycling nutrients. Just as gardeners love horse manure, so fish shit is great for seaweed and phytoplankton, the engines of production in the sea. They liberate nutrients to the legions of detritus feeders and animals such as corals, mollusks, and sponges that feed on floating particles. A recent study based on computer-simulated removals of fish species from food webs in tropical freshwater lakes and streams showed how some species play disproportionately important roles.[19] Inconveniently for us, the most important species are often those favored by fishermen, because they tend to be large, or abundant, or both. When the simulated extinctions followed the sequence of importance of species in catches, the loss of nutrient recycling function was far greater than when species were removed at random.

In these simulations, when species with similar ecological roles were allowed to increase in abundance following the removal of others, they were often able to compensate for the roles played by their vanished colleagues. Diversity is a key to resilience. Where there are lots of species and much overlap in what they do, there is more scope for some of them to disappear before losses trigger catastrophe. The trouble with fisheries is that they are often unselective and take out several species at a time. And when one declines, fishers switch to others, so the scope for compensation is limited. This is certainly what I have seen on Caribbean coral reefs, where fewer and smaller species are left behind as fishing intensifies. The dainty little parrotfish that pluck seaweed from Jamaica's ransacked reefs are all that are left after fishermen have taken their fill. They cannot possibly control the runaway seaweed growth that is overwhelming coral there, while the dense groups of bulky parrotfish that exist on lightly fished reefs easily strip seaweed back to a green baize film.

It saddens me to see the loss of beauty and wild energy of the sea. Tangled forests of mangrove that teemed with life have been replaced by the bare embankments of shrimp ponds. Verdant green meadows of sea grass have withered or been buried under blanketing mud. The labyrinthine creeks of salt marsh where once you could leave the modern world behind have been drained and built over time after time. These are just the places that are easy to see, but we find similar losses in less accessible corners. Cold water coral reefs in the North Sea were already half destroyed by trawlers when they were first discovered in the 1990s. The deep reefs off Miami where liveried grouper once held court in palaces of coral are now mostly rubble, their trawled ruins inhabited by tiny fish and worms. At some level these losses discomfort every one of us. But the disintegration of untamed nature reaches beyond aesthetics. It is undermining the ability of the oceans to sustain human needs and well-being.

CHAPTER 15

Ecosystems at Your Service

To people who allege that we
Incline to overrate the sea
I answer, "We do not;
Apart from being colored blue,
It has its uses not a few;
I cannot think what we should do
If ever 'the deep did rot.'"

—Sir Owen Seaman, 1861–1936[1]

My graduate student Ruth Thurstan spent the summer of 2010 sampling seabed sediments around Britain, wrestling with coring equipment and a sometimes restive team who didn't share her enthusiasm for 5:00 A.M. starts. They pounded coring tubes into the bottom, raising clouds of mud that thickened as they worked until they had to communicate by touch alone. Once the cores were removed, she split them from top to bottom, and half went to be dated using radioisotopes while she spent the next six months sifting fragments of shell, spine, and test from the other half. These remains help trace the transformation of the bottom from an exuberance of oysters, horse mussels, and clams to silty scarcity. The change happened when trawlers and dredgers scraped the bottom clean in pursuit of fish and oysters. They sundered the layer of invertebrate life that once crusted the bottoms of estuaries and great expanses of continental shelves. Over years, decades, and centuries, the trawlers pulverized and buried creatures that once enlivened the seabed. Ironically, oysters introduced to Europe from North America to replenish exhausted beds set loose a disease that wiped out the remaining natives.

British fishers immediately regretted the loss of the invertebrate crust over the bottom, because it destroyed their source of bait for longline fisheries for cod, haddock, turbot, and the like. Some also saw that it would undermine the productivity of the fishing itself, since a breakdown in habitat would compromise living conditions. The British surgeon and naturalist John Bellamy predicted as much in 1843:

> [T]he employment of the Trawl, however, during a long series of years, must assuredly act with the greatest prejudice toward these [fishy] races. Dragged along with force over considerable areas of marine bottom, it tears away promiscuously, hosts of the inferior beings there resident, besides bringing destruction on multitudes of smaller fishes, *the whole of which, be it observed, are the appointed diet of those edible species sought after as human food*. . . . An interference with the economical arrangement of Creation, of such magnitude, and of such long duration, will hereafter bring its fruits in a perceptible diminution of those articles of consumption for which we at present seem to have so great necessity.

Like all good Victorians, Bellamy continued in poetic vein,

> [W]e might suppose the species thus injured by an abstraction of their food, to utter some such expressions as the following:

> *Nay, take my life, and all, pardon not that:*
> *You take my house, when you do take the prop*

That doth sustain my house; you take my life,
When you do take the means whereby I live.[2]

Bellamy's warning and those of many less poetic scientists after him went unheeded, and the trawlers obliterated habitats formed over thousands of years, not just around Britain but around the world. More than 150 years later, Bellamy's prediction has been proved true. A study of fish on southwest British trawling grounds showed that several species are skinny and stunted compared to those in less trawled areas, because they cannot find enough food.[3] Had there been any places left that trawlers had not worked over, the contrast would have been more obvious still. Productivity has fallen not just as a result of direct removal of the plants and animals, but because simple habitats offer fewer living spaces and sustain less life than complex ones. The citadels and piazzas, avenues and apartments of these underwater worlds have crumbled, and their flattening, not just by trawlers and dredges but by the panoply of stresses I have described, has left room for only a fraction of their former inhabitants.

European seas are far less productive than they once were. The fact that the UK bottom trawl fleet lands only half the fish today that it did when records began in 1889, despite a massive increase in fishing power, says all we need to know about how we have squandered natural capital.[4] People in wealthy nations have so far been sheltered from the impacts of overexploitation. They are rich enough to buy fish from parts of the world where there are still some to be caught. But when fish stocks collapse in the developing world, desperation follows. Nowhere has this been more potently illustrated than in the Philippines. This country should be the world's underwater wonderland. It lies at the heart of the most diverse sea on the planet, one crammed full of a dazzling variety of fish, coral, and almost every other form of life. Filipinos have lived by and from the sea for thousands of years. They depend on it, getting some 70 percent of their animal protein from seafood. In recent years, though, their explosive population growth has upped demand far beyond the capacity of the sea to supply. The result is a cycle of overfishing that has become ever more frantic with each passing year; the most extreme manifestation in the world has got to be the practice of *pa-aling*.

When times were better, Filipinos fished using traditional methods, such as traps, nets, hooks, and spears. A day's fishing would easily supply a family's needs and leave extra to sell. In a familiar pattern, as stocks declined, people devised ways to maintain catches by upping the power of these simple methods. The process of escalation eventually led them to invent *pa-aling*—or drive-net fishing—in which a team of men and boys suffers extreme peril at the limits of human endurance for a few bucketfuls of fish. When reefs have been emptied of life in the shallows, there is nowhere to go but down.

A team of ten or fifteen men and boys will dive down as far as 100 feet, or 130 feet, with a huge ball of net, breathing diesel-tainted air through a length of garden hose supplied from a compressor onboard a boat. Meanwhile, another team on the boat struggles to keep the engine going and all the hoses from turning into a lethal tangle. Underwater, the fishermen unfurl the net into a dome the size of a village church. It is weighed down at the sides and one end with rocks and buoyed at the top with air-filled plastic bottles. The dénouement comes after a break at the surface, as the team swims along the bottom with a scare line to drive fish into the net, trampling and bashing the bottom with rocks in their advance. One drive can clear a place of half its fish and leave the reef in tatters.

The human cost of *pa-aling* is brutal: Divers frequently far exceed safe diving times and are stricken by agonizing bends that can paralyze or kill. Others drown when they are tangled in the net or air supplies are cut off. Captains often exploit child labor in their pursuit of the last fish. This is the last gasp of overfishing. When the fish have been taken from these depths there will be none left within reach of divers. The Philippines should have the world's most spectacular coral reefs. Instead, it serves as a dreadful warning.

The world is living on borrowed time. We can't cheat nature by taking more than is produced indefinitely, no matter how fervently politicians or captains of industry might wish it. Rich nations can outsource production to poor ones, but at some point fish stocks will collapse there too, and then there will be no fish to be had at any price. In essence, what we have done in the last few decades is to mine fish, bringing them in at rates faster than they can replace themselves. Sharks, bluefin tuna, cod, Chilean sea bass—all have declined steeply as a result of excessive fishing. The price that must be paid for today's rapaciousness will be tomorrow's scarcity or, in some places, seas without fish. If we follow the trajectories of loss seen today, that point may be only forty or fifty years away. Fishermen and -women of the future will have a tough time keeping their nets full.

Nature has survived planetary upheavals many times before, and life will survive this one. Natural selection is a powerful force. It has brought us life in all its wonderful guises and leaves us in breathless admiration at how such exquisite adaptations as the eye, or the color-shifting skin of a squid, or our self-aware brain could have come about. Life is endlessly adaptable and can eke an existence in some of the most hostile environments on Earth, like superheated hydrothermal vents under extreme pressure in the complete darkness of the deep sea. The planet has remained habitable for over four billion years; life has endured through times when the oceans turned more acid, when the atmosphere choked on carbon dioxide and methane, when the continents coalesced into parched desert, and when ice ages froze the sea.

But timescale is everything. A starving person will not be spared by the prospect of plentiful food three years hence. His needs must be met now. Evolution can take a very long time. It works in the currencies of life and death, and the ticking hand of evolution's clock counts out time in generations. Each generation gives forth its young to face the vicissitudes of life. As Charles Darwin deduced, those best suited to the conditions prevail to give birth to their own young, who in turn share some of the characteristics that favored their parents. So it is that species can adapt to changing conditions over periods of many generations.

Human stresses on the environment exert potent selective pressure on living organisms, winnowing those less able to cope. The hand of evolution can already be seen in the emergence of slower growth and smaller size at maturity in exploited fish, for example. The trouble is, we are changing the environment at such a breakneck pace it will far exceed most species' ability to evolve adaptations quickly enough. For an animal like a whale, deep-sea fish, or reef coral, whose generations can span decades, there is too little time. Microbes, and many phytoplankton and zooplankton, will produce generations in hours, days, or weeks. There is plenty of scope for them to evolve their way out of trouble. But evolution imposes constraints too. The loss of coral reefs for long stretches of geological history suggests that it isn't easy to find a way to precipitate solid carbonate from seawater when acidity gets too high. So there is probably no get-out-of-jail card to be had here.

Animals, plants, and microbes do everything they can to survive. When conditions turn, they adopt all kinds of strategies to persist, like switching to different foods, going deeper to find cooler water, or discovering pockets of favorable habitat in otherwise hostile terrain. We can draw comfort from the fact that life has been through many crises before. Pretty much everything alive on Earth today made it through the deglaciation that happened from about twenty thousand years to ten thousand years ago. This transition saw the world warm several degrees and included rapid shifts in conditions that would have tested the hardiest. Life has experienced some twenty glacial cycles in the last two million years, so it is well versed in the business of tracking environmental change, and ecosystems have undergone many reassortments of their component species.

Financial products these days always include a caveat in the small print below their headline claims of amazing returns: "Past performance is not necessarily a guide to future performance." There is an important difference between those past climate swings and the present day: us. Species didn't have to cope with the multitude of stresses they face today. We really are sailing into the unknown.

We are beginning to count the human cost of altered oceans in currencies other than food. In the far distant past, when Roman emperors ruled and North America was a patchwork of tribal territories, estuaries and coastal seas were clearer and bluer than they are today. In estuaries like America's Chesapeake Bay, where oyster reefs lined channels and inlets, they could strain the entire volume of water in a few days, a rate that has fallen a hundredfold today with the decline of these shellfish. Species on the seabed of open coasts did much the same. The massed ranks of mollusk, sponge, and coral drew nutrients and mud from the water and locked them away in shell and sediment. We need them more than ever today to help deal with the wastes that pour from our industries and towns, streets, and fields.

In Seattle, coastal residents have suffered growing discomfort and health problems since the 1980s, due to heaps of sea lettuce rotting on their beaches. The lettuce produces the toxic gas hydrogen sulfide that wafts from the beach to cause eyes to stream, sore throats, and in some cases breathing difficulties. It may be significant that hospitalizations from asthma in Washington State are higher than the U.S. average and have increased in the past decade.[5] However, it still isn't possible to disentangle trends in ocean pollution from increasing traffic pollution or poverty as causes. The seaweed problem is due to nutrients from sewage outfalls and lawn fertilizer washed from the city's gardens into Puget Sound. You may want to think twice next spring before liberally fertilizing your lawn. Similar problems are popping up in other parts of the world. Residents in the Western Australian town of Busselton saw their beach transform from one of the best in the country to an "environmental nightmare" within the space of a decade, as heaps of rotting seaweed piled up.

In France, the effects of intensive pig farming in Brittany hit the headlines in 2009 when Vincent Petit was exercising his retired racehorse on the beach at Saint-Michel-en-Grève. Without warning he and the horse were sucked into a pit of rotting seaweed up to the horse's shoulders; the horse died within minutes and Petit lost consciousness. He was saved by a bulldozer operator working nearby who happened to see him go under and dragged him from the hole. For years great heaps of *Ulva*—the French equivalent of Puget Sound's sea lettuce—had washed ashore and gathered in piles, liberally fertilized by runoff from pig farms. Sometimes the sun dried their upper surfaces, so they became impermeable lids that trapped poisonous gas below. Petit's experience raised the profile of marine pollution in France and spurred a municipal cleanup that a few weeks later cost the life of a bulldozer driver who was overwhelmed by hydrogen sulfide. You know things are bad when beaches turn lethal. French authorities have yet to tackle the farm pollution at its source. Until they do, beaches will continue to accumulate stinking seaweed, and the risks it brings.

These stories are shocking but still unusual. Many people remain puzzled by environmental horror stories, because they don't match their daily experience. If things are going to hell, they say, why does my own life feel like it is getting better? According to the Millennium Ecosystem Assessment, a compendious review of the state of the world environment, they are right.[6] Indicators of human well-being are on the rise, even as habitats and species are lost around us. This has been called the environmentalist's paradox. It helps explain some people's forthright skepticism of climate change, and why they have concluded not only that it doesn't exist, but that there is a massive conspiracy among scientists and greenies who peddle this nonsense.

Well-being has certainly improved by several measures: childhood survival, GDP, and level of education have risen steadily the world over, with the exception of sub-Saharan Africa, where they lag badly. These measures don't capture all of the dimensions of well-being, but where aspects of life such as happiness or gender equality have been measured, they bolster the view that life is getting better for many of us. There is one measure, however, that might give even the most ardent climate skeptics pause, and that is the apparent increase in exposure to natural disasters. Catastrophic floods and landslides are made worse by loss of habitats such as marshland and forest. The impacts of Hurricane Katrina in New Orleans were worse than they would have been had the city not been built on drained marshes that had sunk below sea level. When the levees breached, the city became a brackish lake. But while our exposure to disasters has gone up, our ability to deal with them has too. We are richer and better prepared, so the risk of death from natural catastrophe has gone down globally.

There are several possible explanations for the environmentalist's paradox. The first is that food is overwhelmingly the key to human well-being, and food production is up. Production of fish, meat, and cereals has outpaced human population growth globally, and health and life expectancy have benefited. (Growth has not been even, and there are many places where rapid soil erosion or water scarcity have triggered famines.) A second explanation for the paradox is that our inventiveness has spared us the adverse consequences of our altered environment. Today, if production falters in one place due to drought or pestilence, we can bring food from somewhere else. Soil erosion and lost fertility can be countered by chemical fertilizers. Fish can be taken from distant seas to fill larders as local stocks decline.

The final possibility is more ominous: We have not yet felt the real cost of our activities but payback is on its way. Prudent financial managers will tell you to spend your profit and not draw down your capital. But for decades now we have been depleting natural resources at an unsustainable rate and literally eating into our assets. Times are still good, but we will feel the impact of our spendthrift ways when our savings run out. Our astonishing population growth is speeding the arrival of that day, although the payback will be in stages, as some resources will run out before others, and people in some places will have to pay sooner rather than later. The hit has come more quickly for a Filipino fisherman than a supermarket customer in the United States, but one day it will happen to us too. We are like debtors living the high life until the moment our creditors force us into bankruptcy.

Already there are credible estimates that suggest we are using something like one and a half planet's worth of natural resources, based on how much area our present consumption would occupy were extraction and energy use sustainable.[7] The only way this overshoot is possible is by living on our savings—in other words, by depleting stocks of natural resources that have built up over thousands or millions of years. Technological innovations and globalization have so far spared the wealthy, but as environmental debts begin to be paid we will all be forced to confront the costs of our lifestyles and those of previous generations. The 2006 Stern Report on the economic costs of climate change, commissioned by the UK government, estimated that there could be two hundred million climate refugees by 2040.[8] We can see them already in the tide of refugees that flows to Europe today on the run from drought, famine, flood, and despair in places like Somalia, Sudan, and Pakistan. One widely cited report estimated there were already twenty-five million environmental refugees in the 1990s, and the figure was expected to double by 2010.[9]

Human relations with the sea have shifted with the passage of time. For tens of thousands of years our ancestors frequented coasts and estuaries, where they hunted and gathered seafood. Close to the places they lived, the remains of shells and fish bones piled up over millennia tell a story of how they reduced the abundance of some species and altered the structure of ecosystems. But their influence was minuscule against the virgin canvas of a measureless ocean. This was the era of carefree use. Humanity neither cared nor needed to worry about what it took from or added to the sea. Ocean resources were limitless against the scale of human need.

From a human perspective, there was a "golden age" of the sea: when people first began to harness the oceans for their own ends at a large scale but before these uses had compromised their abundance and fecundity. This period came in antiquity in the Mediterranean, when commercial fisheries developed for tuna, marine aquaculture was invented, and large-scale ports and harbors were built. The steep growth of our influence in the last few hundred years, and our impact, led people to realize for the first time that their activities could cause problems, like pollution or overfishing. At first those troubles were local and people's responses, if any, were piecemeal. Often problems such as declining fish catches were ignored or overlooked, as people could still obtain what they needed from further afield. But over time, despite a growing awareness, problems snowballed.

Will we overcome the difficulties we have created? We don't yet know how matters will play out. I can find reasons to worry and reasons for optimism. There is the enormous scale and accelerating rate of change, the bewildering multiplicity of our impacts, and our inexperience in dealing with global problems that are causes for concern. But there are good reasons for hope. People are ingenious and can rise to challenges; we have found some solutions and can devise others. Thus far, I have focused on the problem. In the remainder of this book I will concentrate on what we can do to improve the state of our seas and keep our ocean planet healthy. What is inescapable is that we must change. Continuing on the present course will lead to an environment that is increasingly hostile, both to aquatic life and to our own well-being.

I am an optimist. What gives me hope is the thought that we have never before had so much raw power with which to solve our problems. Of course it is our power over planetary processes that brought the difficulties in the first place. Nonetheless, it puts us in a different position from our less fortunate predecessors, whose civilizations failed to survive their own transformations of the environment.[10] What we do share with past societies is our humanity. We possess qualities honed by the struggle for existence that predispose us to short-sighted and selfish behavior. If we can overcome these evolutionary quirks, then a sustainable future may be possible. The Harvard scientist E. O. Wilson has summed up the challenge: "We have

Paleolithic emotions, Middle Age institutions, and God-like technologies." No wonder we struggle to shift from overuse to sustainable uses of natural resources. If we are to survive, we will have to replace our present ways of organizing society, forged over thousands of years, with new means, forged in less than a century. The longer we delay action to reduce stress on sea life, the harder it will become to reverse the unwelcome changes we have unleashed. It is easier to fall into the pit of environmental damnation than to climb back out. The risk is that this backward slide will be tough to reverse. We treasure and love our seas and coasts. They are places to have fun, relax, and go for inspiration. None of these pleasures can be guaranteed, however, unless we recognize and address the dangers now faced by marine life—and, through them, the dangers we face.

How will it feel for our children's children to live with changes we have imposed upon them? To get an idea of what the future has in store, look no farther than Iceland, with its hardy people scratching a living on a scrap of barren volcanic rock.[11] It wasn't always this bare. When Scandinavian seafarers discovered Iceland they found a country covered in soil, shrubs, and trees. Overgrazing by their livestock destroyed the fragile vegetation, and the soil was lost to the sea, leaving the gray desert we know today. Until a few years ago nobody there realized their island had once been green. But now they can look over the monotonous rocky landscape and into their imaginations, stroll through woodland and heath, and breathe the scent of rowan flowers in spring. If we fail to control climate change and find ways to live that are less demanding of our planet, I doubt that our descendants will so easily forgive us the problems we bequeath to them.

PART 2

Changing Course

CHAPTER 16

Farming the Sea

The sea is inexhaustible, and there can never be a general and simultaneous depopulation. The ocean fisheries will always be copious and easy, and their yield will be greater as the ocean becomes more familiar and the methods employed more perfect.

—Marcel Herubel, French marine biologist[1]

Marcel Herubel wrote these lines in 1912. He was a gifted thinker. He was well aware that fisheries could be overexploited locally, but like many others of his time, he could not imagine that humanity could ever dent the productivity of oceans on a global scale. Now landings statistics collected by the Food and Agriculture Organization of the United Nations (FAO) show that we have attained the unimaginable. After a thousand years of virtually uninterrupted increase the underlying trend of global wild-fish catches has been down since 1988.[2]

The picture looks worse if we set fish landings against the explosive growth in human population. Wild fish available per person has fallen by more than a quarter since it peaked in 1970. If fish had been equally available to every citizen of the planet, and all of the catch had been eaten directly—rather than fed to livestock—everyone could have eaten 6.5 ounces of fish per week in 1970; today that has shrunk to 4.75 ounces per week.[3]

Since at least the early twentieth century, governments have exhorted us to eat more fish to be healthy and slim. Some governments publish guidelines on how much we should eat. New Zealand and Greece think people should eat 17.5 ounces or more per week, while Canada and Austria think 5 ounces is fine. The United States recommends people eat 12 ounces per week. Averaged across the world, we are told we should eat about 9 ounces of fish and shellfish every week, an amount nearly double the availability from wild fish catches. There isn't enough wild fish to go around.

Ours is an ingenious species, and we love technological fixes. If there are no longer enough wild fish in the sea to feed us, surely it makes sense to farm the oceans? Aquaculture, the production of fish and shellfish in captivity, is a no-brainer. Many now talk of a blue revolution in aquaculture to match the much vaunted green revolution in agriculture. Farmed fish and shellfish already make up around 46 percent of the fish sold for human consumption worldwide.[4] So is aquaculture the answer?

Aquaculture has a long history. The earliest representation of fish being kept in ponds was painted on the wall of an Egyptian tomb in the Nile Valley four thousand to forty-five hundred years ago.[5] It shows a man seated beside an artificial tank with a central drainage channel, fishing for tilapia with a rod and two hooked lines. Floating plants shelter the fish, while in the background men pick fruit from trees that may have been irrigated by used pond water. Behind the angler, a woman receives the fish.

At about the same time (give or take five hundred years), aquaculture is thought to have developed in China, where freshwater common carp were reared in ponds using fry collected from rivers and lakes. The first manual of fish culture was written in 475 BCE by the Chinese author Fan Li, who is better known today for his popular guide to the rules and pitfalls of doing business. The Chinese refined their methods over the centuries into a sophisticated system of polyculture, raising several other species of carp with different diets, mainly in ponds, but also in rice paddies and lakes. There is a possibly apocryphal story that this diversification was a direct response to an edict by the seventh-century emperor Li Shimin, who declared that nobody was allowed to keep common carp because the fish shared his name. The different species used in polyculture systems eat algae, insects, food waste, and even sewage. They benefit the rice by keeping the mud well aerated and the weeds and pests in check.

In Europe, the Romans, whose appetite for fresh fish is the stuff of legend, took aquaculture to the sea over two thousand years ago.[6] Fish ponds were mainly built using a Roman invention, hydraulic concrete, which sets underwater. They have proved incredibly resistant to wave and tide—so much so that many remain intact and almost appear ready to receive fish today. These ponds were not intended to feed the masses. They were expensive to construct and costly to keep stocked. Lucullus, perhaps one of the greatest fish gourmands, reputedly spent more on his fish ponds than he did on his coastal villa.[7] He cut a tunnel through a hill to bring water from the sea for his fish.

Roman ponds were remarkably sophisticated. Some were built into dining rooms to form the centerpiece for banquets. Proximity to the fish they ate was important to Romans. For one dish the fish were slaughtered at the table, and guests would enjoy the color changes of the dying animal. Some ponds had viewing platforms, while others had fishing towers from which an angler could pursue his lunch. Many ponds had ceramic pots built into their sides to provide hiding places for eels, which were highly prized at the time. (Some eels were kept as pets and adorned with gold jewelry inserted into body piercings.[8]) Coastal ponds were often built around natural coves or grottos and were connected to the open sea by channels closed with grates. Waves and tides renewed the water over the tops of the walls and through these grates. Like all previous attempts at aquaculture, Roman fish ponds were really just ranches. Fish were taken from the wild to be stored and grown in captivity, thus ensuring a ready supply no matter what the weather or season.

When Captain Cook first reached Hawaii in 1778 he found an extensive network of several hundred fish ponds.[9] Some were enormous, with rock walls over half a mile long, up to seven feet high, and thirty-six feet thick at the base. In Cook's day, freshwater and marine ponds together supplied the people with an estimated one thousand tons of fish per year. Hawaiian aquaculture is believed to have originated a thousand years ago and contained what might be called an entire ecosystem. Larval and juvenile fish entered through grates that opened to the sea, and grew in size within until they were trapped and periodically harvested.

Necessity has long been a wellspring of creativity. In the Middle Ages, between the eleventh and thirteenth centuries, the avalanche of soil washed off lands newly plowed for agriculture in Europe triggered the crisis in freshwater fish supply that I described in Chapter 2. At the same time, sturgeon, salmon, and whitefish found their migration routes newly blocked by mill dams. The fall in supply was set against an increase in demand from a larger and more urbanized population. Medieval Europeans had come to value fish more highly due to the spread of Christianity, which prohibited believers from eating the meat of quadrupeds on certain days. The result was the development of pond culture for carp sponsored by monasteries and the aristocracy.

European freshwater carp culture was more sophisticated than Roman sea-fish ranches, because it encompassed the entire life cycle, from egg to adult. It was not until the nineteenth century that efforts began in earnest to close the life cycle of marine fish in captivity. Again, necessity provided the impetus. Salmon runs in European rivers had for centuries been in private ownership, with fishing rights mainly reserved for the aristocracy. The Industrial Revolution caused many of those rivers to become so polluted and blocked by dams that salmon numbers plummeted. Fish hatcheries were established and salmon were stripped of their spawn in captivity and their offspring raised to fingerling size before being released into rivers to complete their life cycle in the wild. The growing sophistication of these efforts is attested to by the fact that from 1868 until the early twentieth century salmon and brown trout eggs were taken by ship to New

Zealand for release into rivers and lakes for the sport and sustenance of expatriate Brits.[10] Brown trout got going quickly, but the salmon was troublesome, eventually establishing in the watershed of only one river.

The late-nineteenth-century revolution in sea fishing put wild fish stocks under intense pressure again and stimulated the first aquaculture experiments with plaice and turbot. As with salmon, these efforts concentrated on spawning fish in captivity to supplement wild populations with large-scale releases of young fish. Marcel Herubel captured the spirit of these pioneering efforts in 1912:[11]

> Piscifacture may legitimately claim to be of social value. It goes further than pisciculture, which limits itself to rearing fish; it endeavors to "manufacture" fish. The first art is a nurse, the second a mother. . . . Piscifacture will restore to the sea what it would have produced had man not multiplied his catches.[12]

For fairly obvious reasons, these well-meaning attempts ended in failure. Juvenile fish suffer extremely high mortality rates in the wild, and few reached adulthood. Even the release of millions of young fish had little impact on the numbers that eventually found their way into fishing nets, especially since captive-reared juveniles tend to be less fit than their wild-spawned brethren. So the direction of aquaculture shifted.[13] A parallel development concentrated on raising wild-spawned animals in captivity from egg to adult. A third strand saw the refinement of methods to raise shellfish such as oysters and mussels. Shellfish mariculture has ancient origins. In the Broughton Archipelago of British Columbia, First Nations people erected rock walls across bays to trap sand to the depths favored by clams, which were an important winter food.[14] Nobody is sure when these walls were first built, but they may go back five thousand years and show just one of the many ways these people transcended the simple hunter-gatherer stereotype to nurture nature to support their needs.

In less than a century aquaculture has evolved from a frustrating practice pursued by mavericks into a highly successful international business. From 1950 onwards, when the FAO began to collect statistics, aquaculture production has expanded from less than one million tons a year to more than fifty-five million tons, just over half of which is from freshwater fish like carp and tilapia. To that we can add another eighteen million tons of seaweed. The growth of aquaculture has outpaced every other form of food production in the last decade or two, except organic (which is rising fast from a low base). Since 1950, it has increased at three times the rate of growth in world meat production, and even outpaced world population growth. So when aquaculture is included, the total amount of fish available to eat per week has risen from 6.9 ounces to 7.2 ounces per person, better than the picture from wild fish catches alone but still a fifth less than the average of health recommendations. Unfortunately, about a third of wild catches are fed to livestock, pets, and aquaculture, leaving just 5.6 ounces per person per week to eat.

While nearly four hundred different species of fish and shellfish are cultivated, a handful dominates production. On the marine side, the major players are mussels, clams, and prawns. Shellfish are cultured all over the world, mainly on ropes strung across bays and mud flats in operations that can cover tens of square miles. Mussels and clams are the mainstays of these farms, but even free-living mollusks like scallops are grown on ropes in bags.

Atlantic salmon is the species most Western consumers associate with aquaculture, but in 2008 salmon farms produced over 1.5 million tons of fish, only about a twelfth of the volume of mollusks and prawns sold that year. Salmon are joined in the top ten farmed fish by a cast of lesser characters, including rainbow trout, seabream, sea bass, Japanese amberjack, barramundi, and the unfortunately named bastard halibut, whose forename is usually dropped by supermarkets! But other highly lucrative species are also now farmed, such as bluefin tuna, groupers, and sturgeon.

In 2008, four out of five fish eaten in China were raised in farms.

Outside China, one in four fish eaten was farmed, up from one in twenty in 1970. It seems that there is hardly a sea loch or fjord in the developed world that does not support, or is not earmarked for, aquaculture. There are parts of Japan and China where you can stand on the beach and see nothing but fish pens or seaweed strings all the way to the horizon. In terms of food production, aquaculture has been the roaring success of recent times. Putting that aside, the industry has many problems. Some are acknowledged even by its most enthusiastic advocates, but others have long lurked in the background and have only recently begun to cause concern.

While filter feeders such as shellfish can feed themselves, the awkward fact is that farmed fish must be fed, and many of the species most prized by consumers in developed nations, like salmon, sea bass, or tuna, eat flesh. People often think of fish farming in terms they are familiar with from the cultivation of cattle or sheep. We raise animals to convert things we can't eat, like grass, into things we can. But this simple logic doesn't apply to aquaculture of predatory fish, which happen to be the ones most Western consumers like best, because they are large and tasty with firm, succulent meat. While some freshwater fish such as carp and tilapia eat a largely vegetarian diet, almost all of the marine fish raised in farms are predators. So it is that we must catch wild fish to feed to captive ones, mainly so-called "forage fish," like anchovies, capelin, herring, and menhaden that live in large schools. These animals would make perfectly good food for us if we ate them directly rather than passing them through salmon or sea bass first. Here, then, is the first of many controversial aspects of fish farming: It can sometimes take several pounds of wild-caught fish to produce just one pound of farmed fish.[15]

Several of the species we farm are ferocious predators and come from very high in the food chain. Bluefin tuna are caught as juveniles in Australia and the Mediterranean and then grown to market size in pens, during which time they eat an eye-watering quantity of wild fish, up to twenty times the weight of flesh they produce. You get an idea of the problem by considering the idea of trophic levels. They are a shorthand way of assigning a species' place in the food chain. Think of them as steps on a ladder that ascends from plants and organic muck at the bottom, to sharp-toothed, beady-eyed carnivores at the top. Plants and detritus are level 1, herbivores and detritus feeders are level 2, meat eaters are level 3, carnivores that eat other meat eaters are level 4. Few food chains go longer than five links because something like 90 percent of the energy passed upward is lost at each step. It gets dissipated sustaining life and producing the next generations. Only about 10 percent builds bodies. This energy loss explains why lions are rare compared to antelope and eagles less abundant than rabbits.

This is a simplified view of a food chain, since it assumes that species occupy just one trophic level. Most animals have a mixed diet, eating foods from more than one rung of the ladder. We can reflect this by giving them intermediate values. Most people eat a mix of meat and vegetables, so the average trophic level of humans is about 3.2. The trophic levels of Atlantic salmon and bluefin tuna are 4.4, which gives them a position on the food web higher than African lions, which come in at less than 4. If you are ever stuck for conversation at a party, you might drop the startling nugget that most domestic cats have a higher trophic level than lions. All that salmon and tuna that Tiddles eats places her well ahead of most big cats in the league of carnivory. Because predators sit much higher in food webs than sheep, goats, or cows, the ratio of what you get back to what you put in is much less favorable.

In the last ten years producers have worked hard to rebalance this equation. Across the industry as a whole, the ratio of wild fish going in to farmed fish coming out has fallen from roughly 1:1 to 0.63:1 (excluding species like mussels that feed themselves from the water).[16] Mostly their motivation is profit rather than concern for the environment, although there are refreshing exceptions. Fishmeal and oil from wild sources is expensive, and prices have risen faster than inflation as demand has grown. About half of the fishmeal and nearly all the fish oil produced globally is now consumed by aquaculture (the rest goes to pigs, hens, and even cows). Since wild fish catches have leveled off, aquaculture will likely swallow all the world's fishmeal by 2050, if not sooner. Those who can squeeze more fish out for every one put in gain a powerful kick to their bottom line. And while there have been improvements in feed conversion ratios, for some species the figures remain bad. Because of the high fish oil content of salmon feed, pressed out of animals like anchoveta, salmon still average about five pounds of wild fish to produce one pound of farmed.

There are several substitutes for wild fish flesh in aquaculture feeds. Some manufacturers blend scraps from processed fish into their feeds, but their oil content is low so it is only an answer for some farmers. Others grow marine worms for fish to eat. The most widely used substitutes are protein and oils from plants, such as soy and canola. The problem is there is a limit to how much plant matter a carnivore can take before its health suffers, and of more immediate concern to consumers, its taste. Moreover, the much vaunted health benefits of eating fish stem from their content of omega-3 oils. There have been many claims over the years about the wonders of fish oil, including that they reduce cholesterol, protect the heart, sharpen the mind, ease depression, lubricate the joints, improve vision, prevent Alzheimer's, and on and on. The protective effects on the heart are well accepted, although many other presumed benefits remain unproven. Omega-3 oils are produced only by aquatic plants, so feeding terrestrial plant substitutes to farmed fish will rob them of their health food cachet. The German agrochemicals giant BASF has an answer in the form of canola plants that have been genetically modified to produce marine oils. These plants have been approved in the United States, but many people remain wary of genetic manipulation, so this alternative is unlikely to satisfy all.

In some places, the struggle to find enough fish to feed aquaculture and farming has taken a troubling turn. Bottom-trawl fisheries can have fiendishly high rates of bycatch of unwanted species, most of which are thrown away dead. In India and Venezuela, a new use has been found for this bycatch.[17] Instead of being chucked over the side, it is dried and converted to fishmeal that then goes to chicken farms and aquaculture. At first this seems like a good idea, but it can keep trawling profitable long after the practice has driven the original, high-value target animals to economic extinction. The one last check on a fishery, which might spare an ocean habitat from being completely destroyed by overfishing, is its inability to turn a profit. If every scrap of life brought up in the trawl can be sold, then the trawlers will continue to drag their nets until there is nothing left at all.

Not all farmed marine fish are as predatory as salmon or tuna, and many shellfish eat plankton, plants, or detritus. Nonetheless, finfish culture consumes more than one pound of wild fish for every pound produced, and across the board, aquaculture swallows two-thirds as much wild fish as it produces. Hardly a solution, then, to problems of overfishing. Why don't we just eat the wild fish directly? This is exactly what many people, celebrity chefs included, have tried to encourage. The main ingredients of fish meal are animals like anchovies, sardines, sprats, herring, and blue whiting, fish whose flesh simply drips healthy oil. The problem is that most are small and therefore fiddly, inconvenient, and full of fine bones that wedge between your teeth or catch in your throat on the way down. In a world of ready meals and convenience foods in developed countries it is hard to persuade people to get intimate with heads, tails, and slime when the alternatives are plump, bone-free fillets from farmed stock. Supermarkets love farmed fish too, because they can guarantee supplies of even-size, uniform quality fish fillets no matter what the season or weather. Farmed fish are definitely here to stay.

There is concern that fishing for "forage" fish to sustain Western appetites for predatory fish robs the world's poor of their staple diet. These small fish made up over half of the fish protein eaten in a sample of thirty-six developing countries from Africa, Asia, and other parts of the world.[18] In something like half of African nations, fish protein makes up a quarter to over a half of the total animal protein eaten. Industrial fishing by other countries in African waters under access agreements threatens fish supplies and has already driven up demand for bushmeat, to the detriment of wildlife on land.[19]

One alternative to these fish is krill, a finger-sized, bright carmine shrimp that lives in polar seas. Estimates suggest that Antarctic waters harbor enough krill to sustain all of the world's aquaculture with little difficulty. Already, hundreds of thousands of tons of krill are caught annually to make into fishmeal.[20] Nothing is that simple, sadly. Fish, penguins, whales, and seals depend on krill and industrial fishing could harm them if it dents krill stocks. And krill are already on the way down as a result of climate change. Krill survive over the winter by grazing algae attached to the underside of floating ice. If the ice melts, krill supplies will crash.

Aquaculture also suffers problems of husbandry. Farmed fish, like any animals kept at high density, are troubled by diseases and parasites. To combat infection and infestation, most farmers dose animals in ponds and pens with antibiotics, pesticides, and fungicides. Since pens are mostly open to the sea, these chemicals spread pollution far beyond the bounds of farms. Nor are these problems confined to captive fish. Take salmon farms, for example, that are mostly located in estuaries that lead directly to rivers inhabited by wild fish. Adult salmon in the wild rarely come into contact with juveniles, because they either die after spawning or return to sea before the young hatch. Now, young salmon must run a gauntlet of salmon farms on their migration to sea, bringing them into contact with the diseases and parasites afflicting captive adults. Sea lice, for instance, feed on salmon skin and muscle. The lesions they create don't normally kill adult fish, but they can kill the young who wouldn't normally have encountered them, at least not in such densities. In recent decades wild salmon runs in rivers that empty past farms have declined to a far greater degree than in those that don't. Estimates put the fraction of wild salmon smolts killed by sea lice from farms at up to 95 percent in one part of British Columbia.[21] In the Atlantic, salmon are ghosts of their former abundance in the wild, so such killing rates could wipe them out entirely. Since each river has its own salmon stock that has adapted to local conditions, every disappearance is a step on the road to extinction.

Farmed animals suffer from more aggressive illnesses than their wild brethren. These diseases have emerged as a result of the unnaturally high densities at which animals are kept, and their prevalence is encouraged by stressful conditions within pens and genetic uniformity of the stock. Infectious salmon anemia seems to be a new strain of a virus closely allied to influenza that has adapted to exploit the high densities of susceptible fish in salmon farms.[22] Before the advent of pen culture, there had only been a single report of this virus in the wild. When an outbreak occurs it sweeps through captive fish and many countries now require all the fish in affected farms to be destroyed. Ironically, the industry itself has been the main vector for disease spread. Outbreaks in Scotland in the 1990s were closely linked to visits between farms by boats that carried salmon stock.

Infectious salmon anemia can be carried by fish that have recovered from the worst effects of the illness. It can also be transferred from fish to fish by sea lice, which raises serious concerns for the viability of wild fish stocks that have come into contact with infected farms. The disease has now spread to the United States, Canada, and Chile. It was first detected in Chile in 2007, and at the time of writing had halved Atlantic salmon production there.

In shrimps, another disease has jumped from pond to pond all the way across the Pacific. White spot syndrome virus emerged in China in the 1990s, then spread to Japan and Southeast Asia. Affected shrimps become lethargic and discolored, but in truth there is hardly time to notice these symptoms since the virus can kill the entire stock of an affected farm within a few days. By the mid-1990s, the virus had nearly eliminated Chinese shrimp output, and by the late 1990s, it had made it to North and South America. Amid all the death and destruction, white spot syndrome virus has at least had one positive effect. Before the disease emerged, shrimp farming often relied on capture of wild juveniles to stock the pens. Now the industry has mainly moved to spawning its own shrimps from certified disease-free broodstock. Pens also used to exchange highly polluted water with the open sea, but there has been a shift toward greater separation between pens and the sea to minimize possible spread of infection.

Isolation of farms from the open sea is a good idea for many reasons. When faced with disease, farmers have three choices: prevent, treat, or risk losing it all. Options to prevent disease include vaccination, prophylaxis, or disinfection. Vaccination is clean and clinical but only available for a few species and diseases, and often only in developed countries. Prophylaxis usually involves feeding animals antibiotics, while disinfection can mean dosing ponds with toxic compounds such as malachite green, which is known to cause cancer in people. There is growing disquiet about the use of antibiotics in aquaculture. Chile uses over one hundred tons of quinolones every year, mostly on fish farms.[23] These antibiotics are important in human medicine, but such heavy use risks the development of antibiotic resistance in microbes. Bacteria tested from sediment and water samples near fish and shrimp farms have begun to develop such resistance, and some can shrug off several antibiotics at once.

You might think that there is little chance of marine bugs causing trouble for us, but bacteria are able to swap genetic material, and genes for resistance have already made the leap from sea to land, and from bugs that affect animals to ones that infect people. A cholera outbreak in South America in the 1990s was of a strain that had picked up antibiotic resistance from contact with a bacterium that owed its enhanced resistance to the heavy use of drugs in Ecuadorian shrimp farms. The sometimes fatal gut bacterium *E. coli* has also acquired antibiotic resistance via aquaculture. For some human pathogens, exposure to antibiotics in aquaculture can be direct. Asian carp ponds are often praised as a model of sustainability because they raise fish that eat vegetables, dirt, or insects, and they recycle human wastes. In Vietnam, some cities rely on vast aquaculture complexes to treat their sewage, which might give you pause the next time you spot some plump fillets of basa, a type of catfish, in the supermarket.[24]

So human pathogens swirl around ponds where they can gain antibiotic resistance before being passed back to us when fish are harvested.

Many developed countries have recognized the dangers of antibiotics and have banned their use. In Norway, most salmon are vaccinated against problem infections. Things that happen in Asia or South America or Africa can seem comfortingly remote to many of us. But this is a globalized world and what happens in one place soon spreads to others, as outbreaks of bird and swine flu show. More than 90 percent of world aquaculture production comes from developing countries, so until the problems are solved there, little progress has been made. In Asia, where most of the world's farmed fish are grown, farms routinely use industrial quantities of antibiotics, pesticides, disinfectants, and antifungals.

Toxic chemicals released to the sea are a worry, but they highlight another of the ways in which aquaculture does damage. Where pens exchange water with the open sea, as most do, the wastes from excess food and feces simply washes away. If there are few farms and plenty of water to slosh past them, we might see little impact and easily live with the consequences. But farms are often put in sheltered bays, estuaries, and fjords where water exchange is limited, and the growth of fish farming means that some places are packed with pens, each of which can contain tens or hundreds of thousands of animals. That adds up to a huge amount of waste. By 2000, Scotland's salmon farms released as much nitrogen—one of the main fertilizing nutrients that fuel plankton blooms—as is contained in the untreated sewage of 3.2 million people, over 60 percent of Scotland's population.[25] Output of phosphorus, the other major plant nutrient, was equivalent to the waste produced by 9.4 million people. In the early years of the twenty-first century, Chinese shrimp ponds released forty-seven billion tons of effluent into coastal seas, compared to over four billion tons of industrial effluent and sewage from land-based sources.[26] Shrimp farms produce mind-boggling amounts of nutrient-rich sludge: in intensive culture, tens to hundreds of tons per hectare for every shrimp crop.[27] These farms have doubtless played a critical role in the problems of oxygen depletion, harmful algal blooms, and fish kills that have plagued regions like China's Bohai Sea. Blooms in China now affect an area bigger than the Mississippi dead zone. Toxic algae provide a strong motivation to find cleaner ways to farm fish, since they can wipe out stocks and profits within hours when they strike. In Philippine bays so crowded with cages and pens there scarcely seems any open water, since 1999 yearly mass kills of milkfish and tilapia have destroyed hundreds of thousands of tons of stock. Such problems not only threaten production, they pose real health concerns to consumers.

If the Bohai Sea is not the most polluted sea on the planet, it must be close. It lies downstream of the heartland of the Chinese economic miracle. This transformation of fortunes, like the past Industrial Revolutions of Europe and America, has been bought at the price of severe pollution. Industrial effluents mingle with sewage discharges from tens of millions of homes. Maps produced by the Chinese government show almost every part of the Bohai coast suffers from excessive levels of contaminants like lead, chromium, mercury, and arsenic alongside the fertilizing nutrients.[28] Off the city of Tianjin, water quality is so bad that most falls into the pollution category of "offensive in color, smell, and taste." This pollution threatens the burgeoning aquaculture industry. So filthy has the water become that there is widespread use of antifungal and antibacterial compounds to keep the animals alive, like nitrofuran, a known carcinogen banned throughout much of the Western world. So farmed animals are contaminated in multiple ways—with DDT, flame retardants, and mercury from the water, and with prophlylactic chemicals applied by farmers. As China has become a major exporter of farmed fish and prawns, these problems are a concern for consumers far beyond the shores of the Bohai.

In previous chapters I talked about how aquaculture has contributed to the spread of invasive species and diseases around the world. This "biological pollution" is considered by some to be among the worst effects of fish farming, because it is usually irreversible by the time the problem is noticed. Atlantic salmon have already established themselves in several rivers in British Columbia. Huge Pacific oysters, three times the size of man's hand, are now embedded in northern European seas after they were introduced to replace native oysters. In the Gulf of Maine, imports of European oysters probably brought in the invasive seasquirts and fingerlike *Codium*, a vibrant green seaweed, that competes aggressively for space with native fauna. No one knows how these or hundreds of other alien species incubating in new environments will turn out; some will definitely go rogue. But there is another form of release that is just as irreversible and scares some people even more: the use of genetically modified fish.

Like any form of farming, aquaculture has its high technology enthusiasts who can spot ways to make better animals. At the time of writing (2011), the first transgenic salmon was nearing approval in the United States.[29] Called Aquadvantage, this Atlantic salmon has been given a chinook salmon growth hormone gene, and another from ocean pout that prevents the growth hormone from being switched off in winter. So these fish pile on weight year-round to reach market size twice as fast as normal salmon. Protesters argue that escaped animals will interbreed with wild and set free Frankenstein fish. But the company that makes them says it is highly unlikely, since 99 percent of fish have an extra copy of the salmon genome to ensure infertility, and they will be grown in closed containers on land. Protesters do have a point, though, since reassurances from the biotech industry have several times proved less robust than claimed. In any case, this salmon will have to clear regulatory hurdles outside the United States to gain a foothold on the market since most salmon farming is in Canada, Chile, and Europe. But despite passionate resistance among environmentalists, it seems inevitable that in the long run aquaculture will embrace genetically engineered fish.

The explosion of marine fish farms across the world has transformed coasts, estuaries, and deltas in dozens of countries. China has pursued one of the most aggressive aquaculture development programs of recent years. In 2003, marine aquaculture covered nearly six thousand square miles, about half of China's twenty-thousand-mile coastline.[30] Most of these farms were carved out of mangrove forests, mud flats, salt marshes, and sea grass beds. Satellite photographs of the Bohai Sea coast, one of the most intensively farmed regions, show the toll taken. Straight-edged ponds incised in blue and turquoise pack the coast to depths of a couple of miles inland, and crawl seaward across mud flats. The Philippines and Vietnam have lost three quarters of their mangrove forests in the last few decades, half of them to aquaculture. Sadly, many of these ponds have been abandoned. When mangrove soils are exposed to the air, they become acidic. The acid leaches into pond water together with toxic quantities of aluminum, so the ponds cannot be used unless they are lined.

Mangroves and salt marshes are self-repairing buffers that defend coasts against storm and flood. If they remain healthy, there is a good chance they could also ameliorate the worst effects of sea-level rise by trapping sediment and building upward. In many places aquaculture has not only removed this benefit, it has caused the land to sink by sucking freshwater from belowground to create brackish ponds for shrimp and milkfish.

Ironically, the loss of these habitats starves fish farms of two things they need most: clean water and animals to stock their ponds. Coastal wetlands draw nutrients and pollution from the water that washes through them. They are also nurseries for a huge variety of fish and shellfish. Although disease has forced many shrimp farmers to switch to hatchery-raised fry, countless farms throughout the tropics still depend on wild sources to stock their ponds, despite its catastrophic environmental cost.

Tiger prawns are much favored for their large size and rapid growth, but they represent a tiny fraction of wild shrimp fry. Most wild prawn fry are caught by night lighting in shallow water under a moonless sky. Within minutes, netters are surrounded by a confusing buzz of hundreds or thousands of tiny animals, fish and fry, all drawn to the light like moths to a flame. For every individual tiger prawn caught in Malaysia and the Philippines, several hundred fry of other species are wasted. Half of the hundreds of millions of tiger prawns grown in Bangladesh are from the wild. The waste doesn't stop at other shrimp species either. In one study, scientists found that every tiger prawn fry collected cost the lives of up to hundreds of finfish fry and over a thousand other animals that live in the plankton.[31]

The wetland grab in developing countries has robbed the poor of the common lands from which they once eked a living by fishing and gleaning. Aquaculture provides jobs and income for some of the displaced. But the divide between rich and poor has grown, and their quality of life suffers with every waterway blocked, every forest tree felled, and every pint of pond sludge pumped into the sea. In several countries, such as Bangladesh, Thailand, and Honduras, some of those bold enough to protest the injustices have been murdered to quash opposition.

So far aquaculture has sheltered most of us from the effects of over-fishing. Throughout the developed world supermarket shelves creak under the weight of an ever expanding range of farmed fish and shell-fish. There is something for everyone if you can afford to pay: stur-geon caviar, bluefin tuna sashimi, and oysters on the half shell tempt high-end diners, while there are salmon and shrimp for the masses. And yet, the amazing expansion of fish farming since 1950 has come at a terrible cost to coasts, wetlands, and shallow seas. Governments hell-bent on foreign exchange or job creation have encouraged aqua-culture heedless of warnings. It is not hard to see that it can't go on like this.

The drawbacks of aquaculture are such that one might question whether it really is the solution to overfishing. Wouldn't it be better to protect fish in their natural habitat? If we were to manage wild fisheries well, we might be able to increase supplies from the open sea by a third to a half. (I will come back later to how we can do this.)

But a 50 percent increase falls far short of the needs of nine billion hungry people expected by 2050. So if the world aspires to a healthy diet of fish, aquaculture will be essential. Like any kind of farming, there are better and worse ways of doing it. The present blue revolution will need to turn blue-green for aquaculture to become a net contributor to human well-being. What would it take to do this?

Better farming practice comes in many forms. In some countries, shrimps are grown in ponds at low enough densities that nature can feed them without the need for supplementary food. But such ponds take up more space than intensive farms, at a greater cost to wetlands and coastal ecosystems. In the Philippines, the value of mangrove forests is now becoming recognized thanks to the tireless campaigning of scientist and activist Jurgenne Primavera. She says, based on decades of research, that shrimp ponds should not exceed a quarter of the area of mangrove forest if we are to preserve the ecological function of coasts. Her work has led efforts to restore forests and move ponds behind sheltering buffers of trees, and she was hailed a hero of the environment by *Time* magazine in 2008.[32] (Shellfish and seaweeds can be also be cultured in natural mangrove and salt-marsh channels.) At the other end of the spectrum, experimental farms in Belize have gone for high-technology, superintensive methods. There shrimps are raised under covered raceways and fed on biofloc, clumps of microbes formed around starch grains and sprinkled into the water. This reduces the need for expensive feed and helps recycle nutrients from shrimp excreta, which reduces sludge production.

We may need to say good-bye to the succulent predators favored by Western consumers and rich Asians alike. Marine aquaculture will have to learn instead from the ancient art of carp polyculture practiced in the freshwater ponds and rice paddies of Asia. We need to find species that grow well together so that one will clean up the waste produced by another. Some fish eat seaweed or detritus and can be raised more sustainably than those that crave fish flesh. Mullet grub around on the seabed for their food and might reduce pollution problems if farmed together with more predatory fish. Likewise, sea cucumbers are considered a delicacy throughout much of Southeast Asia and have been seriously overfished in the wild. These animals are a bit like vacuum cleaners with a hole at one end where detritus goes in and a bum at the other where something that looks nearly identical comes out. Years ago, before I was aware of seafood problems, I tasted sea cucumber soup. I have to admit I'm no connoisseur and it felt a bit like chewing on a rubber band. But the Chinese and Japanese love sea cucumbers and it could make sense to grow them beneath pens of fish to help recycle their waste.

The industry will need to work hard to raise standards and improve sustainability. I have met many fish farmers who are committed to doing just that. With their energy and enthusiasm, aquaculture could indeed help feed the world. But there are challenges ahead. Growing shellfish has always been touted as one of the most environmentally friendly ways to produce seafood. Mussels, clams, and scallops feed on plankton and other organic matter filtered from the water around them. They don't need to be fed wild-caught fish and can improve water quality. But there is a catch. They depend on their carbonate shells, and life is going to get much tougher for them, and would-be aquaculturists, as the seas become more acidic. If you are fond of mussels and scallops, you may want to think hard about what we can do to reduce carbon dioxide emissions.

CHAPTER 17

The Great Cleanup

There is an old adage that "dilution is the solution to pollution." Follow that logic, and you would think that there is no better place to dispose of waste than in the sea. But even the sea can reach saturation. Over the last twenty years, estuaries, bays, gulfs, and enclosed seas have been overwhelmed by the volume and variety of waste, both of the kinds you can see, like plastic, and the kinds you can't, like toxic chemicals and fertilizers.

In the past nature could be relied on to cleanse our effluents as they passed through coastal marsh and mangrove. Once at sea they were filtered out by the gaping mouths of millions of suspension feeders that packed the seabed or swam in shoals so large they darkened the water for tens or hundreds of miles. But as the sea has been stripped of its consumers, and its coasts cleared of the wetlands that once lined creeks and waterways, so its power to regenerate waste has waned. Dilution is no longer the solution. Today the emphasis in pollution control must swing to reduction, prevention, and, in some cases, removal.

For persistent pollutants like heavy metals and PCBs, reduction and prevention are essential. There has been some progress over the last two decades, but past generations have left us a toxic legacy that we must pay for and every year we are adding more. In Europe and North America bans have long been imposed on the use of the most toxic and lingering compounds like PCBs and organochlorine pesticides like DDT. These measures have begun to reduce the burden of pollutants in the Baltic Sea and the Gulf of Maine, but DDT and its toxic breakdown products are still with us and these bans will have limited global impact until all of the developing world follows suit. Left to their own devices many of these chemicals bind to particles of silt and clay suspended in the water. When sediments sink to the bottom the toxins are buried, and in an ideal world that is where they would stay. Unfortunately, storms and tides mix sediments back into the water column and circulate the toxins back into the food web to haunt us again. And there is also a human force mixing mud and its toxic hitchhikers back into the water: bottom trawling and dredging.

It amazes me that more has not been made of the way that trawlers and dredgers kick up mud from the seabed. Trawlers fishing shallow water are visible from space because of the "contrails" of mud that stream from their nets. The Gulf of Maine is one of the most heavily trawled seas on the planet. Oceanographers have known for years that a thick and permanent layer of suspended mud overlies much of the seabed, but they weren't sure why it was there. When worms that normally live within seabed sediments showed up in traps set 80 feet above the bottom, the mystery was solved.[1] Bottom trawlers churned mud, worms, and pretty much everything else into the water above. Parts of the Gulf of Maine contain high levels of persistent pollutants washed downstream from factories and heavy industry of New England. Here and elsewhere, trawlers are revisiting upon us the ghosts of chemical industries past.

It seems only recently to have dawned upon many developed nations that the oceans are a common pool shared by all, and that the pollutants released by one nation can swiftly slosh to the shores of another, even thousands of miles away. Plastics are easily seen, but pollutants like mercury or DDT, harder to detect and easily overlooked, have spread to every corner of the globe. Bizarrely, many developed countries still permit companies to manufacture and sell toxic compounds like DDT overseas. DDT itself is permitted for control of mosquitoes that spread malaria, and three to four thousand tons of it are sprayed inside people's houses in Africa and Asia every year. In India and North Korea, people still use DDT illegally for general agriculture. Malaria is terrible, but can't we come up with a less toxic alternative?

Like many others, this problem requires a global solution. The Stockholm Convention on Persistent Organic Pollutants came into force in 2004 and at the time of writing 176 of the world's 195 nations were party to it.[2] Currently it commits signatories to eliminate seventeen of the most toxic and accumulative chemicals, and limit production and use of several others, but we still have a long way to go. There are still too many countries where highly toxic pollutants are used and ultimately wash into the sea.

Much of the mercury emitted, although toxic, falls outside the umbrella of international conventions, or even national laws. Perhaps because of past successes in controlling mercury releases from industry, it is coal burning and waste incineration, not chemical use by manufacturers, that releases the bulk of mercury today (although precious metal mining is a big contributor in some places). Mercury gives us another excellent reason, alongside greenhouse gas reduction, to move away from coal as a source of power. Given the rising mercury contamination of marine predators and the role of methyl mercury as an insidious poison, I don't think this problem has yet got the attention it deserves. Following close on mercury's heels, pharmaceuticals are emerging as another class of pollutants that needs to be taken more seriously. Again, the only way to clean up pharmaceuticals is to prevent them getting to the sea in the first place. Sophisticated scrubbing technologies are being trialled on sewage treatment plants in the European Union, but they are expensive and energy intensive. The search is on for cheaper and more environmentally friendly alternatives.[3]

Despite my concerns, I am encouraged by recent efforts to deal with pollution. Clean air and water laws are continuously introduced and upgraded, and these directly benefit marine life and water quality. The tide even seems to be turning against plastics. By one estimate nearly six million plastic bits and pieces are discharged into the sea every day; if we include microplastic particles like the ones we find in facial scrubs that number rises to billions. In a fine example of the way in which an individual can make a difference, a UK wildlife camerawoman named Rebecca Hoskins became concerned about the impacts of plastic bags on marine turtles who mistake them for jellyfish. She persuaded her home village of Modbury in England to do away with plastic bags, beginning a fight back against the tyranny of plastic waste that has since led to a countrywide reduction in plastic bag use by UK supermarkets. In the United States, first California, then other states and cities, all the way to Washington, D.C., started charging customers five cents per plastic bag—the impact has been quick and impressive. The backlash against plastic is also reaching the developing world. In 2005, the Indian state of Maharashtra reached its breaking point when plastic refuse choked their streets, creeks, and storm drains, and the authorities banned plastic bags. South Africa similarly banned the thinnest plastic bags in 2003; they had become so common blowing around the countryside that they had been dubbed the country's "national flower."

For several decades now, environmental groups the world over have mounted beach cleanups, often to great effect. The trash piles heaped up at the end of the day are horrific testaments to how messy we are. Recently, efforts have moved offshore to tackle garbage at sea and on the seabed. In Europe a project called "fishing for litter" aims to get the fishing industry to bag up the junk they catch in their nets and bring it ashore for disposal, reversing their previous practice of shoveling it back over the side with the bycatch. Were this approach to be applied more widely, it could make a big difference to the amount of trash lying at fishable depths. The European Commission announced in 2011 that it would trial a scheme to pay fishers to catch plastic in the Mediterranean, Europe's filthiest sea.[4] The Mediterranean has earned this epithet because it is surrounded by populous countries visited by millions of beachgoers and is almost completely enclosed, so junk has accumulated. The state of Hawaii also has a program to collect lost and discarded fishing gear that is then burned to produce power for the islands.

What has shocked the world into taking plastic more seriously are the great ocean garbage patches circling endlessly on voyages without destinations. Unfortunately, these are not of much interest to fishermen. They are short on critical nutrients and have low levels of plankton productivity, so there aren't many fish. Unless we do something, these places will continue to suck in trash like oceanic black holes. The Great Pacific Garbage Patch is said to be twice the size of Texas: enormous at nearly 440,000 square miles, but not beyond the bounds of control. In a way, the ocean gyres do us a favor by concentrating floating garbage into places where we could gather it up without bankrupting ourselves. Twenty boats equipped with hundred-yard-wide skimmers could sweep the gyre in the North Pacific in five years if they worked round the clock. Such boats don't yet exist but more conventional cleanups have been attempted at smaller scales.

This would only work for stuff that floats, and only for fragments above a certain minimum size. To save the albatrosses we would have to fish up everything larger than a bottle cap, which would be a technical challenge. Is it worth going to so much trouble for a bunch of silly birds? The answer is yes. Even if you can't care less about the albatross, self-interest might compel you to do something. Plastics crumble into ever smaller fragments over time, picking up toxins and passing them up the food chain, potentially ending up on our plate as we enjoy a tuna sandwich or sushi lunch. They take hundreds, in some cases thousands, of years to degrade, so unless we clean up the ocean garbage patches, they will continue to accumulate and break down until we stop making our plastic so durable. Fishing for litter will make little difference in the long run, however, unless it is married with efforts to stop littering in the first place.

The vast majority of stuff picked up by beach cleanups is local. At a remote beach in Brazil, collectors found bleach and detergent bottles, hypodermic syringes, and a host of other things washed downstream from towns and villages by a nearby river. In Majorca, beach trash is dominated by cigarette butts, plastic water and soda bottles, and tampon applicators. Litter peaks in summer, when the island is busiest. A former student who went to work in Asia wrote to me of her horror at the huge quantities of plastics and other garbage that choked the banks of rivers in Nepal. She asked a villager if it was ever taken away. He said, "Yes, yes . . . when the monsoon rains come, it is all taken away." She met this detritus again when she sailed across the Bay of Bengal, where junk carried from the Himalayas piles up in the ocean. Local efforts to tackle garbage would make a big difference, and in some places skimmers have been strung across rivers to catch things before they head to sea. But skimmers are really a last resort, and create problems of their own like blocking the passage of boats.

It would be far more effective to tackle pollution at its source. In the case of plastics, the three r's apply: reduce, reuse, recycle. To them we can add recover and redesign.[5] By recover we mean energy recovery using waste plastic to generate power, as in Hawaii. But burning produces other kinds of pollution, so it is better to recycle where possible. Anyone who has wrestled three layers of plastic from their box lunch knows manufacturers could do much more to reduce their use of plastic and design recyclability into their products. In a promising development, some new plastics are made from biodegradable materials like corn starch. More problematically, others are designed to crumble rather than biodegrade. They break down when exposed to ultraviolet light from the sun. That isn't much good in a landfill or out at sea, where floating plastic bags are quickly covered in algae that block much of the light and prevent breakdown.[6] Setting this difficulty aside, making a bag that shreds into small pieces will only thicken the broth of microplastic particles circulating in the sea.

Preventive measures will ultimately do more to reduce the problem of plastics than laws that are impossible to enforce on the open sea. Despite the fact that it is illegal to dump plastics from ships, the amount of rubbish picked up from British beaches in cleanups sponsored by the Marine Conservation Society increased 88 percent between 1994 and 2010, much of it jetsam from boats, and nearly two-thirds of it plastic.[7] In South Australia, a ban on dumping from fishing vessels failed to reduce wildlife entanglement.[8] The number of sea lions from Kangaroo Island wrapped in bait box packing straps increased sixfold and more than twice as many fur seals were trapped. The results of a similar scheme in South Georgia in the southern Atlantic were more encouraging, as incidence of seal entanglement halved.

Just before the Beijing Olympics, a flotilla of small boats set forth every day into a green sea near Qingdao. For weeks their crews forked great heaps of seaweed into the boats, piling it up like marine haystacks. Qingdao had become so polluted by nutrients from sewage, agriculture, and industrial runoff that thick masses of seaweed obstructed the passage of boats. The Olympic sailors would have recorded some pretty feeble times without their efforts.

It is relatively easy to regulate sewage or factory effluents that can be traced back to their sources and much harder to regulate nutrients washed to sea from farms and city streets, or blown offshore from plowed fields. Nonetheless success is possible, as an unintended experiment with the Black Sea proved.[9] By the early 1980s, fertilizers bled from the plains of the southern Soviet Union had turned the Black Sea into a green soup with few fish and an alarming number of jellyfish. The collapse of communism triggered a slump in fertilizer use, as state-controlled farms ran out of money to buy agrochemicals. Water quality has significantly improved since the early 1990s, with fewer and less intense plankton blooms and less oxygen depletion in coastal waters. Better water quality seems to have made life harder for the invasive comb jellyfish *Mnemiopsis*. Its population has collapsed and fisheries for small open-water fish have recovered. The Black Sea experience offers hope for other enclosed coastal seas that now suffer from pollution-induced dead zones, jellyfish explosions, and harmful algal blooms.

Nutrient pollution has choked the Baltic Sea, surrounded as it is by fourteen industrialized countries with farms that rely on agrochemicals, heavy industries that produce copious discharges, and a

combined population of eighty-five million whose fertile sewage effluent ends up downstream. As I explained in chapter 8, this modest
inlet of the North Atlantic is shallow, with an awkward circulation
that leaves puddles of dense, salty water to stagnate near the bottom.
Over the years there have been many plans to restore the Baltic to
health, but none has yet worked. Indeed, as living standards improve
in Eastern European countries, their downstream impact on the Baltic grows worse. In desperation, perhaps, engineers have begun to ask
whether it is possible to fix the Baltic by brute force, or technological
brilliance as they might prefer to call it.[10]

In recent years, dead zones have swallowed up to twenty-three
thousand square miles of the Baltic Sea. Because of its difficult circulation, thick layers of undigested organic matter accumulate in its
basins in water that has been drained of all oxygen. If we could just
stir some oxygen from the surface back into these lifeless waters, life
could return and take care of the heavy burden of waste. That might
happen if we could bulldoze Denmark out of the way to improve
circulation, but the Danish wouldn't approve. So instead, how can we
mix oxygen from surface waters to the seabed? Engineers have calculated that it would take 2.2 million to 6.6 million tons of oxygen a
year to resuscitate the dead zones. It is hard to imagine a figure like
this, but easier if you think of it in more concrete terms: 6.6 million
tons of oxygen would fill fifty-five thousand railroad cars!

One scheme foresees a sea of wind turbines that would drive
pumps to flush surface water to the bottom. Another idea is to install
vast arrays of vertical pipes. Waves washing over them would push
aerated water down with the help of gravity to exit near the seabed.
Another idea is to throw in some alum or powdered limestone to lock
up the phosphorus so that plankton blooms will run out of fuel, but
the volume is daunting and the addition would contravene the London Dumping Convention. There is now so much phosphorus in Baltic
sediments from fertilizer and sewage runoff that there have even been
suggestions it should be mined. Some have proposed we should intensively fish sprats, which eat zooplankton, to spare the zooplankton

that graze on phytoplankton blooms. I am sure this would be popular with the fishing industry, but it hardly seems a persuasive solution, as it would create as many problems as it would solve. In fact, close scrutiny shows none of these ideas to be realistic or viable right now, so it is back to Plan A: reduce nutrient pollution flowing into the Baltic. Countries around the Baltic have committed to a reduction of 50 percent, but it will take strenuous effort to meet this target.

Bottom trawling multiplies the problems. Organic matter and nutrients can get buried on the bottom so they are no longer available to power blooms of plankton or poisonous algae. Trawlers churn sediments to release those nutrients back into the water so that plankton blooms on and on. A small trawler fitted with two twenty-five-foot-wide nets and a chain that cuts an inch into the seabed can raise approximately two thousand tons of sediment per hour of trawling, of which over two hundred tons will remain in suspension for days.[11]

Why do we do it? Why do we foul and defile our own living space? To take one recent example, in the 1990s China embarked upon a program of economic expansion based on coal-fired power stations in full knowledge that they would soon choke on their own prosperity. Perhaps they didn't anticipate just how fast that would happen or with what suffocating vehemence the pollution clouds would envelop them. Throughout the world, rivers and beaches are clogged with the jetsam of civilization, air and water filled with the noxious hallmarks of progress. I think our minds are still stranded in the past. For most of history, the number of people on the planet has been small and the unoccupied space seemed almost limitless. We used natural materials that when discarded soon crumbled back into water or soil to feed new life. Today we are proliferating faster than the adaptive capacity of our biological natures. Before the twentieth century, nobody lived long enough to witness a doubling of the world's population. It is a mark of just how much the rate of growth of human societies has accelerated that some people alive today have witnessed not just one doubling, but two.[12] Adapt we must, because the tide of progress is not about to recede any time soon.

Although cleansing the oceans is a task bigger than any other cleanup attempted in history, pollution control is perhaps among the easiest of the oceans' problems for which to gain political will. People place health concerns high on their political to-do lists, especially in democracies. Action will only follow public awareness and demand, but the latter is on the increase. However far we get, and there is a long way to go, pollution control and prevention will always remain a work in progress. New pollutants are constantly introduced as by-products of our inventiveness and industry. We must be ever watchful.

CHAPTER 18

Can We Cool Our Warming World?

Wouldn't it be wonderful if the world had a thermostat that could be reset a couple of degrees? How many of our problems would disappear! In one sense, it does, for greenhouse gases such as carbon dioxide and methane act as a control. Add more to the atmosphere, and you dial up the heat. Remove them, and we could regain our cool. Technologists and governments, while searching for ways to reduce emissions, are also on the hunt for engineered fixes to our climate problem.

The planet has ways of dealing with high levels of atmospheric carbon dioxide. The greenhouse worlds of the past did eventually cool through reactions that removed carbon dioxide and methane from the atmosphere. The carbonic acid formed when carbon dioxide dissolves in the oceans is neutralized in a reaction that converts carbonate to bicarbonate. This reduces the amount of dissolved carbonate and so expands the layer of deep water that is undersaturated with carbonate toward the surface. This exposes billions of tons of carbonate sediments on the seabed to corrosive water, which can dissolve them. As the carbonate dissolves, it buffers the effects of acidification, so the oceans keep taking carbon dioxide from the atmosphere. The weathering of silicate rocks above water also takes up carbon dioxide. The awkward problem for us is that these feedbacks take tens or hundreds of thousands of years. In this respect, the effects of fossil fuel burning could outlast pollution from nuclear energy. (After ten thousand years, mixed nuclear waste becomes about three thousand times less radioactive, whereas the effects of a massive release of carbon on world temperatures will have declined far less, to between a half and a quarter of the peak.[1]) Conditions in the sea took one hundred thousand years to return to normal after the Paleocene-Eocene spike in world temperatures and ocean acidity 55 million years ago, and probably a million years after the Permian-Triassic greenhouse episode 251 million years ago. We can't wait that long. This means we must prevent the release of further carbon dioxide and find ways to take some back.

There is an emerging global consensus that we should restrict the rise in temperatures from greenhouse gases to four degrees Fahrenheit at most, which means we cannot go above about 450 parts per million of carbon dioxide in the atmosphere. Beyond that the risk of dangerous climatic upheaval is deemed too high. But even that target leaves us exposed to a small chance that we will pass an unknown tipping point and suddenly find ourselves on a headlong course toward catastrophe. To date the world has concentrated on ways to reduce emissions. These include greater efficiency in energy use and tapping more green energy from renewable sources. The oceans already play their role in the fight against climate change, as the biggest and most reliable sink for carbon dioxide, but maybe could do more. They cover more than two thirds of the planet and are, for the most part, uninhabited by people, so they offer considerable opportunities for engineered solutions.

Hydropower and wind have long provided us with energy, but there are severe constraints to their expansion on land. Most major rivers have been dammed, at least in the developed world, and new megadam projects are highly controversial because of the harm they cause to people and the environment. Wind farms are sprouting around us but are often bitterly opposed by those who feel they disfigure the landscape. The restless sea, out there and once offshore not in anyone's back yard, offers a less obtrusive world of opportunity. Because the sea is pretty much flat and covers 71 percent of the planet, nearly 90 percent of the world's wind energy is offshore.[2]

Already wind farms dot the horizon around many countries, and huge blocks of shallow sea are being licensed for development. In one extraordinary case of fractured thinking, the UK Crown Estate, which owns the seabed, licensed Britain's entire share of the Dogger Bank, a shallow hill in the middle of the North Sea, for wind farm development. The Dogger Bank also happens to be one of the richest fishing grounds, but it seems nobody thought to ask for the fishermen's views on the plan, nor the conservationists, who have now declared the place a Special Area of Conservation. Wind farms have yet to be built there, and it looks like there will be tense discussions when that stage is reached. Completed and proposed wind farms will soon grow to cover more than 20 percent of the UK's territorial waters, which extend up to twelve nautical miles offshore. They are expected to supply about twenty-five gigawatts of electricity by 2020 (a gigawatt is one billion watts), roughly 7 percent of projected demand. If all of the world's accessible offshore wind energy were tapped, it could supply around five thousand terawatts (a terawatt is one thousand billion watts), which is equivalent to about one third of the world's current energy use.

Plans are forging ahead for wind farms in U.S. waters with the first likely to be a 130-turbine farm in Nantucket Sound. As in many countries, the U.S. path to offshore windpower has not been without controversy. Critics contend that wind farms are bad for wildlife, such as birds that fly into the turbines. However, the evidence from Holland, where offshore wind farms have been in use since 2006, points the other way.[3] Turbine footings created new habitat for creatures like mussels, anemones, and hydroids, and shelter for schools of fish. Since the Dutch farm was put off-limits to fishing, it may in time come to act as a protected area of sorts. While some birds avoided the farm, like common scoters and gannets, others were attracted, like the great cormorant. Harbor porpoises were heard calling more often within the wind farm than outside, perhaps because they could find more to eat there. The potential for offshore wind energy has also been recognized in China with the first farm completed in 2009.

Developers there have plans for many more, given the great length of the Chinese coast. This move into clean energy is welcome for a country that has relied so heavily on coal up to now.

Waves and tides afford other possibilities for power generation whose potential has barely been touched. Two notable exceptions are a hydropower barrage across the Rance estuary in France, which has been in use since 1966, and a tidal turbine at the mouth of Strangford Lough in Northern Ireland, which has two fifty-foot diameter blades. Tidal dams have existed since at least the Middle Ages, when they were used to power coastal corn mills. But our capacity to take advantage of tidal power has grown beyond the wildest dreams of the monks and millers of long ago. Today plans have been drafted for colossal schemes like the Severn Barrage in southwest England. At its most ambitious, this scheme would bridge eleven miles of the Severn Estuary which has, at forty-nine feet, the second greatest tidal range in the world. It could supply over 10 percent of the UK's electricity needs in 2020, but comes with a large environmental price tag. Like dams further upstream, barrages change water levels. Tidal range would be reduced and the area of rich salt marsh and mudflat habitats vital to hundreds of thousands of migratory birds and commercial fish would fall. So far, the environmental costs of the Severn Barrage have been perceived as too high and the project was shelved (again) in 2010 while eight new nuclear power stations were given the go-ahead. But as the costs of climate change become clearer, resistance to tidal barrages may wash away. The pace already seems to be picking up. There has been a small tidal generator in Canada's Bay of Fundy since 1984, and in 2006 a demonstration project was installed at Race Rocks in Vancouver Island. South Korea has several tidal power stations and ambitious plans for more, while China has installed one scheme and the first Indian station is planned for 2013.

Wave power schemes are more palatable to conservationists. To stand at the edge of a cliff, stiff to the wind and watch the waves hurl themselves ashore, and feel them rumble through your body is to know the power of the sea. The immensity of the energy is palpable.

But many engineering challenges must be overcome before waves will be viable competitors to wind and tide. Many ways have been devised to harness their power. In one, waves rise and fall in an air-filled tube to push air through a turbine. Another involves flexible floating sausage strings whose bending pressurizes hydraulic fluid to drive motors that feed a generator. There are several prototypes off the European coast, one named Pelamis, after the seasnake (a better name than "sausage"). A third method relies on waves breaking over a wall to fill a reservoir which then drains through a turbine. None of these methods is commercially attractive as yet.

Generators might also tap the power of strong currents to turn underwater rotors, following the principle of tidal barrages. The mighty Gulf Stream, as it pours through the narrows between Florida and the Bahamas, might be an ideal location, or the Kuroshio Current as it races past Japan. Such generators would have to be moored in deep water, probably one hundred feet or more below the surface, to avoid causing a shipping hazard. To date, none have been built.

Progress toward restraining emissions to keep them below the target of 450 parts per million of atmospheric carbon dioxide has been desperately slow. As of 2011, emissions were rising in line with the worst case scenario of the IPCC. With every failed negotiation the likelihood rises that we will overshoot. An increasingly vocal group of scientists has begun to think the unthinkable. If we can alter the planetary climate by filling the atmosphere with carbon dioxide, perhaps we can find another way to dial down the heat. Some are now looking for ways of removing carbon dioxide from the atmosphere or finding methods to cool the planet. Such "technofixes" remain highly controversial, and there is a real risk that tinkering with the planetary thermostat could do more harm than good. Ken Caldeira, the American scientist who was among the first to point out the risks of ocean acidification, put it well in his testimony to a British Parliamentary Committee in 2008:

Only fools find joy in the prospect of climate engineering. It's foolish to think that the risk of significant climate damage can be denied or wished away. Perhaps we can depend on the transcendent human capacity for self-sacrifice when faced with unprecedented shared, long-term risk, and therefore can depend on future reductions in greenhouse gas emissions. But just in case, we'd better have a plan.[4]

Carbon dioxide levels may already be too high for the most sensitive habitats on Earth and the most delicately balanced parts of the world's climate system. I keep coming back to coral reefs. They have suffered badly from global change already and their very survival is now in doubt because of the one-two punch of warming and acidification. In the summer of 2009 I spent a day with a distinguished group of experts to talk about what we can do to save coral reefs. Late on a sunny London afternoon, David Attenborough and the Australian scientist Charlie Veron presented our conclusions. They told the waiting press that coral reefs as we know and love them are probably doomed unless we can reduce carbon dioxide levels to less than 350 parts per million—about 10 percent less than where we are now and 100 parts per million less than target levels negotiated for unsuccessfully in Copenhagen in 2010. If these magnificent habitats are to survive, we will have to find a way to suck some of the carbon dioxide we have already released out of the atmosphere.[5] Our only hope is carbon sequestration.

There is no shortage of ideas of how to dial back our carbon footprint, some madcap, others serious. Every meeting I go to these days seems to attract a collection of entrepreneurs touting their schemes for climate geoengineering. Glossy brochures are handed out, breathless talks given, and promotional videos screened. There is a smell of money in the air, and these people are in hot pursuit of cash from carbon sequestration. At the moment they are pursuing money from companies who want to offset their carbon footprint, but government cash may follow in the future if binding emissions targets are negotiated. It is hard to fault their enthusiasm, but most of these schemes seem to have been poorly thought through. Engineering fantasy often seems to trump common sense and suppress the desire to test whether the machinery on offer can possibly live up to the hype, or whether it is even sensible to use it in the first place.

One company plans to break down the two-layered structure of our tropical oceans where warm, low nutrient water floats on top of cooler, nutrient rich water below. Their idea is to speed up carbon dioxide removal by getting nutrients back to the surface, where there is more light, to stimulate plankton growth. Their scheme would deploy vertical pumps powered by waves to suck water from a few hundred meters down and release it near the surface. To make a difference you would have to seed the oceans with millions of these pumps (remember, there are $$ in their eyes). As the company rightly points out, global warming will increase the stability of these ocean layers by heating the surface more. Their plan, they explain, could help to feed the world, since more productive oceans produce more fish. The problem is most of the carbon taken up under this scheme would simply be released again when the plankton and fish in the surface waters died. In most places only a tiny fraction of the extra production would be taken out of circulation for the long term. The effect would probably be piffling. Producing electricity from their wave generators might be a more reliable means of reducing carbon emissions. The company believes that their pumps could be used to stir up dead zones, or cool coral reefs threatened by bleaching. Since hurricanes form in places with high sea temperatures and are sustained by drawing heat from the sea, they have also suggested that their upwelling pumps could reduce the frequency and intensity of tropical cyclones. Their entrepreneurial fire burns bright.

Other means of making the oceans more fertile and productive have been proposed. Far from coasts, enormous swathes of the sea are not short of the nutrients that usually limit plankton growth—

nitrogen and phosphorus—but they have very low productivity nonetheless. Their problem is a lack of the micronutrients that are needed in trace quantities to create the building blocks of life. Iron is one essential micronutrient, whose value to life was established billions of years ago in iron-rich water as metabolic pathways evolved. Mostly the sea gets its iron from river runoff or windblown dust, like the red dirt that blasts from the Sahara across the North Atlantic. Places like the southeastern Pacific and mid-Indian oceans are short of other nutrients essential for life, but about a quarter of the oceans are limited by iron alone. The whole of the Southern Ocean around Antarctica is chock full of nitrogen and phosphorus but far from any source of iron.[6] Likewise, the tropical eastern Pacific has plenty of nutrients but little iron. Perhaps, some have speculated, plankton production could be given a push if iron were added.

One of the first to notice the link between plankton growth, iron, and climate was the U.S. oceanographer John Martin, from Moss Landing Marine Labs in California.[7] In the late 1980s he noticed an inverse relationship between the amount of iron in deep-sea sediment layers and the amount of carbon dioxide in the atmosphere at the time those sediments settled to the seabed. Periods of low atmospheric carbon dioxide corresponded to ice ages while higher levels were present in the mild periods between glacials. Martin thought that iron was the key to low carbon dioxide levels, because it boosted plankton growth that pulled carbon from the atmosphere. The bodies of dead plankton and fish carried some of that carbon to the deep sea. In an experiment in the tropical eastern Pacific in 1993, Martin proved that a soluble iron compound mixed into the sea could trigger a plankton bloom. In the flush of enthusiasm that follows success in science, Martin made a quip that became famous: "Give me half a tanker of iron, and I'll give you an ice age." Luckily not!

Entrepreneurs and other freelance enthusiasts have since taken a leap of the imagination and proposed that ocean fertilization on a massive scale could rid the atmosphere of excess carbon dioxide. A niggling flaw in their plans is that nobody is sure where carbon dioxide absorbed by plankton will end up. Indeed, many scientists now believe that the great majority of extra carbon dioxide absorbed will simply be rereleased into surface waters as the plants and animals decompose. Surface waters are pretty good at recycling nutrients through microbial action before they sink to depths beyond the reach of mixing by winter storms. How deep that is varies from place to place and depends on the violence of winter weather and vigor of local currents, but in most places it is somewhere between three hundred and three thousand feet. Until carbon gets beyond this mixing zone, it hasn't really been taken out of the atmospheric system.

As of now, there have been twelve attempts to fertilize the sea with iron. Most caused a pleasing spike in phytoplankton growth but offered little evidence of carbon export to the deep sea. Early on, several companies set up shop and began to pull in venture capital money on the grounds that they could sell "carbon offsets" from artificial ocean fertilization. One company, Planktos, even got around to doing its own fertilization off Hawaii, dumping a paint-based iron compound into the sea along a thirty-mile track from the rock star Neil Young's antique yacht. While Young might have been pleased to have offset his massive carbon footprint, the findings of this "experiment" were never made public.

The idea of dumping iron to offset our carbon consumption is highly questionable, since fertilization contravenes the terms of the London Dumping Convention. In recent years two United Nations Conventions (on biodiversity and marine pollution) have come out against ocean fertilization. Planktos folded in 2008, and its founder bitterly attacked environmentalists for what he saw as a hate campaign. There are still others in the game, such as Climos and the Ocean Nourishment Corporation (who propose to fertilize with urea as a nitrogen source in areas not limited by iron), but enthusiasm for this approach is fading. Projections suggest that even the most optimistic scenarios could only increase ocean uptake of carbon dioxide by about a ninth of present emissions. It turns out that John Martin's half tanker of iron was way off the mark. To trigger an ice age you would have to continuously fertilize the entire Southern Ocean, all eight million square miles of it. If you stopped that fertilization, things would swiftly bounce back to where they were.

Another possible downside of fertilization is the effect on deep water oxygen. As I have already explained, when dead plankton sinks beneath the mixed surface layer of the sea it falls into a world where oxygen is scarce. Rotting plankton will use up precious oxygen and could enlarge the area of the ocean where there is too little to sustain anything but the simplest life-forms.

A different approach to tackling our climate problem is to extract carbon dioxide from the atmosphere and pump it into the deep sea. Below about ten thousand feet deep the high pressure and low temperatures turn carbon dioxide into a liquid denser than seawater. The idea is to inject liquid carbon dioxide through long pipes into ocean trenches or pool it on the seabed. While these ultradeep trenches would conveniently hold lakes of carbon dioxide, each sustains its own distinctive community of life, with many species found nowhere else, and it would likely cause extinctions. Experiments with small puddles of carbon dioxide placed on the deep-sea bed off the California coast suggest that the approach is feasible, but again this solution has fallen foul of environmentalists. Video footage of puddles show fish swimming nearby apparently unharmed, but capping large areas of ocean sediments with liquid carbon dioxide would certainly kill most of the animals that live there. There are also concerns that carbon dioxide stored in this state would find its way back to the atmosphere. Calculations suggest that only about 6 percent would mix back into the water over two hundred years (though up to 20 percent more would return later), but the technique would still buy us precious time to reduce emissions. It would, however, enhance the problem of acidification in the deep sea, so this idea is not very popular either.

A more realistic and practical solution to carbon dioxide disposal can perhaps be found in active or spent oil wells. One Norwegian oil company has extracted carbon dioxide from natural gas and pumped it back into wells below the North Sea since 1996.[8] The method was introduced to reduce emissions taxes and transport expenses and to increase yields from the well rather than to reduce emissions of a greenhouse gas. This concept, known as carbon capture and storage, has become a cornerstone of the much vaunted "clean coal" technology recently proposed for the next generation of power stations. There is little chance that we will abandon fossil fuels in time to avoid serious climate change, so we urgently need to find ways to generate power at less cost to the natural environment. There are several ways to capture carbon dioxide at the point of emission. The gas can then be pressurized, piped as a liquid, and forced back underground. Active or spent oil and gas fields seem to offer a means to lock away carbon dioxide for thousands to millions of years. Water from salty aquifers deep beneath the Earth's surface could also be displaced by carbon dioxide (the excess water would leak back into the sea). Such carbon capture and storage is now a central plank of international negotiations to avoid climate change. It could save up to 20 percent in emissions.[9] Unfortunately, current underinvestment in the technologies required means we will probably overshoot emissions targets and so risk dangerous climate change.

There are still other ways in which the oceans can help us engineer a cooler climate that remain the stuff of daydreams. One idea is to enhance the weathering of silicate rocks by exposing ground-up rock to hot carbon dioxide and water, and either storing the reaction products as solid rock or dispersing it throughout the oceans. Silicates make up most of the rock in the Earth's crust and include sand, clays, quartz, olivine, and diatomaceous earth, a substance made of the ancient remains of planktonic diatoms whose shells were built of silicon. Of these, olivine is especially suitable. Ocean dispersal of the carbonate reaction product has the advantage that more carbon dioxide is used up, and it could help reduce acidification because of the bicarbonate produced when it dissolves. The huge expense and disruptive impacts of mining the silicate rocks is a potential killer, regardless of any scruples about possible harmful effects on marine life.

Another scheme foresees a future in which floating platforms out at sea blast columns of salty vapor into the atmosphere to trigger the formation of denser and brighter low-level clouds.[10] The extra cloud cover would increase the Earth's albedo, or reflectivity, and help bounce some solar heat back into space. As well as keeping temperatures down, these clouds would help prevent and might even reverse polar ice melt. The idea sparked controversy almost as soon as it was first proposed. Some contended that if clouds were seeded in the Atlantic, it might dry out the Amazon, while others suggested that rainfall would go up, provoking floods and landslides. The problem is that many of these ideas for climate reengineering inhabit the margins of our understanding. Predicting their effects is almost impossible, and there is a serious risk of unexpected consequences.

A lower-cost and perhaps less intrusive solution may be to place microscopic particles into the stratosphere, high in the Earth's atmosphere, where they would form an aerosol and bounce some solar radiation back into space.[11] Huge volcanic eruptions do this sporadically and reduce temperatures fast. Mount Pinatubo in the Philippines erupted in 1991 and released a vast cloud of ash high into the atmosphere. It caused the biggest injection of sulfur dioxide into the atmosphere of any twentieth-century eruption and depressed global temperatures by about one degree Fahrenheit for the following two years. In engineered cooling schemes, sulfur dioxide or hydrogen sulfide gases would be released that would then react to create sulfate aerosols of the right particle size. Others say we should release sulfuric acid itself at high altitude for a better effect. Some have suggested that the world's commercial airline fleet could be used to seed sulfuric acid far above the land. While this would wrap the earth in a reflective blanket, it would do nothing to reduce either the buildup of greenhouse gases or ocean acidification. The same is true of cloud brightening. Sun-blocking strategies may be cheap and quick, but they are far from perfect.

High-altitude sulfate particles could also slow the repair of the ozone hole, which has gradually been mending since we reduced emissions of ozone destroying CFC gases in the late 1980s. It would also lead to a whitening of the skies, so say good-bye to those gorgeous cloudless mornings with heavens of brilliant azure. On the other hand, things will liven up in the evenings as the particles scatter light to produce sunsets of rich vermilion and carmine. Climate control schemes would have to radically transform the oceans to make any kind of difference to carbon dioxide levels in the atmosphere. They would introduce a whole slew of new human impacts on the sea and so are highly controversial. And yet, as Charlie Veron, the Australian coral expert put it, "when your house is on fire, you don't worry about wetting the wallpaper to put it out." As evidence of climate change mounts, our appetite for geoengineering fixes might grow.

One problem with all geoengineering schemes is that their effects are somewhat unpredictable. The Earth's climate system is complex, and we struggle to reproduce the nuances of its behavior in simulations. Doubtless, not everyone would benefit from climate geoengineering. Filling the atmosphere with sulfate aerosols to reflect sunshine back into space would reduce rainfall, since solar energy drives evaporation and cloud formation. Some places experienced a severe drought after Mount Pinatubo erupted, so the ethics of deliberately simulating the effects of an eruption are questionable. The fact that there will be some losers in geoengineering need not disqualify the use of such technology outright. We vaccinate children against disease knowing that a small minority will be harmed by the vaccines, because the great majority will benefit. However, the distribution of benefits has to be overwhelmingly skewed toward the positive to justify a risky intervention. A sensible guiding principle is the basic creed of medical ethics, "First do no harm."

Most geoengineering schemes have so far foundered on the shoals of skepticism and fear. The costs may be enormous and the perceived benefits too uncertain to warrant full-scale investment. But some technologies have been tried and tested, and even make commercial sense, such as stuffing carbon dioxide back into the oil and gas fields from which it was released. If this process were allied to an efficient system for extracting carbon dioxide from the air, it could begin to push the climate change genie back into the bottle from which we have called it. Middle Eastern countries often fret about what to do once their oil runs out. With all that sunshine, empty desert, and those spent oil fields, they have the energy, space, and carbon dioxide storage capacity to shift from being the purveyors of climate change to planetary saviors.

Before we embark on risky experiments that could backfire and are likely to be difficult to reverse, it is worth remembering that there is another method of carbon dioxide removal from the atmosphere that doesn't involve any risk of making things worse and can be achieved using technologies tested over millions of years. Healthy, natural ecosystems are carbon sinks that draw carbon dioxide from the atmosphere and sequester it in sediments, peat, and carbonate rock.[12] It has been estimated that every year healthy salt marshes, mangrove, and sea grass beds collectively remove carbon dioxide equivalent to half the emissions of the world's transport network (which totaled about 13.5 percent of global emissions in 2000).[13] Despite covering only one two-hundredth of the area of the world's terrestrial vegetation, these habitats remove a comparable amount of carbon from the atmosphere. This makes them some of the most intensive carbon sinks on the planet.[14]

And yet, just when we need them most, these habitats are being lost at an alarming rate of 2 percent to 7 percent a year. If we were to halt our losses today, we might have the same impact as the 10 percent reduction in emissions that would be required to stabilize warming at four degrees Fahrenheit or less. If we embarked upon large-scale habitat restoration efforts, like those undertaken for Vietnam's majestic mangrove swamps following the wide-scale use of defoliants during the war, they could contribute more. So if you can do just one thing, protect the salt marshes and mangrove swamps! We limit our options if we think only in terms of global emissions. I hope you will by now agree with me that a compelling case can be made for protecting, nurturing, and rebuilding ocean life to assist our safe passage through the climate crisis ahead and to safeguard the planet for our children's children.

CHAPTER 19

A New Deal for the Oceans

I had lunch not long ago with a high-ranking official from the European Union's fisheries directorate. We chatted amiably about politics and fish and why the Common Fisheries Policy was so feckless. Over coffee he became reflective. He said he remembered paddling the tide line of a beach near his home in Spain as a boy: "Every so often I would step on a flatfish and feel it wriggle away in a puff of sand and indignation. Now when I go there with my children, we never see them anymore." It's a familiar story, and one I hear more and more often everywhere I go. Unwittingly, he had described a once common method of fishing that dates back to prehistoric times. Our early ancestors would wade the tide line, spear in hand, ready to pounce when they disturbed hidden fish. There are still places with enough flatfish to stumble across one here and there, but as life dwindles around us, it is depressing to consider how easily forgotten such simple pleasures are.

Throughout this book I have touched on the many different ways in which people have changed the oceans since their arrival on the scene. Like it or not, we are stressing and killing marine life, making it harder for all manner of species to sustain and reproduce themselves. The struggle for existence has become a lot tougher.

It baffles me that conservation bodies don't pursue more energetically the recovery and rebuilding of nature. Instead, they often wage battles to hold the line, with some wins but frequent losses. Each skirmish for this or that species is fought with passion and vigor, but while we are winning individual battles, overall we are losing the war. We need to defend nature if we want to protect our own long-term interests. Nature conservation is too often perceived as a luxury, a view that has become embedded in attitudes and policies. Many people think of the oceans as a remote and incidental feature of our world. Their importance is felt in a physical sense, but people don't realize how much we all owe to *life* in the sea. Human well-being and economic prosperity cannot be unhitched from nature. Climate change is exposing the folly of our neglect for the ecological under-pinnings of life. Humanity is in retreat all over the world. Once productive landscapes have lost their soils, fertile valleys have become too salty to grow crops, lakes and rivers are drying up, groundwater has been drained far beyond the reach of roots, and now the sea is rising again to reclaim some of the world's most fertile and populous lands. In the oceans, rich fishing grounds have been emptied and life choked from estuaries, bays, and vast areas of the seas. Not since the last ice age has there been less living space for people. Our scant regard for nature means that we have shrunk the ability of the planet to support us just when, with seven billion mouths to feed, we most need to expand it. Contrary to what some might imagine, enlarging the planetary life support system doesn't mean pressing nature into an ever smaller area.

The solution to our present difficulties can only come from adopting strategies that rebuild nature's vitality and fecundity. Refurbishing the natural world is not an antidote to all ills facing the sea, but it is as close to a cure-all as we are going to find. This doesn't mean that we all have to become vegan or give up holidays abroad. Nor does it mean we have to turn off our computers and return to the technologies of yesteryear. We just have to become smarter about how we think about and use nature.

What we need in times of great planetary stress is a vibrant bio-sphere. The biosphere—that part of the planet that supports life—is sustained by four elements: energy, raw materials, the variety of life, and abundance. Let's take energy and raw materials for granted for a moment and consider life's variety and abundance. Both are fundamental to the sustenance of a habitable planet. Together they constitute the engine of the biosphere. Life's variety makes up its component parts, but the speed at which the engine runs depends on its abundance. In many places, wildlife has been so depleted that ecological processes now only just tick over. Keeping a few water-filtering oysters around will not maintain good water quality in estuaries; for that we need dense beds of filter-feeding invertebrates and aquatic vegetation.

To date wildlife conservation has concentrated on sustaining variety and has neglected abundance. The world's wonders would be less today without the efforts of thousands of people who since the nineteenth century have expended incredible energy to protect rare species and save wildlife on the brink of oblivion. Many creatures are now so rare that they depend on intensive care for survival. In some places entire nature reserves have been established to protect a single species—even a single individual. We must never give up such work, but now that nature is in retreat all around us, we have to shift our focus and give greater attention to abundance. Conservation of course is all about increasing numbers; the more common a species is, the less its extinction risk. But we have to set our sights higher, because extinction prevention and life as a museum will not keep the oceans healthy and sustain humanity through the difficult times ahead.

There are two bywords for resilience in the natural world: diversity and abundance. More enlightened economists refer to them as key elements of natural capital. Diversity begets flexibility and adaptability. Greater diversity enhances capacity to respond to change in positive ways; to keep the machine of life running, if you like. Abundance dictates how fast the machine runs. If humanity is to survive into the future we must come to appreciate that we cannot prosper without revving the engine of life back into a higher gear, so that life's processes run swift and sure: the uptake and locking away of carbon; water filtration and detoxification; the construction of biological structures like reefs, marshes, and forests that resist sea-level rise and protect coasts; indeed, the production of oxygen itself. If life were a multinational company, many of its subsidiaries would have gone under by now as a result of lost productivity. The whole business would be at risk of failure. When banks or countries fail we are quick to step in to prop them up, recognizing that we cannot afford a systemic breakdown. We are reaching that point with nature. It is time that we apply a little shock therapy to our environment. So allow me to propose a New Deal for the oceans, a bold and ambitious plan to rebuild and sustain the abundance of life at sea. This New Deal should know no boundaries and should apply to every ecosystem and region on this planet.

Where should we start? People love technological solutions. If nature is going bust, then surely we can build our way out of trouble. We can throw up seawalls to hold back the ocean, farm fish in ponds, mix up nutrients artificially in barren areas to boost productivity, even plant plastic sea grass (God forbid!) to re-create nursery habitats for fish and prawns.[1] To many, the answer to improving fisheries is simple: build an artificial reef and they will come. Better still, fill the sea with artificial reefs. Scuba divers have long sought out wrecks where glittering shoals of tiny fish attract animals with bigger mouths. A wreck adds a third dimension to a flat seabed and has the added benefit of discouraging trawlers and dredgers from risking their gear nearby. Anglers love wrecks too. There is always the chance of enticing some grizzled giant from among their rusting pipes and panels. But a nagging question remains: Do they simply concentrate existing fish or do they add living space and so increase the amount of fish a habitat can support?

This debate has bounced around the world of fisheries for decades without any conclusive answer. Perhaps that is because, like so many questions, there is no definitive answer. Sometimes they do, sometimes they don't. But even if artificial reefs did provide some uplift in fish stocks, is that a good enough reason to fill the sea with junk? Many of the most ardent advocates of artificial reefs seem to have large quantities of bulky waste to dispose of: tires, power station ash, railway cars, ships, and oil rigs. I once saw a newspaper photograph of three garbage trucks being pushed off the side of a boat in Thailand. The caption read, without any hint of irony, that this was a scheme promoted by Queen Sirikit to "rescue the sea from years of overfishing and pollution." Ever since I have carried an image of these trucks chugging around the bottom of the Gulf of Thailand collecting garbage!

The Gulf of Mexico must be one of the most junk-filled seas in the world. All of the states bordering the gulf have artificial reef programs and appear to be dedicated to getting rid of trash under the guise of environmental protection. Here is what the State of Alabama Department of Conservation and Natural Resources has to say:

> In 1993, the U. S. military in addressing the need to de-militarize obsolete battle tanks realized that immersion in seawater was an acceptable method. The idea of using these obsolete materials to create artificial reefs was born. . . . As plans developed the full extent of the impact that this program will have on the reef fisheries and associated economy of Alabama became obvious. The conservative estimate for the life span of the tanks is 50 years as artificial reefs. The potential economic impact of these tanks as artificial reefs during this time is millions of dollars. Even this conservative estimate far outweighs any other method of removing these tanks from military service. It is an outstanding and creative way to convert swords to plowshares.[2]

I love the language: "immersion in seawater was an acceptable method" of disposal. It sounds so much better than "offshore dumping." Southern states now use the sea as the repository of choice for all manner of waste, including old highway bridges, oil and gas platforms, mine spoil, school buses, you name it.[3] The states of Alabama and Florida have gone further than others in that they permit individuals and companies to dump in designated areas. Only the people involved know the precise locations where they have sunk their junk, so they will benefit most from possible improvements in fishing. Mostly these new "reefs" consist of old vehicles or bits of concrete. I don't want to suggest that creating artificial reefs has no merit. They keep bottom trawlers out, and there is some evidence that recreational catches of red snapper are higher near the reefs than in areas of open seabed. But I have to say the idea makes me deeply uncomfortable. There are better ways to recover productivity than to dump our discarded trucks and bridges in the sea.

In Japan, artificial reef creation has been married with aquaculture in "sea ranching," whereby areas of coast are used as semienclosed fish ranches within which habitats are "enhanced" with concrete reef structures. Fish are fed at certain places and times, and in some cases music is piped into the sea so they come to associate the music with feedings. I can't help but think that rather than being a sensible way to enhance fish production this is just another way in which Japan offers subsidies to its politically well-connected concrete industry. The majority of fishery subsidies in Japan go to the cement industry to fund construction of harbors and wharves.

The natural world is far better at generating the services ecosystems provide than we are at engineering them. Unlike real reefs, Florida and Alabama's underwater scrap yards disintegrate within five or ten years of disposal to litter the bottom with old car parts and crumbling panels. Seawalls cost millions of dollars per mile to build and have to be regularly maintained at great expense, whereas salt marshes, mangroves, and coral reefs protect the coast far better and look after themselves. Fish ponds produce one or a few varieties of fish, usually with large subsidies from wild nature in the form of feed, clean water, land, and waste disposal. The mangroves, salt marshes, and tidal flats they replace are nurseries to dozens of commercially important species that grow and disperse to sustain fisheries more widely. In New Zealand, tests with plastic sea grass showed that it attracted a considerable variety and abundance of fish. Many estuaries in New Zealand have lost nearly all their natural sea grass since the 1960s. But real sea grass is far better at the job, with the added benefits of oxygenating the water, filtering wastes, binding sediment, capturing carbon, and providing food for a host of animals, from turtles to snails. Natural habitats are solar-powered and will continue to deliver their multiple benefits as long as the sun shines and we take care not to harm them. So why are we finding plastic and concrete solutions rather than preserving and protecting our natural resources?

There are many ecological services that we cannot artificially re-create or which would simply be too expensive or impractical to reengineer. It would be impossible to create a high-capacity filtration system that worked as well as the carpet of invertebrates that once covered much of the seabed. Stirring oxygen into dead zones would be an enormous undertaking. And no man-made reef will ever come close to the richness of a healthy estuary. Restoration efforts are most worthwhile when they work with nature. In America's Delaware Bay, intensive efforts to reestablish living oyster reefs are beginning to pay off, as transplanted oysters are seeding the bay with their own off-spring. And restored beds provide the foundation for rich communi-ties of other invertebrates and fish that the natural beds of old did. In the Philippines, replanted mangroves help protect coasts from erosion while also creating nurseries for wild fish and shrimp that later on fill the nets and baskets of fishers. In Britain, farmland long ago reclaimed from the sea is being reflooded to encourage salt-marsh growth and replace mud flats that are disappearing under sea-level rise.

Although I spend much of my life behind a desk, wild nature inspires and renews me. I carry the memory of it with me wherever I go, and I return to the well whenever I can. A few years ago I was lucky enough to witness the elemental struggle between two great creatures of the ocean in the Galápagos Islands. I was with a small group in a dingy when some sixty feet away we spotted a hunting orca plunging through the face of a wave as it cornered a green turtle against a hidden reef. White foam streamed over its dark flank lend-ing it a frightful beauty. Another orca crested beside it, then lashed its tail as the king of predators snatched the turtle, and flung it across the surface of the sea. The orcas seemed to play with the turtle after that, rising and diving, arching their fins against a backdrop of volcanic slopes, black on black. It was a timeless moment: the millennial struggle for survival played out over and over in these waters. Storm petrels flitted above in excited flocks, while far overhead frigate birds circled and waited, hoping for a share of the spoils. Nature offers us things that no artifice could ever create. If we give it a chance, it can heal many of its wounds.

I am not foolish enough to think that we can wind back the clock to some primordial condition when our species first walked out of Africa. Change is inevitable. Species rise and fall; ecosystems shift and warp with time. But if our biosphere is to weather the storms ahead, then we must strengthen the twin pillars of life's variety and abundance.

How can we deliver this New Deal for the oceans? Marine reserves are a critical part of the solution. Placing areas off limits to exploitation and excessive use has proven time and again to be a powerful tool to protect and enhance life at sea. But if reserves are to make a real difference, there must be lots of them in extensive networks that are well enforced. They cannot simply safeguard a smattering of beauty spots. We should think of these interconnected safe havens as the new foundations underpinning everything we do in the sea. Fortunately, there is a growing appetite for this kind of protection in the world at large.

The mountains of Arran rise up from Scotland's Firth of Clyde in a ragged ridge of brown and gold. The island is known as the "sleeping warrior" for its profile from the mainland. Its warrior spirit must infuse some of Arran's five thousand inhabitants, for it was here that the first battle against overfishing and degradation of the Firth of Clyde was fought and eventually won.

In 1995 a small group of islanders got together to campaign for the protection of a short stretch of coast in Lamlash Bay. They were

shocked by how, within a few decades, the bay had been nearly emptied of fish and its fragile seabed habitats destroyed by scallop dredging. For years they were rebuffed by fisheries managers, government
officials, and politicians. Even Scotland's own nature conservation
agency wouldn't support them, because they felt that the bay was not
special enough to warrant protection. The islanders refused to give up.
Over the years their campaign group, the Community of Arran Seabed Trust, grew to include half of Arran's population.[4] Votes count,
and political support for the protected area followed. Through dogged
persistence and a conviction that the destruction of the Firth of Clyde
should not be ignored, the people of Arran achieved their goal. In
2008 Lamlash Bay became Scotland's first marine protected area to
be closed to all fishing. A small community had taken the future of
the sea into their own hands.

Arran's faith in the regenerative powers of the sea quickly proved
well-founded. Within two years of protection habitats inside the
reserve began to take on a more lush and healthy appearance, and
juvenile scallops had become more numerous under protection than
in nearby fishing grounds.[5] Other benefits, like recovery of big fish,
will take longer to accrue, but reserves elsewhere show they will come
if protection is maintained.

Across the world, other communities have trodden parallel paths
to Arran's. In the Philippines, a few villages troubled by the growing
scarcity of fish, established small protected areas in the 1970s to help
rehabilitate reefs and fisheries. Those areas were studied by researchers
from Silliman University under the direction of Angel Alcala, an
affable and determined man who later became a government minis-
ter, and Garry Russ, a scientist at Australia's James Cook University.
Over time, fish and corals rebounded in the reserves. A change of
leadership in one village led to the Sumilon reserve being reopened to
fishing in 1984. Within a year stocks fell by half, and the connection
between protection and benefit was firmly established. By way of
contrast, support for the reserve at Apo Island remained strong and
fish stocks there kept on going up, year after year.[6] Twenty-five years

of sustained protection has since increased Apo's stocks eleven times over.

Word spread about these islands of protection in what are otherwise some of the most intensely overfished seas on the planet (remember *pa-aling*). With the encouragement of Alcala and others keen to revivify Philippine seas, groups of fishers who lived near reserves traveled the country to share their experiences. Other villages copied their approach. The Philippines now has hundreds of community-managed protected areas. Even though the fishers involved were desperately poor, they realized that if they didn't make some short-term sacrifices, they would have no fish to eat in the future.

In Fiji, local chiefs have for centuries jealously guarded patches of the coral reef exploited by their communities, always keeping some areas off limits to fishing (called *tabu,* which is where our word taboo comes from). In recent years, their efforts have been harnessed in the creation of a national network of community-managed reserves. Fiji has committed to the protection of at least 30 percent of its seas by 2020, and with nearly two hundred protected areas already, it looks likely to reach this target long before then. Big fish are still easy to come by in Fiji, and the islands have some of the most exciting shark dives in the world. Examples of effective protection are multiplying across the world. The Hol Chan reserve in Belize, Cabo Pulmo in Mexico, Ras Mohammed in Egypt, Exuma Cays Land and Sea Park in the Bahamas, Tsitsikamma in South Africa, the list grows longer month by month, all thronging with life and testament to the spectacular effects of protection. Time and again, when people realize what they have lost, and understand that careful stewardship of their resources will bring some of their riches back, this realization kindles enthusiasm for action.

Marine protected areas represent an indispensable building block of our New Deal for the oceans. In the last century and a half we have expanded our exploitation of the oceans from a narrow strip around the coasts to the deepest and most remote seas. In the early nineteenth century, the untapped sea exceeded the areas that were fished by at least a hundred to one. Today that position is reversed. This expansion, like our earlier spread to every corner on land, has left far less space for wild nature. The difference between land and sea is that we depend on wild nature to supply the majority of our seafood. To ensure continuity of this food before it is too late, we must set up parks at sea, and lots of them. If well protected, they could harness nature's resilience to produce a resurgence of life. As Apo Island proved, exploited species can increase by five-, ten-, even twentyfold within a decade of protection, and sometimes in just a few years. After eleven years of protection in Spain's Cabo de Palos reserve, dusky grouper, a popular Mediterranean eating fish, had leaped in abundance by over fortyfold, while in Florida's Dry Tortugas Ecological Reserve, mutton snapper rebounded by three times in only four years.[7]

Watching reserves refill with life over a span of years, as I have many times in the Caribbean, is an uplifting experience. Caves, ledges, and overhangs repopulate with animals that had formerly been scarce, while the blue waters above thrill again with the flash and strike of predators and prey. After an absence of ten years I recently revisited a small marine reserve at Molasses Reef in the Florida Keys National Marine Sanctuary, which has been protected from fishing since the 1970s. I was delighted to find many more fish than on my first visit, including several kinds that are usually absent from exploited reefs. Three-foot-long blue parrotfish grazed the bottom in amiable groups, their pale flanks a pleasing contrast to the indigo shades of the midnight parrotfish mingling with them. I even spotted a thick-set Nassau grouper between two coral heads, once the mainstay of Caribbean fish dinners, it is now missing from much of its former range.

Refuges like these cannot solve all ills. They will not address the root causes of global change, nor reduce pollution, but they can alleviate some of the effects of these problems. This is how: Marine reserves—places off limits to all exploitation—have proven to be spectacularly good breeding grounds. This is a simple consequence of two things: increased abundance and extended lifespan of the animals they contain. Early on in this book I explained that overfishing had caused a collapse in fish and shellfish stocks because it culled the big, old, fat females that were the engines of egg production. A big fish can produce hundreds of times more young than a small fish, and many marine creatures keep on getting bigger as they grow older and remain fertile most of their lives. So having craggy and ancient giants around is a brilliant way to ensure masses of offspring. If you think of a marine reserve as a fountain of microscopic eggs and larvae pouring into the sea, you won't be far off. As populations recover and individual inhabitants grow larger, the flow increases from a trickle to a stream to a gush. Long-established reserves offer up countless millions of eggs, larvae, seeds, spores, planulae, or whatever other means their inhabitants use to reproduce. Like dandelion seeds caught in the wind, currents carry offspring away from their parents for distances of feet to hundreds of miles, depending on the species. So the new life spreads far beyond the limits of the protected habitat.

Stresses like climate change and pollution reduce survival rates and make reproduction harder; by increasing the flow of offspring, protected reserves can mitigate these effects. Bigger populations that reproduce vigorously will be much better able to battle through the battery of stresses that we throw at them, tipping the balance between life and death back in their favor.

Protected areas do things that fishery managers cannot achieve without them. The Leigh Marine Reserve in New Zealand is just three miles across, and yet it produces more young snapper, a burly gray fish much loved by fishermen, than sixty miles of fished coast. For some animals in long-established reserves the multiple is even greater and protected areas outproduce fishing grounds 100 to 1.

Marine reserves can boost the resilience of fish populations and fisheries in another way too. One of the effects of fishing, if you recall, is that it increases the variability of fish stocks, which induces swings in the fortunes of fishers. This happens when fishing drives down populations and shortens lifespans so far fewer offspring are produced. Reserves reverse this effect by securing a steadier and richer flow of offspring.[8] If you fish for a living, that difference matters.

When the stream of new life flows fast, there will always be something to catch. When it is weak and intermittent, catches are unreliable.

Reserves can also bring back habitats swept away by trawls or dredges, and reverse habitat collapses induced by plagues of grazers like sea urchins let loose by the removal of their predators. If you could fast-forward through the years after a reserve was created, you would see kelp forests rise from the seabed, sea grasses thicken and spread, sediments disappear under crusts of invertebrates, like oysters and sponges, that elbow their way above the bottom. The gradual rebuilding of these underwater metropolises increases the capacity of the sea to sustain life. There are more ways to make a living and more places to hide.

There probably aren't too many people who, if given the choice, would want the fish on their plate to come at the price of extinction for other species. But that is the unseen cost of many fisheries today. Most people are unaware that some of the species that show up on the fishmonger's slab simply cannot sustain productive fisheries in the long run. They grow and reproduce too slowly. Most sharks and the bigger skates and rays fall into this category. So does almost everything caught more than sixteen hundred feet down, deep-sea beasts like Chilean sea bass, orange roughy, or roundnose grenadier. They are caught because they are there, and when they are gone they disappear from markets. If you are old enough, you will remember the sudden appearance of orange roughy in supermarkets in the 1980s and 1990s. Their firm white fillets were similar enough to cod and haddock to find a ready market. Just as swiftly they vanished early this century as their fisheries collapsed, the seamounts from which they were pulled left empty. There are good reasons why we farm animals that are highly productive and feed low in the food web, like chickens and cows, rather than bears or cougars. But it is the bears and cougars of the sea that we have grown used to eating.

Many exploited species became rare so long ago that we can no longer remember a time when they were abundant: angel sharks and giant skates in Northern Europe, bluefin tuna in the North Sea (even Iceland), halibut in the North Atlantic, goliath groupers in most of the Caribbean, rosefish in the Gulf of Maine. Celebrity chefs and environmental organizations sometimes confuse variety with sustainability. They exhort us to eat less cod, for instance, to ease pressure on an overfished species, while urging alternatives like ling or wolffish. Here a little historical perspective may help us judge the wisdom of these substitutions. Both ling and wolffish are impressive beasts. Fully mature they can grow longer than a man is tall. The wolffish has a face like a seasoned rugby player and teeth to match, their gappy grins and broken canines testament to a lifetime of hard knocks gained crunching clams and urchins. We are not the first generation to appreciate wolffish steaks: they were much admired in the nineteenth and early twentieth century. Trawl fishing records from the UK show that since 1889 catches of wolffish per unit of fishing power have fallen by 96 percent.[9] Ling catches have fallen by the same amount. My student Ruth Thurstan once gave a presentation in which she showed a photograph of fish landings made at the UK port of Grimsby in the 1890s. Six-foot ling were piled on the dock like logs. A fisherman in the audience rebuked her for confusing tuna with the trawl catches she had mentioned. But he was the one who was confused. He had never seen ling so large or abundant, so he mistook them for tuna. Neither ling nor wolffish is a sustainable alternative to traditional favorites. Today they are rarities that were once common. So if you want to find a substitute for cod, a better option would be something that is still common, like haddock, whiting, or pollock.

Marine reserves offer a way out of the dilemma of whether to fish hard to benefit from high-production, high-turnover species like scallops, prawns, and plaice, or to fish lightly to prevent the disappearance of the big and vulnerable. At the moment, species like the "no-longer-common skate" hang on in scraps of rough ground beyond the reach of trawl and dredge. But these de facto refuges are shrinking under the onslaught of new technologies and fishers chasing better catches as traditional grounds fail. Scuba divers in the west of Scotland report scallop dredgers scraping the bottom right up to the edge of submerged reefs, guided by sophisticated underwater sensing equipment. They lift tons of the sediment that settles over the reefs and smothers life on the rocks so their impact extends even into places of apparent safety. Reserves provide refuges where fragile and vulnerable animals and plants can thrive without hindering fishing operations. Intensive fishing and susceptible wildlife can coexist, but not in the same places. In this respect the sea is no different from land. Sensitive animals and plants that cannot cope in landscapes dominated by industrial agriculture find safety in protected woodlands, heaths, and prairies. We just haven't got our heads around the simple fact that we need to set aside parks in the sea. We can't easily see what is going on down there, but many modern fishing technologies are equivalent to clear-cutting and monocrop farming rolled into one.

With so much in their favor, why is it that marine protected areas are so often opposed by the group that stands to gain most from them— the fishing industry? In 2010 England embarked upon a grand project to create a national network of marine protected areas to redress over a century of accumulated damage to marine life, and to fulfill its commitment made in 2002 at the World Summit on Sustainable Development. Like other coastal nations of the world, the UK promised to create a national network of marine protected areas by 2012. Sadly, many within the fishing industry have chosen to oppose this. Lurid headlines and exaggerated claims have filled the fishing press: NO TAKE ZONES ARE THREAT TO FISHING from *Fishing News* is typical.[10] In reality, the biggest threat to fishing is fishing itself. Overexploitation is putting people out of business, not protected areas. All the evidence suggests that protection keeps fishers at work. In developing nations, marine reserves have been proven to boost the food security of people who live nearby.[11] In the Mediterranean, artisanal fishers, the very lifeblood of coastal culture and community, make a better living in the waters near marine reserves than those in places where there are none.[12]

Three quarters of the United States haddock catch is caught within three miles of the boundaries of large protected areas on Georges Bank, east of Cape Cod, and in the Gulf of Maine.[13] Lobsters have bounced back to seven times their previous abundance in a tiny scrap of protected area that was declared in 2003 next to England's Lundy Island, and the reserve is spilling young lobsters into surrounding fishing grounds.[14] Every year the examples multiply, and yet some people seem to stick stubbornly to views that science has proven to be unfounded.[15] I hear the same arguments against protected areas that I heard fifteen years ago, about how they presage the collapse of the fishing industry and will eviscerate thriving ports. But ports are closing because a lack of protection means that every year there are fewer fish. Here is just one example local to me: The port of Whitby on the Yorkshire coast was a bustling center of fishing in the mid–twentieth century and long before that. Old photographs show the harbor crammed with boats and dockside stacked high with boxes full of fish. In 2010 there were ten boats left, all for sale.

One old chestnut that pops up again and again is how marine reserves will not provide any benefit to mobile fish like cod, which are the mainstays of cool water fisheries in Europe, North America, and Japan. This view is based on a misunderstanding of how they live their lives. In recent years cod have been fitted with tags that track their movements. Many of them are remarkably sedentary and for months on end hardly move at all. Some stay put pretty much all year round. There is a telling example of what protection can do for cod in the convoluted entrance to the Baltic Sea.[16]

Between Copenhagen in Denmark and Malmö in Sweden there is a short and narrow strait called the Öresund. It varies from two-and-a-half to twenty-eight miles wide and is the main route for ships in and out of the Baltic. Because of its importance to shipping, Öresund has been closed to mobile fishing gears like trawls and seine nets since 1932. Catches in the Öresund are made mainly by fixed nets. To the north, a region ten times its size, called the Kattegat, is fished by bottom trawls. The fortunes of their respective fisheries in the last thirty-five years could not be more different. In the late 1970s, fishers trawled sixteen thousand to twenty-two thousand tons of cod from the Kattegat while gillnetters in Öresund took about two thousand. In 2008, the Kattegat yielded 495 tons of cod, compared to 2,350 tons from the Öresund. Research catches made in Öresund that year showed cod there to be fifteen to forty times more abundant than in Kattegat, and much bigger too. Some of the cod pulled from these waters today rival the behemoths of old. Similar size differences were found in several other kinds of fish, including lemon sole, haddock, plaice, and whiting. The Öresund and its fisheries have remained in great shape because they are protected from the most destructive kinds of fishing. Despite the fact that it covers just 780 square miles, and the coasts that border the strait are home to nearly four million people, marine life flourishes.

Fisheries benefit from complexity. As it transpired, not all cod are the migrants we took them for, in endless transit over thousands of miles between nursery, feeding, and breeding areas. Populations like those in the Öresund, or the Firth of Clyde for that matter, are locally adapted and more sedentary. Indiscriminate fishing has laid waste to this population diversity, and in the process it has undermined the capacity of populations, and the fisheries that depend on them, to continue to thrive as conditions change. This adaptability becomes ever more valuable as the world around us changes.

Alaska's Bristol Bay salmon fishery is a perfect example of how population variability can work to stabilize fisheries.[17] Bristol Bay is 250 miles by nearly 200 wide and opens into the eastern Bering Sea. It is fed by dozens of salmon rivers and streams, each of which supports a population of sockeye salmon adapted to local conditions. Salmon stocks vary in size and productivity as the climate shifts. Under any given set of conditions some fare poorly while others thrive. Fifty years of salmon fishing records show ups and downs, but collectively salmon production has remained strong, much stronger and less variable than it would have been if the bay had only supported a single sockeye population.

This has been dubbed the "portfolio effect." Just as investors spread their risk by keeping a variety of stocks, having a variety of differently adapted fish populations imparts stability to ecosystems and fisheries. George Rose, who works out of Newfoundland's Memorial University and saw Canada's cod collapse firsthand, knows more about cod than almost anyone alive. Even as fishermen scooped up the last of the great cod shoals in the early 1990s, he found several populations local to different parts of the coast that appeared unaffected. He describes nature as being like an orchestra in which not all the instruments play loudly at once: "You have the horns playing for a bit, then the strings come in."[18]

The same effect is apparent across species. Every animal and plant makes a living in a slightly different way. Each one responds differently when conditions change; each reacts to stress from human activities in a dissimilar way. Something that one species might barely notice, or easily overcome, could fell another. Where there is a broad portfolio of species to draw upon, ecosystems are much more likely to continue to function as the seas change. Simplified ecosystems of the kind that now predominate in many places have little of this resilience. Their variety has been winnowed to a handful of species, like prawns and scallops. We will not be able to restore life to these places by tinkering with mesh sizes or fishing gears alone. Marine reserves are essential to defend and promote diversity and thereby help both wildlife and the fishing industry weather turbulent times ahead.

One thing that is seldom appreciated about marine reserves is that strong protection is needed for strong benefits. Early on in this book I described the phenomenon of "fishing down the food web," as top predators are depleted and fisheries turn to smaller animals lower down the food web. Effects of exploitation on marine habitats parallel this trend, with progressive loss of structure-forming animals and plants; complex habitats crumble into simple ones. To be truly effective, marine protected areas must push ecosystems back up the slope of collapse. How far they can go—in other words how much recovery is possible—depends in large part on how strongly they are protected. A little protection, such as a small reduction in fishing pressure, will nudge things back slightly. Modest protection, such as a halt to scallop dredging, while continuing the use of trawls targeting fish, will be met by modest benefits. Animals of middling size will probably do well with middling protection like this, but the big beasts of old will not return. Only strong protection—removal of all exploitation and alleviation of other sources of harm—will drive ecosystems back toward their most prolific and productive. Because protected areas generate controversy and often inspire highly vocal opposition, politicians often seek compromise not in whether to create them in the first place, but in how much protection they will be given. The result is that the seas are full of protected areas that aren't really protected at all. Many of Europe's Special Areas of Conservation are like this (they should really be called Suspicious Absence of Conservation). In a way "paper parks" are worse than no protection because they give the illusion of conservation without the reality. America, whose politicians are no less weak-willed, suffers the same problem. Some of America's National Marine Sanctuaries still offer little real protection from fishing, although times are changing there for the better.

Marine protected areas will only contribute meaningfully to raising the variety and abundance of life if they spread and multiply. There is little value in protecting a few beauty spots to a high level if we continue business as usual in the rest. The United Nations Convention on Biological Diversity set a target in 1997 of 10 percent of the oceans protected by 2012. At the time of writing, the tail end of 2011, we had only made it to 1.6 percent, so the target date has been pushed back to 2020. Is this ambitious enough? An honest scientist will say no.

Looking at the question of how much to protect from a wide range of angles gives remarkably concordant answers. Studies have looked at how much coverage is needed to sustain or maximize fisheries production, to protect populations of all species in more than one place (to avoid losses from random catastrophes), to prevent the loss of genetic diversity, and to ensure protected areas are large and numerous enough and sufficiently interconnected to sustain themselves. Across the board the answer is that we need to protect double-digit percentages of the sea to achieve these goals, not single figures. The middle of the range of answers is about 35 percent,[19] so the 10 percent target proposed by the UN is too low. In my view we need to protect a third of the oceans from direct exploitation and harm, and manage the rest much better than we do today. Hitting the UN target means a sixfold increase over the next decade. I accept that it is unrealistic at this stage to expect more. But this target only represents a waypoint on the path to a more sustainable future, not the endpoint.

CHAPTER 20

Life Renewed

I often come across people who think that we can't afford to cut
back fishing when every day there are more mouths to feed. But
simple math tells you that restocking our seas makes economic
sense. Think of it this way: If you have a million fish in the sea and
can catch 20 percent of them every year without depleting the stock,
that stock would give you two hundred thousand fish a year. Now
imagine that you nurtured your fish and gave them a chance to grow
so that you had five million. Your 20 percent would come to one million a year. The interest rate on your capital is the same, but the yield
is much bigger. And with fish more abundant they would be easier to
catch, so you would need fewer boats and each would cost less to run.

Wishful thinking? Not really. In 1889, there were ten to fifteen
times as many large bottom-living fish like cod, haddock, and halibut
in the seas around the UK as there are today. As I explained, a late-
nineteenth-century fleet, made up mostly of open-decked sailing
boats, landed twice as many bottom-dwelling fish as we do today,
with seventeen times less effort. Less fishing, more fish, bigger catches,
food for more people. A World Bank report aptly titled *The Sunken
Billions*[1] highlighted the madness of overfishing when it calculated
that major fish stocks of the world would produce 40 percent more if
we fished them less. It sounds paradoxical—fish less to catch more—
but that is the simple message. In Europe, the benefits of fishing less
could be even higher, perhaps 60 percent, since stocks there are in
such a poor state.[2]

There are people out there who insist that all is well with the seas and that those who say otherwise are just bleeding-heart greenies or scientists who have drunk too deep from the cup of environmentalism. Foremost among them is Ray Hilborn, a well-respected fisheries scientist from the University of Washington in Seattle. The flavor of his argument can be garnered from the title of a *New York Times* op-ed he wrote in 2011: "Let us eat fish." In it he claims, "On average, fish stocks worldwide appear to be stable, and in the United States they are rebuilding, in many cases at a rapid rate." Hilborn has been on a one-man mission to face down "doom-mongers," as he calls them, tramping the world in search of headlines. He finds a ready audience— who wouldn't rather hear that we're doing just fine? But Hilborn's reading of the state of fisheries is highly selective, and he seems immune to any amount of historical evidence of decline. For sure some fisheries are doing well in some parts of the world (and after recent reforms, many are on the increase again in America after years of mismanagement).[3] On a narrow reading this is good news, but these productive fisheries often come at the expense of a host of other species, as is the case of scallop dredge fisheries, which thrive in the open, sandy, predator-free bottoms that are left when other fisheries have carted off the rest. As Canada's cod shows, pulling down fishing rates (Hilborn's main metric of success) doesn't always lead to recovery. Hilborn seems to place more faith in his mathematical models than he does on evidence, like someone who insists on going out in the rain without an umbrella because the forecast said it would be sunny. I admire his optimism, but his relentlessly upbeat reading of the state of the ocean life is Panglossian.

Overfishing has been called one of the biggest soluble environmental problems in the world. We know what must be done and it takes no more than a sentence to say it. We have to fish less, waste less, use less destructive methods to catch what we take, and provide safe havens where fish can reach their full reproductive potential, habitats can flourish, and vulnerable species can be protected. So why hasn't it happened?

Every year Europe's political leaders set ludicrous fishing quotas that over the last twenty-five years have averaged a third higher than the safe catch levels recommended to them by their own scientists.[4] The folly of this approach is obvious. A forester who cuts down more trees than he plants will soon run out of wood, while a farmer who sends ten more cows to market every year than he produces will run out of livestock. There is a simple omission in the complex political calculus of quota setting: You can't cheat nature out of more than it produces. Europe is running out of fish in large part because politicians have chosen to ignore what scientists are telling them they need to do to get the most from the sea. In serving the short-term interests of the fishing industry or their own narrow electoral goals, our political leaders are condemning the industry to death in the medium-to-long term.

The same broken reasoning and flaw in collective decision making has poisoned the management of fisheries on the high seas. Away from the coasts in international waters, catches are regulated by Regional Fisheries Management Organizations. These bodies consist of representatives from all of the nations with an interest in fishing the area in question. Instead of following scientific advice, these political appointees often pursue their perceived national interest at the expense of common sense. The car crash of bluefin tuna management is a perfect example of all that is wrong with the system. Bluefin tuna are hugely valuable (in part because they are now rare) so nations are reluctant to stop fishing them, even when scientists are virtually unanimous in saying that this is what must happen for the stock to recover. Countries are so busy competing with one another for a slice of the pie that they seem blind to the fact that every year the pie is smaller. Prince Charles summed up the folly of this behavior beautifully: "It seems one thing to destroy a species out of ignorance; but it is totally another to destroy it in the full knowledge of what we are doing."[5]

The perversity of political decision making is mirrored within the industry itself. Both are manifestations of "the tragedy of the commons." Fishermen will try to take as much as they can to maximize their own gains, even though they would all be better off in the long run if they exercised restraint. According to economists, people are more willing to look after what they own, and so the idea of "catch shares" was born.

In New Zealand, Iceland, and some of North America's fisheries, the fishing industry has been given shares in the take. For example, the New England groundfishery for haddock, flounder, and other bottom-living fish was converted to catch-share management in 2010, while the Canadian halibut fishery has been managed this way for over twenty years. Own 10 percent of the shares, and you will be able to take 10 percent of a given year's catch. Catch shares have the advantage of ending the race to fish. Prior to catch shares, the seasonal opening of Canada's halibut fishery had been shortened year after year to the point where the entire catch had to be landed in just six days. Catch shares have become very popular with some environmental organizations, who see them as a remedy for all fishing's ills, and with many of the philanthropists who support them. But I think they are oversold. I worry that giving away something that all of us own—the resources of the sea—to a few will set us up for problems in the future. I am not alone. One of their most prominent critics is Daniel Bromley, a professor of applied economics at the University of Wisconsin in the United States. In a stinging polemic against catch shares, Bromley abandoned the neutral language of academia:

> The universal policy response to this [fishery management] failure seems to consist of nothing more imaginative than the free gifting to the commercial fishing sector of permanent endowments of income and wealth under the utopian claims associated with [catch shares]. . . . This . . . can only compound the tragedies of past malfeasance by the dangerous endorsement of this bundle of confusions, contrivances, and deceits.[6]

We seem to have blind faith that new private owners will look after fisheries properly and see to their health and preservation. Their incentive is that their shares will increase in value as fish stocks recover to more healthy levels. But these catch shares often go beyond this and can be bought and sold, with the result that small family businesses have sold out to larger firms and ownership of the fisheries has become vested in fewer and larger hands. Within one year of the New England catch share scheme coming into use, the Gloucester fleet had lost 20 percent of its boats. Before catch shares 80 percent of the fishery value was landed by 68 percent of the boats; afterward it was landed by 20 percent. There is little evidence that the new owners of catch shares feel any responsibility for long-term stewardship of the seas—or indeed that catch shares offer up any wider environmental benefit.[7] Having given away public property, it will cost society dear to get it back if we change our minds. Imagine how you would feel if your government gave away all the national forests to industry and then twenty years later used your taxes to buy chunks back to turn them into nature reserves or return them to the public amenities they once were.

Seth Macinko, a colleague of Bromley's from the University of Rhode Island, pointing to how politicians usually ignore the scientific advice they are given, says, "Catch shares are seen as a solution to the problem of fisheries management, but we haven't tried management yet!"[8] In my view, some of the downsides to catch shares could be solved if they were leased to the industry for set periods and the leases were sold to the highest bidders, much as governments do with other public franchises, like running rail networks. Even then, on their own, catch shares cannot solve the problems of overfishing. They don't protect habitats from destructive fishing practices, they don't stop other wildlife from being taken as bycatch, and they won't save species like groupers or sharks that disappear from places intensively fished for more resilient animals.[9] We need enlightened and effective public management of fisheries for that, and a willingness to establish protected areas to safeguard the fragile and vulnerable.

People often ask me, "What can I do to help?" One place to start is to avoid eating fish that are overexploited in the wild or taken using methods that harm other wildlife. It is easier said than done. In the winter of 2011, I was invited to dinner at one of London's top seafood restaurants in the heart of Mayfair by some wealthy philanthropists to talk about ending overfishing. My fellow guests included government ministers, the heads of conservation groups, journalists, media tycoons, and business leaders. Needless to say, our hosts had gone to a great deal of trouble to produce a menu of tantalizing delicacy and exemplary sustainability. Then came embarrassment served with the very first course. The menu said "hand-dived Loch Fyne scallops" but the three half shells on my plate spoke of an altogether rougher journey from sea to plate. Their edges were worn and chipped from tumbling in the belly of a scallop dredge. It was nothing like the porcelain fragility of the frilled shell edge shown to me by a Scottish scallop diver who offered up a handful of freshly caught scallops. My hosts had been duped. Despite their best efforts, even a high-class restaurant with experts in seafood sourcing couldn't do the right thing. So what hope does a confused customer have contemplating the fish counter at the supermarket and trying to figure out what to get for dinner?

You may have heard of the Marine Stewardship Council, an organization dedicated to helping you choose seafood from sustainable fisheries. It has a blue and white logo in the shape of a fish with a check mark for a tail that implores you to take this packet home rather than some other dubious fish whose background cannot be trusted. Rupert Howes, the chief executive of the organization, certainly hopes you know the logo but you would be in good company if you did not recognize it. Three quarters of all shoppers quizzed in 2010 didn't know what it was. In gaining visibility for the Marine Stewardship Council, Rupert faces a dilemma: The label will make little difference to seafood buying habits if people cannot find any certified fish, but fisheries have to pass tough standards before the label is awarded. They are judged against three criteria: Is the population healthy? Does the fishery harm other wildlife? Is it well managed? From what I have described in this book, you won't be surprised to hear that not many fisheries achieve these standards. So some fisheries are being certified on a promise to get better, which makes environmentalists uneasy. As of 2011, the Marine Stewardship Council's label was carried by more than a hundred fisheries that together made up 12 percent of the world's reported catch. The organization now has visibility, and the fishing industry is falling over itself to gain accreditation after big suppliers such as Walmart, McDonald's, and the German Kaufland chain have pledged only to sell certified fish (McDonald's has already fulfilled its commitment). Holland has promised to sell only fish that carry the logo "as soon as possible." If others follow suit, it could make a big difference.

There is a price to this success. An increasing number of fisheries are being certified that cannot possibly be regarded as doing little harm to other wildlife or habitats. Ask any sponge that has been shredded by a scallop dredge, or juvenile skate carried away by a prawn trawl. Where fisheries for scallops or prawns seem reasonably healthy and well managed for the target species, it seems that the certifiers have simply shrugged their shoulders about wider environmental damage. The report for a Scottish prawn fishery acknowledged that prawns could not be caught without damaging bycatch species and the seabed but concluded that since the seabed had been damaged like this for some time, there was no harm in carrying on. Here is an example of the fish bycatch from four days of prawn trawling off the west of Scotland: to land 28,300 prawns, 12,500 fish from 38 species were caught, and all were thrown away, the vast majority of them dead.[10] The destruction of seabed life went unrecorded.

The trouble is that some of the most robust fisheries today are for species that have high population growth rates and can live in heavily degraded environments. The peculiar fact is that it is possible to have "sustainable overfishing" in which resilient species manage to hang in there after others have long departed. Scallops and prawns, for instance, have benefited from the removal of most of their predators by overfishing. Cod, angel sharks, giant skates, and a bevy of others once crunched their way through legions of scallops and prawns. What remains are fisheries of last resort. They are all we have left when everything else is gone. Scallops and prawns are the last commercially viable fisheries in Scotland's Firth of Clyde (and even then, by 2011 the prawns had become so small from overexploitation the fishermen called them "beetles"). I was left breathless with indignation when a leading environmental organization teamed up with the prawn trawl fishery there to help them get Marine Stewardship Council certification. I asked the chief executive of this organization whether he would support the destruction of rain forest if it could be shown that the palm oil that caused deforestation was produced sustainably. In his enthusiasm to find industry-led solutions to overfishing, he had forgotten that his core business was to be nature's champion.

Does that mean the Marine Stewardship Council logo cannot be trusted? While its reputation is dented, I wouldn't go that far. It is a pretty solid guide to fisheries that are well managed and where the target animal is still reasonably abundant. But that is as far as it goes. Until fisheries of last resort and those that use destructive methods like dredging stop being certified, you cannot be sure of the wider harm that is being done. There is a simple way out of this bind—the MSC needs to embrace the idea of "mitigation." In my opinion there is only one circumstance in which fisheries that use trawls or dredges should be certified and that is if managers establish extensive protected areas alongside fishing areas so that fragile and less resilient species can find refuge. I recently had an awkward conversation with Rupert Howes and Canada's Daniel Pauly, who shares my concerns about the MSC. I put this suggestion to Rupert, but he said the MSC was not in the business of recovering overfished species. Good grief! Given the depleted state of so many of the world's fisheries, it should be. The MSC needs to think beyond sustainability in a narrow fisheries sense, because it isn't good enough. In light of the many pressures the sea faces, we need to manage the sea for resilience, which means "sustainability +." What that entails is the promotion of increased abundance of life across the board, not the survival of the dregs left behind after we have plundered the rest. This shift in approach would multiply the MSC's ability to roll back the harm done by fishing many times over. Then we could eat dredge-caught scallops with a clearer conscience.

So to get back to the stumped customer standing by the fish counter, here is my advice: Try to avoid prawns or scallops and other bottom feeders fished up by dredgers and trawlers, such as plaice, cod, and hake. If you want to eat such fish, try to find them line caught: handlines and trolls (hook and line towed from a boat) have less bycatch than longlines. Eat low in the food web, so favor smaller fish like anchovies, herring, and sardines over big predators like Chilean sea bass, swordfish, and large tunas (you will be doing yourself a favor as these predators also concentrate more toxins). If you can't give up tuna, choose pole and line caught animals that have virtually zero bycatch. "Dolphin-friendly" versions alone may not be very dolphin-friendly, since tuna are often caught with purse seines, walls of net that surround and stress dolphins and snare sharks, turtles, and other wildlife. Farm-raised fish and prawns often come at a high environmental cost in destroyed habitat and wild fish turned into feed. Vegetarian fish like tilapia and carp are better than predators like salmon and sea bass. Organic is better too, since your fish will have been dosed with fewer chemicals. I offer more detailed advice in appendix 1.

Can the oceans continue to feed us? Will marine plankton survive their acid future well enough to support the rest of life at sea? Can we keep dead zones from expanding further? The exponential growth in human population, together with our ever increasing aspirations for material comfort, is taking us into the unknown. The population explosion over the last half century means that rates of change ahead will be greater than anything we have experienced up to now. The only thing we can know for certain is that problems will get worse, perhaps much worse, before they get better. Greenhouse gases already emitted commit us to unavoidable climate change that could push some species and ecosystems to the limit of their endurance, and perhaps beyond them in the case of coral reefs. If we are to avoid the more disastrous of foreseeable futures, we must use every means at our disposal to lower stress and boost the abundance and variety of life in the sea. It could just buy nature enough time for us to stabilize our own population, transition to energy sources other than fossil fuels, and to find ways of living within the limits of a finite planet.

For that we need therapies complementary to protected areas and lowered fishing intensities. Stress, as any sufferer will testify, can affect every aspect of life. Stressed people are more tired, suffer more illness, and are less able to cope with everyday life. Chronic stress can eventually lead to psychological or physical breakdown. Multiple stresses in the sea have analogous effects, triggering lower fertility, higher rates of injury, disease, and parasite infection. Stressed ecosystems are more susceptible to invasive species. Because ecosystems are a product of the interactions among the animals and plants from which they are constituted, the influence of impacts reverberates. If our underwater worlds are to cope with the vicissitudes of climate change, we will have to do everything we can to minimize other stresses. Research in Australia suggests that coral reefs could cope with a couple of degrees more global warming before throwing in the towel if the water that bathes them is clean, clear of sediment, and free of nutrient pollution.[11] We can't perform miracles overnight, but some countries have understood that it behooves them to do everything humanly possible to protect their natural resources. In 2010, in recognition of the compounding effects of multiple stresses, the Thai government closed eighteen popular dive sites when coral reefs bleached due to seawater warming.[12]

Many stresses affecting marine life have local origins so local remedies can be effective. Coral reefs illustrate graphically what is at stake, and what the options are. The Stern Report of 2007 made it clear that dealing with the problems climate change may bring could cost a great deal more than tackling greenhouse gas emissions at source.[13] You can build seawalls to hold back the ocean, but they are massively expensive, they don't work as well as natural habitats, and their costs put them beyond the reach of most of the countries that depend on coral reefs for their survival.

I recently came across a study of historic sea levels in the Maldives in which the authors found that sea levels had been twenty or twenty-four inches higher than now on a couple of occasions in the past few thousand years.[14] They concluded that "there seems no longer to be any reason to condemn the Maldives to become flooded in the near future." This is encouraging perhaps, but before Maldivians take too much comfort from the news, they need to keep in mind that today's coral reefs face a fusillade of other stresses that didn't exist in the past. The Maldives has led on the world stage in negotiations to curb greenhouse gas emissions. Their voice will have to be multiplied to be heard—we need a coalition of "wet-footed" nations from across the world who, by dint of geography, are on the front lines of global change. These negotiations have proven extraordinarily difficult and will take time to conclude, perhaps decades. But time is in short supply for coral countries. Perhaps the only thing that could buy them enough time to survive the transition to a world of reduced emissions will be to raise their standards of coral reef protection to the very highest levels. It may not work but it is the only thing that might.

Many people I talk to are pessimistic about the future, feeling that human nature cannot be changed; we will just carry on depleting and despoiling. It is easy to feel overwhelmed, but it would be wrong to give up and do nothing. Mahatma Gandhi once said, "Be the change you want to see in the world." If we all shoulder some responsibility for improving the oceans, at whatever level of influence we can muster, then change will come. Small effects multiplied millions of times make a great force.

The future is an unknown and sometimes frightening place. When I was a child the Cold War was at its height, and I grew up fearing that nuclear annihilation was just around the corner. In preparation for a postapocalyptic future, I devoured books on survival and self-sufficiency in the wild. Young people growing up today fear a slow-motion apocalypse. The creeping gears of climate change and ever greater burden of a growing population mean that future worlds will be different in ways that defy accurate prediction. Today whole societies must hone their survival skills if they are to prosper. We have to work out how to live within our means, and there isn't much time to do it.

Yet we are adaptable and resilient. Our success is testament enough. In the industrial age, so far, human resilience and adaptability have been directed toward maintaining the use or extraction of our natural capital in the face of dwindling resources. In the case of fishing this has included increased fishing intensity and power, invention of new gears and methods, and periodic switches to less favored species. What we need now is to redirect our energies toward maintaining and rebuilding our resources. Prepare for change, and you have less to fear. It is our good fortune that there are so many ways to rebuild the variety and abundance of life. We have the means but do we have the will? For if we are to refurbish our planet, we need to reinvent our relationship with the natural world.

From the perspective of society as a whole, the best policies are those that maximize long-term benefits and minimize costs. But politicians are human. They tend to serve the interests of those who most ardently court their favor, and weigh short-term interests far more heavily than long-term considerations. I constantly encounter politicians who view nature conservation as an extravagance that affluent societies can afford to indulge in a little, but no more. Perhaps this is why we so often see policies that pay lip service to nature rather than recognize it as indispensable to human welfare. Their logic is that economies compete with nature for space and nature should yield to economic interest. We have to reverse that logic. More space for nature means more space for us. Our interests overlap far more than most people recognize.

CHAPTER 21

Saving the Giants of the Sea

After a lecture in Glasgow one evening, I was collared by a grouchy Scot. Fixing me with an unfriendly stare, he spat out, "Whit are we goin tae do aboot all them gray seals then? They're eatin moor fush than we catch!" My talk had been punctuated with old pictures of monster catches and bleak graphs of fish landings over the last century, whose lines plunged like jagged alpine slopes toward zero. This man had picked out what for me is one of the few bright spots in a sea of decline: the fact that gray seal numbers around the UK have been on the increase since we stopped hunting them in 1970. So too have common seals. Seabirds like gannets and guillemots also bounced back after we stopped eating them in the late nineteenth century.

What this proves is that conservation works. If we don't kill animals they live longer and become more common. Gray whales in the eastern Pacific have made a comeback from near extinction in the 1930s to a population of twenty thousand today. Blue whales are slowly fighting back in Antarctica, the Atlantic, and the Pacific. Several species of fur seals have struggled away from the precipice of extinction in the last century, and Kemp's ridley turtles in the Gulf of Mexico have multiplied out of the risk zone. Florida's goliath groupers are mounting a comeback after fishing for them was banned in 1990. There is much joy to be had from these examples, for which the credit is widely shared.

Many ocean giants are quintessential world travelers so protecting them is complex. In the aftermath of the Japanese tsunami of March 2011, Fukushima's nuclear disaster brought some of those connections to public attention in America. Albacore tuna winter in the seas off Japan before they head for California five thousand miles away, where they support valuable fisheries. Northern bluefin tuna make the same journey but more sporadically. Suddenly Fukushima didn't seem so far away. Humpback whales migrate thousands of miles from chilly polar waters in both southern and northern hemispheres to warm water breeding grounds in places like the Caribbean and the Mozambique Channel between Africa and Madagascar. Gray whales enliven the seas all the way along the west coast of North America as they shuttle between breeding lagoons in Mexico and summer feeding grounds in the Bering Sea. White sharks and whale sharks swim thousands of miles between feeding and breeding grounds. Some albatross circle Antarctica many times in their lives. Protected areas are often dismissed as a waste of time for such nomads. The truth, as you probably already know from migrant birds, is that put in the right places, they can do a lot to help even the most peripatetic species.

Hanifaru, a coral ring on the eastern edge of Baa Atoll in the Maldives, is home to one of the great spectacles of nature. A ledge of coral traps and concentrates plankton as water pours in and out with the tide. When conditions of tide and plankton are right, dozens or even hundreds of manta rays glide in to feed. On a recent dive I was lucky enough to meet them.

Nothing prepares you for the first encounter with a manta ray. As I swam over the lagoon, the white sand bottom dropped away and the water darkened. Dense swarms of copepods coursed around me, causing my skin to tingle as they zinged against my body. I had never seen plankton this thick before. To a manta, it was an irresistible soup. Ahead of me, a large crescent shape loomed from the murk, followed by another, then another. Six mantas in all, head to tail, their broad oval mouths gaping wide. A pale flash of belly caught my eye and, turning, I saw another ray, mid-cartwheel, curling back on itself to make the most of a rich patch of food. Others appeared as I watched, mesmerized by their effortless grace. Soon twenty rays circled below, around and above me in a swirling silent vortex of planktonic gastronomy. Monolithic and impassive, they seemed oblivious to my presence, and yet not a fin brushed me though I was in their midst. Two giants headed directly for one another but at the last moment veered upward, belly to belly, before turning backward. Another seemed certain to strike me on the downstroke of a wingbeat but delicately withheld the flap until it had passed by. The dance, for that is what it felt like, continued for an hour, until the last of the plankton had washed over the reef sill on the falling tide.

Baa Atoll's mantas are lucky, because they have a champion in the person of Guy Stevens. Guy is in his thirties, tall and lanky with sun-bleached hair that is usually wet from his latest dive. He spent years getting to know these mantas and soon found that every one had a unique pattern of black spots on its white belly. Since his discovery he has recorded over the course of ten thousand observations more than two thousand different rays, half of which visit Hanifaru.

Manta rays are the largest of all stingrays, although they have lost their stings and have quit the seabed for life in open water. They are among the largest fish in the sea, with a wingspan up to twenty-three feet and weight of over two tons. For hundreds, perhaps thousands of years they have inspired awe and fear in those who have seen or imagined them. Few nineteenth-century books on ocean life were complete without a vivid description of some hapless pearl diver falling victim to a manta ray. But these were fanciful tales, as manta rays had more to fear from people than the other way around. Many mantas basking at the surface were speared and hooked for sport by adventurers. Great amusement was had in the early twentieth century by hunters chasing mantas off the Carolinas in America. Sometimes the rays outwitted their pursuers, towing their boats with effortless strength until the lines broke.

Today fishermen have the upper hand. Commercial fisheries have developed for mantas in Indonesia, the Philippines, Sri Lanka, and Mexico, where thousands are landed every year. Elsewhere mantas are threatened by entanglement in longlines, capture in purse seine nets set to surround shoals of fish, and boat strikes. Guy Stevens's efforts to protect mantas led to Hanifaru Bay being declared a marine reserve. Although Maldivians have never hunted mantas, Hanifaru was their favorite place to catch whale sharks (now also protected) that were valued for their livers and fins.

Marine reserves can be useful for highly mobile species by protecting places where they aggregate, breed, and raise their young, and in bottlenecks along their migration routes. Caribbean lemon sharks spend their first two or three years of life in coastal mangrove nurseries—without them the sharks would decline. The Mexican lagoons used by gray whales were nearly the species' downfall when whalers discovered them in the late nineteenth century, but protection in the 1970s helped boost numbers. In many places, salmon runs were virtually wiped out when nets were strung across river mouths as they returned from the sea to spawn. Close regulation of fishing on these migration highways has helped Alaska's salmon to a vibrant resurgence. Turtles haul themselves onto the same nesting beaches time and again over long lives. Judicious protection of these places can help save these animals from harm. But marine reserves offer only part of the protection that the giants of the sea need.

I recently came across a study that described efforts to reduce the bycatch from a longline fishery for mahi-mahi in Costa Rica that made my blood run cold. If you pay any attention to the sustainability of the fish you eat, you may know that mahi-mahi (also called dolphin fish or dorado) is considered a good choice. Mahi-mahi are supple-bodied gold-and-emerald fish spangled with blue that sport amid shoals of tiny fish wherever the sea is warm. They grow fast and reproduce early, so they are resilient to fishing. Retailers like to label their products with enticing words like "line caught in the clear waters of the Pacific," conjuring images of a gallant figure pulling on a rod to swing a single fish into the boat—no bycatch, clear conscience. The reality is usually very different. With the exception of pole-and-line-caught tuna, line-caught mostly means longlines. They can be tens of miles long and carry thousands of hooks. The problem is they catch far more than the target species.

The longline fishery in the study I am talking about was at Playa del Coco, and supplied mahi-mahi to America as well as more locally.[1] The collateral damage from the capture of just 211 mahi-mahi, which took fifty-four longlines with forty-three thousand hooks, was atrocious: 468 olive ridley turtles, 20 green turtles, 408 pelagic stingrays, 47 devil rays (close relatives of the manta), 413 silky sharks, 24 thresher sharks, 13 smooth hammerhead sharks, 6 crocodile sharks, 4 oceanic whitetip sharks, 68 Pacific sailfish, 34 striped marlin, 32 yellowfin tuna, 22 blue marlin, 11 wahoo, 8 swordfish, and 4 ocean sunfish. To capture enough mahi-mahi to provide one lunch for five average-sized office blocks caused carnage in Costa Rican seas. What is even more disturbing is that this longline fishery had already switched to so-called turtle-friendly circle hooks! How friendly is that, hooking nearly five hundred turtles for two hundred mahi-mahi? Circle hook points are bent inward rather than upward as in the standard J-hook, so turtles are less likely to be hooked and fewer are gashed if they swallow the hook. But they often die nonetheless, as they tangle in the lines and drown.

Turtles aside, what about the devil rays, the sharks, and the sunfish, the largest bony fish in the world about which we know almost nothing? The Playa del Coco fishery isn't just bad, it's absurd! It clearly isn't possible for a longline fishery of this kind to be sustainable according to any reasonable meaning of the word. Yet this is just one example of the reckless slaughter that pervades the open sea. How many images do we have to endure of sharks having their fins hacked off, their still living bodies thrown back into the ocean before we do something about it? Next time you sink your teeth into a delicious mahi-mahi sandwich, spare a thought for the ghosts of all those others slaughtered to catch that fish. The ocean's big animals need protection beyond the limits of protected areas. Otherwise it won't be many years before this Costa Rican fishery and others like it close shop as there will be nothing left to take.

While I'm on the subject of good seafood choices, let's return to those "dolphin-friendly" signs I mentioned that most cans of tuna sport these days: You should know they are a little misleading. It all began in the Eastern Pacific, where pods of spotted and spinner dolphins associate with schools of yellowfin tuna. Fishermen found it easier to spot dolphins than tuna, so they began to set their nets around dolphin pods, in the process drowning or mangling in their winches tens to hundreds of thousands of dolphins every year. The United States banned the import of tuna caught using this method in 1990, so the fishermen changed their ways. Many still use dolphins to find tuna, but now they haul their nets in a way that lets them escape, assisted by swimmers who jump into the water to herd them out of the back of the net. Today "only" about five thousand dolphins are killed directly by the fishery every year. Of greater concern is the fact that populations of the two species of dolphins that have been targeted are not recovering. Their reproductive rates are down, which can probably be attributed to stress from monthly near-death experiences at the hands of tuna fishers.[2] Calves may also be separated from their mothers in the confusion of fishing operations. Setting purse seine nets around tuna schools is a highly effective way of catching fish, but like long-lining it inflicts severe collateral damage.[3]

In the past, when migrant animals left their seasonal haunts we could only guess where they went and what they did. For much of the twentieth century, our guesses were little better than the long-held belief, common in previous centuries, that swallows overwintered on the bottom of ponds. Now sophisticated electronic tags allow us to travel with them, often in real time. The sea is giving up its secrets, and we can use this revolutionary new understanding of the oceans' biggest creatures to craft better ways to protect and manage them.

The most ambitious tagging program, "Tagging of Pacific Predators," is run out of Stanford, University of California–Santa Cruz, and the Pacific Fisheries Environmental Lab in California. Since

2000, this team has attached electronic tags to over two thousand animals, including seabirds, whales, sharks, turtles, sea lions, even squid.[4] Tagged leatherback turtles in the eastern Pacific follow familiar paths from their nesting beaches to seasonal feeding grounds. These routes could easily form core areas for seasonal protection from the longline fisheries that have decimated them. Despite the name, the project also tags creatures in other parts of the world. They recently combined data from tagged bluefin tuna in the Gulf of Mexico with information about oceanographic conditions they experienced over the course of their migration to build a picture of preferred bluefin habitat. The findings show how fishermen who target other ocean-going predators like swordfish or yellowfin tuna could avoid places where bluefin are most likely to be present. The study showed that bluefin particularly like a region of the northwestern Gulf that was severely affected by the Deepwater Horizon oil spill. Since bluefin go to the Gulf to breed, this is worrying news.

The four hundred northern right whales that migrate along the east coast of the United States and Canada transit some of the world's busiest shipping lanes, and about one a year is killed by a boat strike. Satellite tags and arrays of underwater hydrophones now show areas where whales concentrate. In Canada, hydrophones listen out for these whales and warn ships to slow down and be more watchful when whales are close. They need to be careful, because calculations suggest that for the northern right whale species, the loss of even two females a year could tip the balance toward extinction.[5]

Shipping is a major problem even in protected areas, as most allow the right of "innocent passage" to boats that is enshrined in the United Nations Convention on the Law of the Sea. Stellwagen Bank National Marine Sanctuary lies just above the hook of Cape Cod off the coast of New England. It was created mainly to protect the humpback whales that visit every summer to gorge on thick shoals of tiny fish and krill that flourish where cool and warm currents meet. I visited Stellwagen on a charter boat in 2001 to see the place for myself. The sea was steel gray with a low, rolling swell, but otherwise

conditions were good, and we soon spotted columns of spray from
blowing whales. There were seven humpbacks, and as we approached
I was thrilled to discover they were bubble-net feeding. In this technique the whales blow a curtain of bubbles around a shoal of fish to
ball them into a tight mass before rising through the middle to engulf
them. All at once a group of mouths burst through the surface in
cavernous liquid yawns, fish and water streaming from their barnacle-
knotted brows. These whales would long ago have learned to tolerate
whale-watching boats, but much bigger and more dangerous vessels
cut through their world, as Stellwagen straddles the shipping lane in
and out of Boston Harbor.

More than thirty-five hundred ships transit through the Stellwagen Sanctuary every year, one every two and a half hours. To lessen
the risk to whales, the shipping channels were narrowed and moved
to avoid areas of highest activity, especially of northern right whales.
The move is thought to have reduced the risk of ship strikes by 80
percent. To the north there has been a similar reorganization of shipping routes in and out of Canada. While welcome, this does little to
reduce the shipping noise that occludes whale calls and breaks down
social groups. When big ships pass through Stellwagen, the sanctuary
is so filled with noise that there is little space left for whales to converse.

I have been shocked by how rapidly populations of the seas' largest and
most majestic animals have declined in my lifetime. Leatherback and
loggerhead turtles, most species of albatross, and almost all species of
shark are down by 75 percent to more than 90 percent since the 1960s.
Perhaps more enraging is how slow the world's response has been, even
when the causes are obvious and reversible. The collapse of Atlantic
bluefin tuna has exposed an institutionalized inability to instigate
measures that will allow the species to recover. Here the will of a
conservation-minded majority of nations is subordinated to the will
of a handful bent on business as usual, regardless of the consequences.
International forums are usually constituted on an egalitarian basis,

where every nation has an equal vote and decisions are made by consensus. If unanimity cannot be achieved, the measure fails. This setup gives excessive power to minorities and allows essential measures to manage the environment to be hijacked by self-interested groups.

A perfect example of this took place at the UN General Assembly in 2006. On the table was a proposal to introduce a global moratorium on deep-sea bottom trawling. This fishing method is highly destructive to vulnerable seabed habitats, and its abolition had gained widespread support. The measure came within a whisker of being passed but was vetoed at the last minute by the Icelandic delegation. A nation of three hundred thousand people stymied the introduction of protection critical to the survival of deep sea life. I am a democrat, but that upset me. At some point we need to draw the line so that small minorities cannot torpedo the fate of the planet. The wildlife and environment of our children's generation will be under even greater pressures than they are today. Although it won't be easy or popular, reform is essential.

Bodies like the United Nations often have a consensus decision-making process that gives equal weight to the view of nations, be they large or small, strong or weak. In many of the bodies I have seen at work, such as Regional Fisheries Management Organizations, the European Union, or the United Nations itself, important conservation measures would have gained approval if decisions had been made on a majority vote rather than needing full consensus. We would have an international moratorium on deep-sea bottom trawling, and Atlantic bluefin tuna would be on the road to recovery if we shifted to a majority vote.

Of course majorities could always be gained by coercion or bribery, so decisions that weigh present greed or obstinacy over long-term responsibility would still be taken—it isn't a cure-all. Here is an example. Horrified at the scale and speed of the decline in bluefin tuna and frustrated by the deadlock in the International Commission for the Conservation of Atlantic Tunas, a coalition of conservation-minded nations and environmental organizations proposed at a meeting in Qatar in 2010 that Atlantic bluefin tuna be added to the list of the Convention on International Trade in Endangered Species (CITES for short). CITES regulates international trade in things like elephant ivory in an effort to ensure endangered species are properly protected. Until 2010 it had never dealt with fish of high commercial importance, as there was a tacit agreement that these species were best left to fishery managers, which is what pro-fishing nations argued in Qatar. It seems strange that other delegations swallowed this idea, since the only reason for bluefin to be proposed was the serial ineptitude of fishery managers. Japan, which consumes more bluefin than any other nation, lobbied hard to keep it off the list. In a move that left conservationists and diplomats steaming with indignation, their delegation served bluefin tuna at a banquet and gave away earrings and necklaces made of red coral, another threatened species considered for the list and later thrown out.

Imagine that a country served up roasted Amur leopard or delicate fripperies of Miyako kingfisher with a Philip Island hibiscus garnish at a conservation conference. It was not a proud moment for Japan, and by these selfish acts they called into question the entire worth of CITES. There comes a point at which the world must rise above national hubris and unite to save the common heritage of humanity for the benefit of our own and future generations. Many felt that CITES was the place to do that; now they are not so sure. The scientific case for protecting bluefin tuna could not have been stronger, but it was voted down. Although majority decision making would shift the odds in favor of conservation, it does not automatically lead to better environmental decisions.

I don't have a perfect answer; both consensus and majority voting have drawbacks. The latter will be anathema to some and unsettling to smaller and poorer countries. What is clear is that we must reinvent the concept of social responsibility for a crowded planet. As the world has filled to capacity, the freeboard to exercise willful self-indulgence or flagrant waste or to produce pollutants that spill into the living space of others has shrunk. The world can no longer afford to accept decision making that puts selfish self-interest and short-termism ahead of consideration for fellow inhabitants of the planet, human or other, living or yet to come. At the global level we need a Charter for Nature—a Charter for Life—akin to the UN Universal Declaration of Human Rights. Critics will no doubt be swift to condemn this as untenable in a world where human needs press ever harder. But a charter for wildlife is really a charter for human life and well-being. As Gaylord Nelson, a U.S. senator and the founder of Earth Day, once said, "The economy is a wholly owned subsidiary of the environment, not the other way around." He meant that we depend on the environment for everything. Governments rarely consider the supporting roles of marine life when they devise management strategies for the sea. That omission is beginning to cost us dearly as oceans are stripped of life, breaking down their capacity to provide things we take for granted, such as protection from storms and clean, healthy water. We see this in the rise of toxic red tides off the coast of the southern United States, fueled by polluted runoff that poison fish and people alike. We see it in the spreading dead zones and in the heaps of decayed seaweed that clog fishing nets in China and the Baltic, and the plagues of jellyfish that close resorts in the Mediterranean.

The laws governing the high seas are weak. Countries must first sign up to the UN Convention on the Law of the Sea to be bound by its regulations (the United States still hasn't signed). They also have to sign up to become members of Regional Fisheries Management Organizations to be tied by their rules. Nations that remain outside these agreements are not breaking any laws if they catch fish or pollute or infringe any other legal measures set out by these bodies, because they are not party to them. It seems mad and it is. It is as if a robber were free to break into houses and take whatever he wanted because he refused to acknowledge the law against theft. This legal loophole has spawned a trade in so-called "flags of convenience," wherein countries like Belize, Panama, and Liberia sell their flags to ships that wish to remain outside the law. Often this is to avoid restrictions on fishing, but it is also to avoid international labor laws, so that crews can be made to work under horrendous, exploitative conditions. You can buy a flag online in minutes for a few thousand dollars.

It is hard to combat activities that are not overtly illegal. One approach which is enjoying some success is for Regional Fisheries Management Organizations to "harass" the boats of nonsignatory countries out of their areas by denying them access to ports. The high seas are enormous, so this can make it uneconomic for boats to operate beyond the law. However, even if conservation measures on the high seas were made legally binding to all, governance would still be left in the hands of organizations that place exploitation far above conservation.

Are there any good models of international collaboration to inspire us? The beginnings of a nature-first approach to management can be seen in the Convention for the Conservation of Antarctic Living Marine Resources. This agreement has been signed by thirty-one countries to date. The overarching aim of the convention is to ensure that marine life remains healthy in the Southern Ocean. Only as much fishing as is compatible with this aim is allowed, and populations of penguins, albatross, and seals, among others, are monitored closely to ensure there are no adverse effects. This body has been pretty successful in kicking out longline boats that were slaughtering albatross by the thousands while fishing for Chilean sea bass. Victory in one place can mean disaster for another, however, as fleets of illegal or unregulated boats will simply move elsewhere. What we need is governance like Antarctica's to be rolled out across all of the high seas and for it to apply to everyone, whether or not they have signed up to international conventions and treaties. The world is too small and crowded to let fleets flagged by irresponsible countries sow destruction wherever they wish.

The UN Convention on the Law of the Sea was conceived and written in the 1960s and 1970s, came into effect in the 1980s, and is badly outdated now. Article 62 gives a flavor of the motivations of those who drafted it:

> The coastal State shall determine its capacity to harvest the living resources of the exclusive economic zone. Where the coastal State does not have the capacity to harvest the entire allowable catch, it shall, through agreements or other arrangements . . . give other States access to the surplus of the allowable catch.[6]

In other words, the default position is that all of the seas must be exploited. In fact, the Law of the Sea still has no provision to create protected areas despite wide recognition that they are needed, even within the United Nations. A couple of years ago I asked a legal expert close to the United Nations when we might expect a change. He rolled his eyes and groaned as if to say, "Not in my lifetime."

We don't have much time to reform our management of the oceans. Big creatures go down hard and fast and come back slowly, if at all. Most have few young and can take decades to over a century to come back. Whales often give birth to a single calf every other year and sometimes fewer. Manta rays in the Maldives appear to pup only once every five years.[7] With the megafauna, there are no quick fixes and populations can get stranded at very low abundance where they simply can't give birth fast enough to compensate for losses. But recoveries do happen. In 2010, for the first time in over a hundred years, possibly much longer, a southern right whale gave birth in the Derwent Estuary near Hobart in Tasmania. In the early nineteenth century, boats had to stay close to shore to get from one part of the bay to another because there were so many whales it was too risky to sail through their midst. By the 1830s there were nine whaling stations near Hobart, but by the 1840s their work was done. They had killed almost every right whale that used the bay. It has taken a very long time for them to come back.

Some regional bodies are moving forward to create high seas protected areas. I was lucky enough recently to work with OSPAR, a regional body that oversees environmental protection of the North

East Atlantic. In 2003, minsters of the fifteen OSPAR countries agreed that by 2010 they would fill their seas with marine protected areas that would create a network of wildlife havens. By 2008 it was clear that there was a massive high seas void in the emerging network, so they asked my team at the University of York to find places that warranted protection and to build a case for each one.[8] The outcome could not have been better. In 2010 OSPAR declared six places in the middle of the North Atlantic as marine protected areas; together they cover 111,000 square miles of high seas. For comparison, this is about the size of Nevada, and bigger than New Zealand or the UK. When I heard the news I was at a conference. There can be few moments of such excitement in a scientific career. I felt like pulling my shirt off and running around the room in a soccer player's goal celebration, but there were rather a lot of people there so I decided against it! Our science was the easy bit, though. It took extraordinary political skill and tenacity by many people to see this through. The politics was complex and often frustrating, as we saw firsthand at a succession of international meetings. We had to revisit some sites several times to alter boundaries, or bolster the case. We also had to drop two sites due to objections from some member states including, to my chagrin, the UK. There were last-minute complications when Iceland and Portugal made claims to the United Nations for huge areas of seabed to become part of their national waters. But in the end we won through.[9] Ours may be the first high-seas protected area network, but another is being built in the Southern Ocean, and island nations of the Pacific have declared no-fishing zones for tuna in pockets of international waters between their exclusive economic zones. The world turns slowly, but at last a future is being built for life at sea.[10]

The nineteenth-century biologist Ernst Haeckel spent much of his time looking at plankton through a microscope. His exquisite illustrations were celebrated and inspired the organic forms of Art Nouveau.[11] There is immense beauty to be found in a thimble of seawater, but tiny animals and plants produce different emotions from the ocean's giants. I once went diving in a kelp forest off California's Channel Islands. On the boat ride from the mainland we came upon two blue whales which captivated us for half an hour as they fed. As one sounded, it threw its tail skyward and rivulets cascaded from a fluke as wide as a two-lane highway. I felt awed and humbled to be so near the largest animal ever to have lived on this planet. Nothing about my dive in the kelp forest could match that experience. On the boat ride back, four hundred Pacific white-sided dolphins erupted from the water around us. The sea filled with leaping and twisting bodies ablaze with the warm glow of the setting sun. The megafauna worked their magic again.

The great beasts of the sea have long inspired wonder and fear. They exert a hold that few other creatures can match. More than this, though, they have come to be symbolic of the state of our seas. If whales, dolphins, sharks, and turtles disappear they will leave the sea incomplete; we lose something more important than mere flesh and blood with their passing. Slowly we have come to appreciate that megafauna are not merely embellishments, they are vital to the natural rhythms of life. Preventing these species from going extinct is not enough. Our aim should be to rebuild their numbers.

What my grumpy Scottish questioner had failed to grasp is that gray seals had only become a problem because of our failure to reverse declines in the fish that both seals and people enjoy. In his view, we should shoot the seals, a logic shared by the Scottish government and disingenuous Japanese whaling officials who argue that their hunt, banned by most other countries, is necessary to spare more fish for people to eat. In the past there was fish aplenty for wildlife and for us. That is the point we need to strive for once again. Culling the oceans' largest inhabitants because they eat fish treats the symptoms, not the disease.

CHAPTER 22

Preparing for the Worst

We are on the cusp of one of the great reorganizations of planetary life. Five times our world has been plunged into turmoil that ended with the extinction of huge swathes of life. Asteroid impact and geologic upheavals that belched forth colossal quantities of lava have been implicated in several of these mass extinctions. Many scientists, myself included, now believe we are at the beginning of a sixth mass extinction, for the extinction rates of plants and animals now run at one hundred to one thousand times the background rate seen in the calm intervals between previous geologic extinction spasms.[1] Since we are the cause this time, it is still possible to avert catastrophe.

This is a remarkable century to be alive. According to the best estimates, the human population will reach a peak in the next ninety years at somewhere between nine and eleven billion people.[2] As fertility rates fall, a run of population growth that has been virtually continuous for thousands of years will come to an end. Grave concern tempers this good news. There are important assumptions built into this prediction which we cannot take for granted. The final population figure is incredibly important to our ability to survive the apex of human population. It depends critically on quickly lowering fertility in developing countries. Demographers are concerned that people simply expect to watch human numbers rise to the peak and then gently fall afterward. But lowering fertility fast enough to stabilize the world population at nine billion will require us to ensure people in developing countries have rapid and secure access to family planning.

The world's religions must rise to this challenge too. Just as the old ways of using resources are no longer tenable, so too religious doctrines forged thousands of years ago must be reinterpreted for this century of global crisis.

Regardless of the specific level at which the human population peaks, demand for resources will be intense, and there is no assurance that the world can feed a population this large. Enormous pressure will be heaped upon the natural world as we scramble for ways to live sustainably at this population level. While the curves of predicted population growth all level off reassuringly between 2050 and 2100, the curve of population multiplied by consumption does not; it just carries on skywards in an exponentially steepening rise.

All those new people added to the planet will not want to scratch a living for a dollar a day. They will aspire to better material standards of living, just like the billion plus people who live in abject poverty today. One of the development goals established by the United Nations at the beginning of this millennium was to end poverty and hunger. If, as I said earlier, the world is already using resources equivalent to the sustainable production of one and a half planets, how much more will a world need with 30 percent to 60 percent more people, the poorest of whom will want a bigger share than is claimed by today's poor? This is what scares the pants off me.

If we just carry on with business as usual, humanity has a bleak and uncertain future. The direst predictions see us on a planet suffering from runaway global warming. Drought and inundation by rising seas will rob the world of its most fertile lands, while wildfire, flood, and tempest will cause upheaval elsewhere. The oceans and lands will hemorrhage species as the sixth great extinction accelerates, leaving us with less and less from which to meet our needs. Civil disorder will rise and conflicts among nations will escalate as people struggle to survive. Refugees will flood the luckier nations as they flee war and famine and rising seas, and might overwhelm them. Standards of living will plummet in both developing and developed nations.

There is a less dismal future ahead if we quickly wean society off fossil fuels and apply every scrap of human ingenuity to find ways to live within the limits imposed by nature and to bring population growth down fast. There are many people who hope for the latter, but they will have to fight their case hard to win against the entrenched opposition and apathetic insouciance of others who fear the inconvenience such a move might impose on their lives or businesses. It is a natural reaction not to want to change long-established habits. As I have shown in this book, our relationship with the seas has developed over more than a hundred thousand years. It is incredibly difficult to shrug off mind-sets that have such deep roots we scarcely question them; like the mantra that there are plenty more fish in the sea, or the belief that a little collateral damage in catching them won't hurt, or that the oceans are a bottomless receptacle for wastes, or can disperse and neutralize all of the toxins we release.

Every generation likes to think of itself as different from its predecessors. For much of the last thousand years this has been true as each generation has brought its creative fire to find new solutions to old problems. Over much of the course of this history we have harnessed nature to our own ends to secure comfort and prosperity. But now that relationship has begun to falter: We have surpassed nature's capacity to sustain our wants.

This change has overtaken us rapidly, so it isn't surprising we are finding it hard to adjust. My grandparents never worried about their world running out of fuel, or space to grow crops, or fertile soil, or fish. It is we who must grapple with a planetary account that is in the red. We must rebuild our inventory or humanity faces bankruptcy. At the heart of the problems is a difficult dilemma: How do developing countries meet their aspirations from a dwindling pool of resources? It is not right for developed nations to dictate that they remain poor, or that they pay a higher price to gain access to energy and infrastructure, which is one of the main reasons climate change negotiations have been fraught and to date unsuccessful. On balance we are most likely to tread a path somewhere between the best and worst extremes, but there is one certainty: Things will get worse before they get better. We must prepare for the difficult times ahead.

In the deep past there were comparable changes in ocean acidity to those we are likely to face in the next hundred years that led to widespread loss of life. Some scientists, such as the Australian coral biologist Charlie Veron, see alarming parallels between those crises and the present. With increased ocean temperatures and acidity, he says, we will not just lose one ecosystem; we will set in motion a domino effect in which we will see one ocean ecosystem fail after another. It could be a replay of what happened 55 million years ago during the runaway global warming of the Paleocene-Eocene Thermal Maximum when carbon dioxide levels, temperature, and ocean acidity spiked ferociously. The planet has recovered from previous extinction crises, but not without devastating upheavals and loss of life. As species fell others rose from the ashes of their empires and ecosystems were remade from a few of the old parts and many new. But it took millions of years.

The geologic record gives us more than enough reasons to terrify us into action. But we cannot stop the chemical transformation of our atmosphere in its tracks any more than we can halt human population growth. Both have enough latent capacity to maintain the momentum of planetary change far into the twenty-second century. The seas are likely to become more hostile to life.

Even if we get serious about cutting greenhouse gas emissions, with more and more mouths to feed the assault on the oceans could simply carry on. More fertilizer and sewage input would increase the frequency of harmful algal blooms, intensify oxygen depletion, create more dead zones, and set the stage for the jellyfish ascendancy. The spread of aquaculture will eat away at natural habitats and aggravate problems of nutrient enrichment. More intense agriculture on degrading soils will flush extra mud into coastal waters, which might help build coastal wetlands like salt marsh and mangrove but would destroy sensitive habitats constructed by invertebrates like corals. Sea level rises to which we are already committed by past emissions will lead to more seawalls and other defenses in a process of coastal hardening that will squeeze out productive habitats like mudflat and marsh. With the disappearance of these vital nurseries, wild fisheries will suffer, and there will be fewer feeding grounds for migratory birds. And if we remain wedded to all the comforts that modern technology can give us and remain as wasteful as we are today, the oceans will continue to accumulate toxic contaminants and the soup of plastic particles will thicken.

It doesn't have to be like this, but we are sleepwalking in that direction. We can change the course of the future by what we do today. There is an old adage, much loved of self-help books, that says "today is the first day of the rest of your life." If we change course by a few degrees now, it will take us to a very different place in fifty years time from where we are headed. Our children will reap the benefits. The great thing about the future is that it has yet to come so there is everything to play for. We don't have to accept our fate. In the few pages left of this book I will explore what we can do to maximize the chances that our transition to sustainable lifestyles can be accomplished without impoverishing nature and destroying ourselves. I have to assume that we will work to control human population growth, switch to clean energy, reduce emissions, and lower pollution of all varieties, because without these shifts there is no question the outcome will be bad; the only uncertainty is when will the end finally come?

Aldo Leopold, one of the fathers of America's environmental movement, said, "To keep every cog and wheel is the first rule of intelligent tinkering." It is a phrase often quoted and for good reason. Even if there are some seemingly redundant parts in the mechanism of nature, we don't know which ones they are. Nature's variety is one of its great assets when it comes to coping with change, and therefore it is one of ours. The portfolio effect I spoke of in fisheries applies much more widely. If an investor holds a diverse portfolio of assets, they can be pretty sure that it will continue to yield dividends no matter what the future brings. At any particular time some investments will do well while others languish. A species that is rare today and seems unimportant may be common tomorrow. Under different conditions, it could become the lynchpin of some vital ecological process. We have little or no capacity to predict which species that might be from the thousands around us.

In 2001 European environment ministers set themselves an ambitious goal: to halt biodiversity loss by 2010. To many people, and probably most of the politicians who set the target, this simply means preventing any more species from going extinct. Although that is important, halting biodiversity loss implies much more. It means stopping further loss of natural habitats, turning around population declines, and ending the piece by piece dismantling of food webs. Europe failed to achieve its goal, of course. Halting the collapse of biodiversity without tackling the drivers of loss is impossible. I would no more have expected a bon viveur with a cellar full of fine wine to give up drinking. Stopping biodiversity loss will take far more than bold words. At the global level, the European target was watered down to "achieve a significant reduction in the current rate of biodiversity loss." We failed even to achieve that.[3] While those targets were welcome, they must be met with a complete change in our attitude to nature. Otherwise we will never achieve them.

Some people are preparing for the worst by stockpiling frozen stores of seeds or genetic material of plants, corals, and other animals. These gene banks are a little too science fiction for me. They could work for plants, and maybe corals, but the approach has great limitations for most animals. Captive-bred animals struggle to cope when released into the wild, even after being raised by nurturing parents. Imagine how difficult it would be to resurrect a species from a tissue sample. There are dreamers out there who yearn to stitch mammoths back together from fragments of DNA deep frozen in the Siberian tundra, but these frozen libraries will do nothing to keep our planetary environment running. I think there are better ways to create arks in which species might tough it out until humanity gets to grips with the climatic convulsion we have set loose.

One approach is to identify the places where habitats and their species are most likely to survive, and then give them powerful protection. During past ice ages, the deterioration in climate forced many species from large parts of their ranges. In some places, conditions remained more clement. They formed refuges where animals and plants hung on. When the ice melted, the survivors spread forth to reclaim lands they had lost. Some people are searching for a modern equivalent of these refuges, especially for embattled habitats like tropical coral reefs. Rod Salm is one such hunter. He works for the Nature Conservancy and heads up its coral program. For the last decade Salm has been on a quest to find places where coral reefs are most likely to weather the warming and increased ocean acidity ahead. Some of the most likely places, it turns out, are ones that have few human inhabitants. George W. Bush protected several of them in the Pacific Ocean in an uncharacteristic frenzy of environmentalism at the tail end of his presidency. Rose Atoll, Palmyra and Kingman Reefs, the Northwest Hawaiian Islands—all these places support more vibrant and intact reefs than almost any that fringe inhabited lands.

The UK created the largest marine reserve in the world in 2010 when it placed the Chagos Archipelago off limits to all fishing. This British Territory rises like a vision of paradise from the middle of the Indian Ocean with its white sand beaches, swaying palms, and vivid turquoise lagoons. It seems an ideal refuge from which reefs around the Indian Ocean rim could be replenished with life in some more agreeable future. Aside from a U.S. military base—actually because of it—the Chagos are uninhabited.[4] Its reefs were badly hit by the 1998 global mass coral bleaching event which killed nine tenths of its coral. But they have since bounced back better than any others in the Indian Ocean.[5] The near absence of any other stresses like pollution, construction or significant exploitation enabled the corals to cope. The Chagos and places like it offer evidence that local efforts to reduce the cocktail of human stresses can make a difference.

One organization that championed the protection of the Chagos was the Alaska-based Pew Global Ocean Legacy program. Their aim is to see some of the world's most intact and iconic marine ecosystems given the highest level of protection. The goal is bold, not least because the places on their target list are vast and protection at this scale has never been attempted before. But they have notched up several successes (with the help of George W. Bush and the UK) and have set their sights on Australia's Coral Sea and New Zealand's Kermadec Islands.[6]

Blue Marine Foundation, a UK-based environmental group, secured the financing for the first five years of protection of the Chagos. Blue plans to mobilize private capital to protect extraordinary and important places at sea wherever they may be. Possibilities include South Georgia and Gough Island in the South Atlantic. Gough is a windswept speck of heath and rock that is inhabited only by albatross, penguin, and a handful of other hardy wildlife (and some monster mice that have tripled in size and taken to a diet of seabird chicks since whalers dropped them off in the nineteenth century). It was discovered by the Portuguese in 1505 and was claimed by the UK in 1938. Gough came with nearly 50,000 square miles of ocean around the island. Imagine what value it would have to wild nature if it were protected. Gough has been described as one of the most important seabird colonies on Earth, and its waters are home to numerous whales, dolphins, fish, and invertebrates.[7] Though our world is crowded, there are still many opportunities to safeguard wild places and create a lasting legacy for future generations.

Other organizations are concentrating on protecting hot spots where a rich and distinctive variety of plants and other creatures can be found, which they believe will be of disproportionately high value in extinction prevention. Conservation International, a powerful U.S.-based group, has identified hot spots on land, such as the Atlantic forests of Brazil, California (for its flowers), Madagascar, and the wonderfully named Succulent Karoo of Southern Africa. They are pouring money into these places to save species before it is too late, with considerable success. Similar places exist at sea, although they have only been mapped for a few habitats and species groups, such as coral reefs and open ocean predators, and a systematic catalog of ocean life is still years away.[8] Other organizations, like the World Wildlife Fund, are targeting more diffuse areas for conservation known as "priority" ecoregions. Their thinking is that life operates at large scales that protection efforts should encompass. Such ecoregions include the Gulf of Alaska, the North Sea, the South Kuroshio Current off Japan, and the Adriatic Sea.

Protecting remote, intact areas like the Chagos will give us a sporting chance of saving life's variety as the globe changes. I'm all for it. But nature at arm's length will do little to satisfy our aesthetic or spiritual needs. Places usually remain intact because they are hard to get to. It is lovely to know wild polar bears stalk seals on distant ice floes. It comforts me to think that there are isolated coral reefs that remain virtually as they were two hundred years ago, like Kingman Reef in the mid-Pacific where sharks patrol in almost unthinkable numbers and thick shoals of fish like burly emperors and sweetlips wait for nightfall to fan out and dig for prey. But we need places nearby too.

Nature close at hand is almost priceless; look at any city park if you doubt the truth. New York's Central Park opens like a green oasis hemmed in by cliffs of stone, concrete, and glass, a place of tranquility and beauty away from the dust and din of traffic-snarled streets. A song thrush that sings from your own rooftop is lovelier by far than the unheard call of a loon on some distant northern lake. Nature nearby is what matters when it comes to benefits like clean and healthy water and air. I saw a study by a worthy group of scientists not long ago that suggested we could enhance the protection of biodiversity if we decommissioned the least cost-effective protected areas, sold off the land and bought up a far greater area of cheaper, more effective land instead.[9] It sounds sensible and at face value the figures add up. There is just one drawback—the places deemed most likely to offer poor value for money were parks close to cities, where land values were sky high. To me those are the places where nature is highly valuable, where it is most accessible for us to enjoy. To decommission them in favor of more distant patches of habitat would amount to withdrawal of the lifeblood of human experience. I once flew into São Paolo in Brazil. On the descent we passed a near endless sprawl of high-rises, apartment buildings, and shanties. The buildings seethed up and over the sides of the low hills like mold. Hardly a breath of open space or green leaf broke a man-made vista that stretched to the horizon and beyond. It was one of the most depressing things I have seen.

In the race to save species, conservationists usually home in on the best remaining fragments of habitat. This is understandable in a world where money is tight. Such areas offer a way to maximize species protected per dollar spent. But in doing so, conservation neglects places less rich and remarkable. Sometimes this is because they are more degraded, but often it is because environment and evolution have not bestowed on them the variety gifted to other places. But society has a stake in low-diversity regions too. In fact, these places are likely to be much more susceptible to loss of important ecosystem services than the highly diverse sites seen as conservation priorities. Caribbean coral reefs possess a handful of coral species that do the heavy lifting of reef construction; West Pacific reefs have dozens. Disease epidemics in the Caribbean have knocked out two of the main reef builders and are now hacking away at the third. As a result, growth of reefs in the Caribbean at best has juddered to a halt, and in most places is in reverse as erosion takes hold. It is easy to see the results underwater. Where bright coral turrets garlanded in multi-hued fish once rose, all too often there is a low, hummocky panorama of green weed pockmarked with sponge and sea fan. The results are visible above water too, as resort beaches wash away and condos fall into the ocean. Local conservation efforts matter, even in places that would add little to the catalog of species protected.

We tend to be skeptical of government promises and targets. But I have been impressed by how great a motivator targets for marine conservation have been. Delegates at the 2002 World Summit on Sustainable Development in South Africa agreed to set aside more marine protected areas and to restore fish stocks to more productive and safer levels by 2015.[10] The latter target will not be met in full; indeed it cannot be, because some fish populations grow too slowly to rebuild in the three years left. Like Europe's plan to halt biodiversity loss, I doubt whether the people who agreed to the targets thought they could ever be achieved by the deadlines set. What these goals have done is to set the hares running, and in their wake there has been an unprecedented burst of effort to protect the oceans. Marine protected areas are proliferating. George W. Bush made a huge splash with his Pacific protected areas and single-handedly added over 30 percent to the total area under protection in the world oceans. That makes him one of the world's greatest marine conservationists! That expression might stick in some people's throats given his rather less glorious environmental record on land. Whatever your feelings about Bush, he left a wonderful legacy in the sea.

Other leaders have emerged who are shaping the global agenda. In 2005, the President of the tiny Pacific state of Palau, Tommy Remengesau, launched the "Micronesia Challenge" in which he urged other states in the sprawling Micronesian archipelago to protect 30 percent of their marine environment and 20 percent of lands by 2020. *Time* magazine lauded him as an Environmental Hero in 2007. But Remengesau was a flawed leader, embroiled in corruption and with rather thin environmental credentials since he built his own house without a permit in the midst of a mangrove forest. Nonetheless, the Micronesia Challenge has gained the support of many of the island nations of the region and inspired countries further afield, like Indonesia and the Caribbean island of Grenada. Island nations have the most to lose from mismanagement of the sea, and their example is one that nations better endowed with land should all follow.

There has been much head scratching over how big and how close protected areas need to be to function optimally.[11] Based on how far animals and plants move and disperse, the consensus is that they should typically be about six to twelve miles across and no more than twenty-five to fifty miles apart. At its most risk prone—six-mile reserves fifty miles apart—such a network would cover one ninth of the sea. At its most risk averse—twelve-mile reserves twenty-five miles apart—it would cover a third. In network building, we would do well to heed the investors' caveat that past performance is not always an indicator of future performance. In warmer seas, eggs and larvae will develop faster, which means they will spend less time swimming or drifting in the open sea before settling on a place to live and won't travel as far as they do now.[12] That means we would be wise to establish networks at the higher end of the protection range. Some nations and states have already started to act on these principles.

In 1999 California started to build a network of coastal protected areas that, when complete, will cover some 20 percent of its seas. (Like all U.S. states, California has control over waters up to three nautical miles out to sea; the rest is managed by the federal government.) The English network I mentioned earlier is slated to cover a quarter of its seas, although not all will be highly protected. Some areas will permit less damaging kinds of fishing to continue, like lobster trapping. France has declared it will protect 20 percent of its waters by 2020 and half of those will exclude all fishing. In many countries, renewable energy may add to overall protection, since wind farms often prohibit fishing to protect cabling. Given the enthusiastic embrace of wind energy, the combined effect could secure 30 percent to 40 percent of coastal waters, which is where we need to be.

Beyond the borders of these future protected areas, we need to scale back the intensity and destructiveness of fishing. This doesn't mean forgoing catches in the long run, although there may be short-term costs. Re-inflating fish stocks from their present nadir will bring increased catches within five to ten years, taken at lower cost with less fuel burned. It does, however, mean phasing out or greatly limiting the footprint of the most damaging fishing gears, like trawls and dredges, and unselective longlines. That move is long overdue. The fourteenth-century English peasants who pleaded with their king to ban the newly invented trawl that "runs so heavily and hardly over the ground when fishing that it destroys the flowers of the land below water,"[13] perfectly understood the dangers the method entailed. In the long run, those who decide to protect and treasure their natural assets are the ones who benefit most. Many of the gains that come from establishing protected areas and banning trawling are most apparent locally. Reinvigorating the sea isn't an altruistic act of self-sacrifice; it makes sense on the grounds of self-interest too.

There has been huge progress in the last fifteen years on some fronts and retreat in others. New high-level commitments to establish protected areas are encouraging and invigorating, and I see incredible energy being thrown behind implementing these initiatives. Protection is at last being extended to some places in international waters far from coasts where the foundations of a high seas reserve network are being laid. In Europe, it was finally acknowledged in 2011 that throwing away dead fish after you have caught them is madness. It is better still to let the ones you don't want live on in the sea by fishing less in the first place and using more selective fishing methods. But against these gains, the tide of human pressure continues to swell. In October 2011, as I was completing this book, we passed the seven billion mark. And in this globalized world, the time between discovery and depletion of a new type of fish or shellfish, or new fishing ground, has shrunk from decades to the space of a few years. The urgency for decisive change is greater than ever before.

The Sea Ahead

A few years ago, Inupiat hunters from the North Slope of Alaska caught a bowhead whale. When they cut up its carcass they found an iron harpoon point buried within its shoulder of a kind that had not been used for more than a hundred years. It turned out that this whale was 130 years old. Other older Bowheads have been caught with stone harpoon points in them which indicate that these animals can live 200 years. I wonder what a 200-year-old whale would make of the changes it has experienced in its lifetime.

In the early nineteenth century, there were perhaps as many as a hundred thousand bowheads around the Bering Strait that separates North America from Asia. The sea would have resounded to their calls as they went about the noisy business of life. For more than half the year, much of their world was frozen. Then in 1848, a whaling boat penetrated these waters and slaughtered them like sheep. The bowhead is a docile creature and killing was easy. A year later, their southern grounds exhausted, more than one hundred and fifty other whaling vessels did the same. In the space of a couple of decades, bowhead numbers were slashed. For every hundred whales before, only one or two were left and the whale chorus fell silent. A hundred years on, those seas began to fill with the noise of ships and then in 1984 the world called off its commercial hunt. By then their seas were pierced with a new sound, the deafening thump of seismic exploration for oil and gas. Sea ice retreated and unfamiliar fish and plankton began to move north. Over long years, ever so slowly, the numbers of bowhead whales began to recover. Now there are over ten thousand in Alaskan waters and their lowing calls once again fill the seas, although this time in competition with the roar of ship engines.

Within the space of a single bowhead lifetime, the world of whales changed forever. We are not quite the Methuselahs they are, but the oldest among us have experienced transformations of similar magnitude. Our world is changing faster than at any time in human history. While our relationship with the oceans has hit a low, there is still time for us to change course. The oceans and life within them will endure, albeit in forms that may be unfamiliar today. The largest hurdle on the road ahead is perhaps not amenable to any technical fix: our humanity, shaped by nearly two hundred thousand years of struggle to survive. It is essential for ocean life and our own that we transform ourselves from being a species that uses up its resources to one that cherishes and nurtures them.

People have a deep emotional connection to the sea. The oceans inspire, thrill, and soothe us. Some think we owe our clever brains and the success they brought to our ancestors' close link to the sea. But our relationship with the sea stretches back through time much further than this: all the way to the origins of life itself. We are creatures of the ocean.

In this book, I have made the case for a wholesale reversal of present trends of wildlife decline and environmental degradation. We are inseparable from the living fabric of this world. We depend on other creatures to fix energy for us, so in the most fundamental sense their lives make our own possible. Our present and future prosperity and happiness requires that we stop taking nature for granted, recognize what it does for us, and accord it the highest importance in human affairs. It is not enough to preserve bits and pieces of nature here and there simply to remind ourselves of what we once had. I am not suggesting that we stop fishing altogether, but it is clear that we can't afford to go on fishing the way we do. The value of the fishing industry, large as it might seem, is dwarfed by its environmental costs of lost function and resilience in marine ecosystems. Many of us are only slowly beginning to recognize that the hurricanes and tornadoes, droughts and floods that have been ravaging the world are connected to our degradation of the environment. We are losing our natural defenses and nature unleashed is a terrible thing. We need to refurbish the oceans so they once again heave with life. Doing this will give the world more food and would reduce or eliminate many other problems of reduced wildlife abundance that now plague us. All of the countless things that the natural world does for us and we take for granted would be delivered at higher levels, with greater security at less cost. Flourishing ecosystems are more resilient to shocks and change, and better placed to cope with chronic stress. Fewer species would require expensive protection. Our quality of life will rise with the abundance of wildlife and the seas can continue to inspire us as well as provide.

The changes I have chronicled in this book are those that most affect the oceans today. But new challenges continue to emerge. The economics of growing resource scarcity will soon make it worthwhile to mine metals from the deep sea. In many places the seabed thousands of feet below the reach of light is studded with fist-size nodules rich in scarce metals, a tempting prize for those who can devise a way to bring them to the surface. Already oil companies have pushed their wells into water more than a mile deep. These activities bring with them new impacts and risks to marine life, as the horrific Deepwater Horizon blowout showed.

The pressure is on for aquaculture to use the latest genetic technologies to boost production from the sea. But we must be careful because our engineered hybrids will escape into the wild and once they have done so will be almost impossible to control. Melting ice sheets may soon open up the poles for use and exploitation, just as acidification begins to rip a hole in the polar carbonate system that could transform life there. And just as we sort out how to neutralize our toxic contaminants, we will continue to invent new problems for ourselves, such as microplastic particles in facial scrubs and pharmaceutical pollution. Fixing ocean problems is not a one-time effort. It will be a continuous process of revision and adjustment.

Some of the backstory in this book is disheartening, and the picture gets worse looking forward if we blithely continue on our present course. But I am hugely encouraged by efforts in the last ten years. People have really noticed the spread of human influence across and beneath the sea, and there are countless efforts underway to redress the harm. I have never seen so much energy or commitment to tackle problems, from the humblest village to the debating halls of the United Nations. This is why I remain an optimist. We can change. We can turn around our impacts on the biosphere. We can live alongside wild nature. The alternative is self-destruction.

Seafood with a Clear Conscience

In developed countries, there has never before been such a great variety of seafood on offer. Globalization serves up the delicacies of shoreline, reef, and open sea from the farthest reaches of the planet to anybody willing to pay. Supermarket counters groan under a cornucopia of shellfish and finfish that please the eye and ravish the palate. I am often asked if it is possible to be a seafood lover and a lover of life in the sea. As an enthusiast of both, I have to answer yes, but it is a qualified yes. Buyers with a conscience must choose carefully, because as I showed in this book, for many species on offer, the price paid by the environment for their capture is far greater than the ticket price.

There are four main problems to look out for in choosing seafood:

- Is the species in trouble in the wild where the animals were caught?
- Does fishing for the species damage ocean habitats?
- Is there a large amount of unwanted bycatch taken with the target species?
- Does the fishery have a problem with discards—generally under-sized animals caught and thrown away because their market value is low?

It is difficult to give blanket advice on what to buy and what to avoid, since fisheries for the same species can have very different impacts depending on how and where the animals are caught. But I will attempt a few rules of thumb here. Avoid large, long-lived species that mature late in life, as they are easily overfished. Examples include skates, sharks, swordfish, marlin, some tunas (like bluefin and bigeye), wolffish, halibut, and sturgeon (caviar). Sharkfin soup is a big no, not just because sharks are in steep decline everywhere, but because the fins are often cut from living animals whose bodies are thrown overboard to suffer lingering death. Avoid all deep-sea species because they are very easily overfished and slow to recover. They tend to be long-lived, slow-growing, and mature late in life. However, added to these problems, deep-sea fishing causes immense habitat damage and has severe bycatch problems. Almost none of the bycatch survives because of extreme pressure and temperature changes experienced on being brought to the surface. Examples include orange roughy, Patagonian toothfish (also known as

Chilean seabass), oreo dories, scabbardfish (espado), grenadiers (hoki), black halibut, redfish, and deep-water prawns.

Many fishing methods have terrible impacts on habitats and other marine life
as I have said throughout this book. Bottom trawls, scallop dredges, and hydraulic
clam dredges tear up or bury fragile marine life growing on the bottom like corals,
seafans, and sponges. Gill nets hang like ghostly walls in the water and drown
thousands of marine mammals and seabirds, as well as catching many worthless
species that are thrown away. Prawn and shrimp fisheries have the worst bycatch
record, with five to ten times the amount of usable catch tossed overboard, sometimes more. Think about how much of a pile those dead and discarded animals
would make on your plate the next time you tuck into a meal of wild-caught shrimp
or prawns! Choose animals caught with minimal damage to the environment.
"Hand-picked," "diver-caught," "hook and line-caught" (although beware as this
can mean longlines; handlines are best), "pole and line," "creel," or "trap-caught"
species are usually good bets. Avoid trawls, dredges, gill nets, longlines, and drift
nets. Schooling species like herring, pilchard, and anchovy are caught using nets
but they are very clean fisheries with almost no bycatch, and are an excellent
choice on health grounds (high oil content and little in the way of toxins since they
eat low in the food chain).

"Dolphin-friendly" tuna brands are not always what they seem. As I explained
in the book, much of the tuna caught in the Eastern Pacific still involves surrounding dolphins with purse-seine nets. Even though few dolphins are killed outright,
animals are stressed and mothers get separated from their young. Pole and line
caught tuna is the best choice; look for it on the tin. The same caveat emptor goes
for fresh tuna steaks in the supermarket, most of which will have been caught with
longlines or purse seines. Many of those purse seines are now set around fish aggre-
gating devices that have been deployed for days or weeks by the catching vessel to
attract tuna. The trouble is they also gather together turtles, sharks, whales, and
dozens of other species that may also be caught and killed.

I covered the many problems of aquaculture in chapter 17 so I will only give a
brief recap here. Suffice to say that many farm-raised fish are certainly not guilt-
free, and they may come loaded with contaminant chemicals used to prevent
disease in overcrowded pens. Some fish farms are stocked with fry or young caught
from the wild, like some prawns and bluefin tuna. Aquaculture also has direct
impacts on the environment including pollution from chemicals, excess feed and
wastes produced, alien species introductions, and habitat destruction to make way
for ponds or cages, not to mention the human rights abuses associated with
unscrupulous businesses in some parts of the world. Shrimp farming has caused
vast areas of mangrove forest to be cleared along tropical coasts. You probably
wouldn't want to eat anything that has come out of particularly filthy places, like
China's Bohai Sea. However, not all aquaculture is bad. Choose fish that feed low
in the food chain, such as tilapia and carp. Avoid predators like salmon, groupers,
and tuna because they will usually have consumed far more wild-caught fish over
their lives than the weight of flesh produced. However, some (usually high-end)
suppliers do produce ethical varieties of these fish fed on things like fish trimmings

or purpose-grown worms. Choose shellfish such as mussels and oysters. Farming shellfish can do favors for coastal water quality, as these animals filter plankton and other organic matter from the water. While organic farms cause less contamination of the environment from use of chemicals, the animals still produce polluting feces and are often still fed on wild-caught fish. Organic is a better choice, however.

Knowing all this is one thing, but applying it is another matter entirely. A problem we all face at the fish counter is a dearth of key information, namely where and how was the fish caught (and sometimes what the type of fish it is in the first place as DNA testing shows a significant fraction are mislabeled, even in reputable stores). Don't be afraid to grill your fishmonger on the provenance of their fish. If this all seems too much to remember when buying fish or ordering dinner, you can download wallet-sized cards with advice on fish to buy and those to avoid, or use cell phone apps from the list below.

One of the pioneering and still best efforts to rate sustainability of different fish species is the Monterey Bay Aquarium Seafood Watch Program, based in California (www.montereybayaquarium.org/cr/seafoodwatch.aspx). It produces wallet cards which list species to avoid and good choices. A U.S. national card and six regional guides are available, several in English and Spanish. There is also a sushi guide (www.montereybayaquarium.org/cr/cr_seafoodwatch/download .aspx).

Now there is also a Monterey Bay Aquarium Seafood Watch Android app that brings you up-to-date recommendations for ocean-friendly seafood and sushi. The newest version, with Project FishMap, lets you share the locations of restaurants and markets where you've found sustainable seafood. As the map grows, you'll also be able to see what others have found near you.

Another U.S.-based organization that recommends sustainable seafood is the Blue Ocean Institute (www.blueocean.org/seafood/seafood-guide). They also produce cell phone and iPhone applications (www.blueocean.org/fishphone).

For the best seafood choices within Europe, the UK Marine Conservation Society's Fishonline (www.fishonline.org) is a great resource.

There are many other guides produced for other parts of the world in local languages. The following Web site provides links to guides from the Netherlands, Belgium, Sweden, Finland, Germany, Switzerland, Canada, Australia, Indonesia, Hong Kong, and South Africa: overfishing.org/pages/guide_to_good_fish.php.

Finally, Fish2Fork (www.fish2fork.com) is a campaigning guide for those who want to eat fish in restaurants sustainably; it was founded by Charles Clover, who wrote *The End of the Line*. Currently it rates restaurants in the UK, United States, Spain, France, and Belgium.

A Word on Eco-labelling

Some fish are provided with eco-labels that are designed to alert consumers to sustainable fish. The most recognizable, and so far only independent label is that of the Marine Stewardship Council (www.msc.org). As I described in the book, fish given the MSC label are supposed to come from fisheries that are exploited in accord with the following principles:

> Principle 1: Is the stock at levels considered to be sustainable over the long term, and is it well managed and monitored?
>
> Principle 2: Does the fishery damage habitats or kill other species, including marine mammals and sea birds?
>
> Principle 3: Are good management procedures in place to regulate fishing and ensure rules are followed?

Although the MSC has been criticized (by myself included), especially with respect to not fully implementing Principle 2, it is still a good guide to the sustainability of the labeled fish itself, and is almost always a better choice than fish without the MSC label.

Finally, Is Seafood Safe?

Fortunately, you can avoid the worst of the problems from contaminant chemicals in seafood the same way you can avoid some of the worst overfishing problems: avoid big predators like tuna, swordfish, and marlin, and eat low in the foodweb by choosing species like herring, anchovy, pilchards, tilapia, and mussels. If you are worried about how to balance the health benefits of fish with the contaminant dark side of seafood, there is an outstanding source of information: the SeaWeb KidSafe Seafood program (www.kidsafeseafood.org). Their Web site contains a host of information on the best choices to avoid contamination and promote sustainable fishing, as well as lots of background on the pollutants involved.

Conservation Charities Working to
Protect Ocean Life

If, having read my book, you would like to support efforts to safeguard life in the sea, or just want to find out more, I can recommend some excellent places to begin. The following is a list of organizations that I have worked with over the years and can personally recommend for their dedication, innovation, and effectiveness. Each has its own approach and priorities. Hopefully, you will find among them a match to your own interests.

Blue Marine Foundation (www.bluemarinefoundation.com)
Blue Marine Foundation grew out of the success of the 2009 movie *The End of the Line*. It was founded by George Duffield and Chris Gorrell Barnes, who produced the film, and is chaired by Charles Clover, who wrote the book of the same name. Blue notched up its first major success in 2010, raising funds to finance management of the newly created Chagos Marine Reserve in the Indian Ocean. They believe that there are many businesses and individuals who would willingly support greater protection for the oceans if only they could find a simple and reliable way to donate to worthy causes. Blue is dedicated to finding excellent projects in need of support, and raising the funds to help them. I have been on the board of directors of Blue since it was founded in 2010.

Client Earth (www.clientearth.org)
Client Earth is a group of idealistic lawyers (they do exist!) who have committed themselves to fighting environmental abuses with the law. Their starting point is in human rights: the right that everyone has to live in a healthy environment. But as they point out, that right is in its legal infancy and is still fragile. As their name suggests, they consider the Earth to be their client and their brief is to defend it. Sometimes that means the creative application of rather arcane legislation, but often it just means ensuring that good laws are used to their full force. And where no suitable laws exist, their aim is to nurture them into being. In the marine realm, Client Earth is working to ensure Europe's Common Fisheries Policy is reformed so that overfishing is ended, to halt the reckless slide to oblivion for bluefin tuna, and to challenge supermarkets over shaky environmental claims made on fish packaging, among other initiatives.

Conservation International (www.conservation.org)

Conservation International (CI) was founded in the United States over twenty years ago around the notion that rather than protect a smattering of places as "relics of the past," we should move toward a future with a healthy environment as the green beating heart of a prosperous society. As such, it has become a champion for indigenous communities as well as for the wildlife and habitats their homelands support. Like FFI, the emphasis was initially land-based, but CI soon developed an active marine program. Like its terrestrial counterpart, CI's efforts in the sea have sought protection from some of the richest habitats on the planet, especially coral reefs like those of southeast Asia's Coral Triangle, the heartland of marine biodiversity. It also coordinates regional initiatives in places like the eastern Pacific.

Fauna and Flora International (www.fauna-flora.org)

Fauna and Flora international (FFI) was founded in the UK in 1903. Early on they focused on protection of big game in Africa where hunting was decimating iconic animals like lions, cheetahs, and leopards. FFI has since expanded worldwide and today has satellite organizations in the United States, Australia, and Singapore that guide extensive regional programs. With over a century of practical conservation behind it, FFI is one of the most experienced organizations around. However, it was only in 2010 that they founded a full-fledged marine program, in recognition of the growing peril faced by life at sea, supported by generous donations from Lisbet Rausing, Peter Baldwin, and Arcadia. The shape of this program is still evolving, but its core emphasis is to support the creation and effective management of marine protected areas. As a council member of FFI since 2007, I have been lucky enough to help FFI launch its efforts seaward.

Greenpeace (www.greenpeace.org)

Greenpeace has its origins in the environmental movement of the late 1960s and from the beginning campaigned on ocean issues from its vessel *Rainbow Warrior*. In recent years it has been a passionate advocate for high seas and deep sea protection in the global commons beyond the limits of national waters. My research group did a report for Greenpeace a few years ago in which we asked what a network of high seas marine protected areas might look like if we were to create it today. It was a flag-waving exercise to highlight the almost complete lack of protection these areas had. I am glad to say that through Greenpeace's efforts, some areas we proposed have now been placed off limits to tuna fishing in the Pacific, and OSPAR has further developed suggestions for North Atlantic MPAs and implemented them in 2010. I always enjoy working with Greenpeace—it has a refreshing idealism coupled with steely determination.

Marine Conservation Institute (www.marine-conservation.org)

The Marine Conservation Institute is a highly respected outfit that is headquartered in the United States but whose reach extends far beyond U.S. borders. In no small part that is due to the enormous energy and boundless vision of its founding president Elliott Norse. Elliott is a master strategist and a tireless campaigner. And although too modest to admit it widely, he and his team have notched up a string of successes, notably in campaigns to establish huge marine protected areas during the Clinton and Bush (George W.) presidencies. The Marine Conservation Institute has dedicated enormous effort to seabed protection from bottom trawling, especially in the deep sea. They were a core partner of the Deep Sea Conservation Coalition that gained UN support for deep sea protection in 2008. Like SeaWeb, MCI is committed to generating and using excellent science in the cause of greater protection for marine life.

The Nature Conservancy (www.nature.org)

The Nature Conservancy was founded in the United States in the 1950s but has since spread its wings internationally. For many years, it devoted most of its funds to buy land to create refuges for threatened habitats and species. The oceans are publicly owned and so this approach cannot be translated directly except in coastal areas where they have used it to great effect to protect wetlands and shores. Farther seaward, the conservancy has been working to develop ways to promote resilience in natural habitats so they can better deal with the slew of stresses I have described in this book. Like Conservation International, it has an especially active program of reef conservation in the Coral Triangle.

The Ocean Conservancy (www.oceanconservancy.org)

The Ocean Conservancy is based in the United States and has historically concentrated on protecting national waters. It has dedicated itself to causes such as reducing plastics and trash that despoil oceans and coasts, and to establishing more and better Marine Protected Areas (MPAs). The Ocean Conservancy was among a small number of players that were instrumental in getting California's Marine Life Protection Act onto the statute books and using it to build a network of MPAs that will, in my view, become a source of national pride over the course of time. They are also campaigning for a fresh approach to environmental protection in the Gulf of Mexico following BP's blowout of 2010. They see opportunity in the aftermath of disaster and propose that some of the damages fund be put toward ending years of environmental neglect and degradation in the Gulf of Mexico. (Remember those seas of trash, I mean artificial reefs, I mentioned!)

Oceana (na.oceana.org)

Oceana claims to be the world's biggest organization dedicated wholly to ocean conservation. It has over half a million members and is active all over the world, especially the United States and Europe. Oceana runs campaigns across a wide spectrum of ocean problems, from overfishing to ocean acidification to mercury pollution. Its successes include efforts to protect the oceans from bottom trawling in Alaska, ban shark finning in Chile, and ban the trade in shark fins in California (and as the saying goes, where California leads, the nation follows). Oceana is one of the few organizations to tackle the pernicious issue of harmful government subsidies that prop up overfishing. It has waged a long campaign to remove subsidies that is not yet won, but the rewards are potentially huge. Withdrawal of subsidies from high seas and deep sea fisheries could eliminate overnight much of the harm these industries perpetrate, since they are only economic with generous taxpayer handouts.

Pew Environment Group (www.pewenvironment.org)
The Pew Environment Group was set up with support from The Pew Charitable Trusts, which has long had an intense interest in the sea. It runs the well-known Pew Fellows Program in Marine Conservation, which supports people in mid-career who are dedicated to promoting greater protection for the oceans (I am lucky enough to be one). Under the tireless leadership of Josh Reichert, Pew has been a powerful advocate for the oceans, establishing the Pew Oceans Commission that reported on the state of the sea in the United States in 2003. Although the focus has traditionally been concentrated in America, increasingly the reach extends outward as it turns to solving problems that respect no borders. Efforts include campaigns to establish very large and highly protected areas, like the northwest Hawaiian Islands, British Indian Ocean Territory (Chagos Islands), and Australia's Coral Sea. Among many other things, they are working hard to put in place strategies to protect sharks. The Pew Environment Group has given strong support to basic scientific research that can inform and underpin conservation.

Rare (www.rareconservation.org)
Rare is an organization that works internationally from the bottom up by promoting a stronger cultural and emotional connection between people and nature. I first came across them years ago when I was studying the effectiveness of marine reserves in St. Lucia. This Caribbean island was where it all began for Rare, with a passionate campaign to save the native parrot, then on the brink of extinction. The parrot was uniquely St. Lucian and the successful campaign to save it tapped into a growing sense of national identity and pride. Since then, Rare has rolled out this approach in more than fifty countries, training local conservation fellows to apply lessons learned from these efforts to achieve rapid and lasting benefits for the environment and thereby enhance quality of life for their communities. In recent years Rare has begun vigorous efforts to increase numbers and effectiveness of marine reserves in places that desperately need them, such as the Philippines.

Sea Shepherd (www.seashepherd.org)
This is the only organization in this list that I have not worked with or experienced personally in some capacity. But for those of you who prefer your conservation to be more confrontational, Sea Shepherd can probably not be bettered. Their mission is to protect wildlife at sea, and there is not much they won't do to get between those intent on killing animals and their quarry. Their most high profile campaign has been against Japanese whaling in Antarctica, but they have also been active in many other places. Among dozens of campaigns, they have helped patrol the Galápagos

Marine Reserve against shark poachers. They have tried to protect Taiji's migrating dolphins (whose plight was brought to international attention by the movie *The Cove*) from slaughter in Japan. And for many years they have fought against Canada's harp seal hunt.

SeaWeb (www.seaweb.org)

SeaWeb's motto is "Leading voices for a healthy ocean." High-quality, accurate scientific knowledge of the state of the sea forms the bedrock for all SeaWeb's programs and campaigns. They take a unique approach to conservation, putting the latest understanding from communications science to work in their efforts to raise awareness of problems at sea, change mind-sets, and put in place measures to protect marine life. Its commitment to basic science has made SeaWeb a trusted source of information by those looking for the facts without the spin that is often imparted by those with agendas to pursue. I mentioned one of those information portals in Appendix 1—Kidsafe Seafood (www.kidsafeseafood.org)—which helps parents gain the health benefits of fish for their children while avoiding the toxic downside posed by compounds like mercury. Over the years I have benefited greatly from one of SeaWeb's services to ocean professionals: a regular e-mail update on the latest scientific research (Marine Science Review: www.seaweb.org/science/MSRnewsletters/msr_archives.php). For those who want to see some of the issues I have covered in a more graphic way, SeaWeb runs a Marine Photobank (www.marinephotobank.org) that contains thousands of incredible photos from all over the world, donated by the photographers and available for use by the media.

Rather than considering the corporate world an enemy, as some conservation bodies do, SeaWeb has sought to engage them, with considerable success. Its flagship program, Seafood Choices, has for years been building better relationships between the seafood industry and conservationists. To achieve this they have enlisted the help of celebrity chefs, restaurateurs, and captains of industry to speak out and encourage others to follow the path toward more sustainable fisheries. Their hugely popular annual "Seafood Summit" provides neutral ground at which conservation and consumption can meet and exchange ideas. SeaWeb's work has broadened over the years from a focus on campaigns to protect species threatened by overfishing, such as swordfish and sturgeon (caviar) to tackling the threats posed by pollution and climate change. As a board member of SeaWeb since 2007, I know firsthand that the SeaWeb team works incredibly hard to foster greater understanding of the importance of the sea to us.

Sylvia Earle Alliance—SEA (www.sylviaearlealliance.org)

Sylvia Earle is the most eminent advocate for ocean conservation in the world. She has had a long and starred career as scientist, oceanographer, explorer, government appointee, and author. Although now in her seventies, she never scoffs at hard work and dedicates greater energy to conservation than most people half her age. She was awarded the TED Prize in 2009, which gives recipients to the chance to make a wish. Hers was that we "use all means at your disposal—Films! Expeditions! The Web! More!—to ignite public support for a global network of marine protected areas, 'hope spots' large enough to save and restore the ocean, the blue heart of the planet." The upshot has been dramatic. Sylvia has done a huge amount to raise awareness of the perils facing life in the sea among people who have significant resources or means to do something about them. She founded SEA as part of her TED Prize and has since raised millions of dollars for marine conservation.

Wildaid (www.wildaid.org)

Wildaid focuses its efforts on reducing demand for wildlife products ("when the buying stops, the killing can too"). It is one of the few conservation groups to have made serious inroads into the Chinese center of Asian seafood consumption. They are best known for their quirky and beautifully made infomercials that urge people to change their eating habits (such as shunning sharkfin soup in China or turtle eggs in Mexico). They enlist celebrities to carry these messages to a wide audience, including Leonardo DiCaprio, Richard Branson, and Chinese basketball ace Yao Ming.

WWF (www.worldwildlife.org [United States], www.wwf.org.uk [UK], www.panda.org [international], and many other national offices)

WWF (World Wildlife Fund in the United States and World Wide Fund for Nature everywhere else) celebrated its fiftieth year of conservation effort in 2011. It is a vast organization that now claims millions of members in more than a hundred countries of the world and has projects in many more. WWF works with partners all the way from village level in places like Africa through national governments to international bodies. It has long taken the view that it must work directly with governments and businesses to promote environmental protection, drawing criticism from those who consider some of these relationships too toxic to be pursued. However, through these initiatives they have often gained privileged access to high-level meetings and forums that have enabled them to score many landmark conservation successes. I served two terms on the national council of WWF-US and have been an ambassador for WWF-UK since 2009.

Forgive me if I haven't mentioned your favorite organization here. There are literally hundreds worldwide. I have concentrated on those working internationally in the above list so my apologies to all those excellent groups working to make things better nationally, like Canada's Living Oceans Society, the Australian Conservation Foundation, the UK's Marine Conservation Society, Mexico's Communidad y Biodiversidad, all of whom I can vouch for personally, as well as countless more.

Notes

Prologue

1. McClenachan, L., "Documenting Loss of Large Trophy Fish from the Florida Keys with Historical Photographs," *Conservation Biology* 23 (2009): 636–43.

Chapter 1: Four and a Half Billion Years

1. Valley, J. W., "A Cool Early Earth?" *Scientific American* (2005): 58–65.
2. We can date the timing of the Earth's formation very precisely, to 4.567 billion years ago, based on the age of the earliest meteorites. Valley, J. W., "Early Earth," *Elements* 2 (2006): 201–4.
3. Sleep, N. H., et al., "Initiation of Clement Surface Conditions on the Earliest Earth," *Proceedings of the National Academy of Sciences* 98 (2001): 3666–72.
4. Fedonkin, M. A., et al., *The Rise of Animals: Evolution and Diversification of the Kingdom Animalia* (Baltimore: The Johns Hopkins University Press, 2007).
5. Strangely, it has been very hard to find evidence for higher and more energetic tides in the geological record. As Harvard professor Andrew Knoll put it to me, "Perhaps we don't know what to look for."
6. Wilde, S. A., et al., "Evidence from Detrital Zircons for the Existence of Continental Crust and Oceans on the Earth 4.4 Gyr Ago," *Nature* 409 (2001): 175–78.
7. Drake, M. J., "Origin of Water in the Terrestrial Planets," *Meteoritics and Planetary Science* 40 (2005): 519–27.
8. Isotopes are variants of elements that have different atomic weights from the original because they contain different numbers of neutrons. They help us distinguish the origins of ocean water, since the isotopic composition of the historical source must be similar to that of present-day oceans.
9. Hartogh, P., et al., "Ocean-like Water in the Jupiter-Family Comet 103P/Hartley 2," *Nature* 478 (2011): 218–20.
10. Schopf, T. J. M., *Paleoceanography* (Cambridge, MA: Harvard University Press, 1980).
11. Koeberl, C., "Impact Processes on the Early Earth," *Elements* 2 (2006): 211–16.
12. Charette, M. A., and W. H. F. Smith, "The Volume of Earth's Ocean," *Oceanography* 23 (2010): 112–14.
13. Hawkesworth, C. J., et al., "The Generation and Evolution of Continental Crust," *Journal of the Geological Society* (London) 167 (2010): 229–48.
14. Snelgrove, P. V. R., *Discoveries of the Census of Marine Life* (Cambridge, UK: Cambridge University Press, 2010).

15. Nutman, A. P., "Antiquity of the Oceans and Continents," *Elements* 2 (2006): 223–27.

16. Hessen, D. O., "Solar Radiation and the Evolution of Life," in E. Bjertness, ed., *Solar Radiation and Human Health* (Oslo: The Norwegian Academy of Science and Letters, 2008), pp. 123–36.

17. Battistuzzi, F. U., et al., "A Genomic Timescale of Prokaryote Evolution: Insights into the Origin of Methanogenesis, Phototrophy, and the Colonization of Land," *BMC Evolutionary Biology* 4 (2004): 44. doi:10.1186/1471-2148-4-44.

18. Some of this warming probably came about through the splitting of methane in the upper atmosphere and the conversion of ethane, another powerful greenhouse gas. Haqq-Misra, J. D., et al., "A Revised, Hazy Methane Greenhouse for the Archean Earth," *Astrobiology* 8 (2008): 1127–37.

19. This lack of physical evidence means that the study of the earliest life is highly controversial and is constantly under review as new evidence emerges.

20. Our best estimates suggest that the oceans of the deep past were tens to hundreds of times less productive than today.

21. Canfield, D. E., et al., "Early Anaerobic Metabolisms," *Philosophical Transactions of the Royal Society B* 361 (2006): 1819–36.

22. Canfield, D. E., "The Early History of Atmospheric Oxygen: Homage to Robert M. Garrels," *Annual Reviews of Earth and Planetary Science* 33 (2005): 1–36.

23. Buick, R., "When Did Oxygenic Photosynthesis Evolve?" *Philosophical Transactions of the Royal Society B* 363 (2008): 2731–43.

24. Anbar, A. D., et al., "A Whiff of Oxygen Before the Great Oxidation Event?" *Science* 317 (2007): 1903–6.

25. Kopp, R. E., et al., "The Paleoproterozoic Snowball Earth: A Climate Disaster Triggered by the Evolution of Oxygenic Photosynthesis," *Proceedings of the National Academy of Sciences* 102 (2005): 11131–36.

26. Williams, D. M., et al., "Low-Latitude Glaciation and Rapid Changes in the Earth's Obliquity Explained by Obliquity-oblateness Feedback," *Nature* 396 (1998): 453–55.

27. Kappler, A. et al., "Deposition of Banded Iron Formations by Anoxygenic Phototrophic Fe(II)-Oxidising Bacteria," *Geology* 33 (2005): 865–68.

28. Anbar, A. D., and A. H. Knoll, "Proterozoic Ocean Chemistry and Evolution: A Bioinorganic Bridge?" *Science* 297 (2002): 1137–42. Sulfide is toxic to oxygen-producing cells, so they were excluded from deeper layers.

29. The earliest oxygen users might have gained it from the breakdown of hydrogen peroxide. Dismukes, G. C., et al., "The Origin of Atmospheric Oxygen on Earth: The Innovation of Oxygenic Photosynthesis," *Proceedings of the National Academy of Sciences* 98 (2001): 2170–75.

30. Gill, B. C., et al., "Geochemical Evidence for Widespread Euxinia in the Later Cambrian Ocean," *Nature* 469 (2011): 80–83.

31. Dahl, T., et al., "Devonian Rise in Atmospheric Oxygen Correlated to the Radiations in Terrestrial Plants and Large Predatory Fish," *Proceedings of the National Academy of Sciences* 107 (2010): 17911–15; Lenton, T. M., "The Role of Land Plants, Phosphorus Weathering and Fire in the Rise and Regulation of Atmospheric Oxygen," *Global Change Biology* 7 (2001): 613–29.

32. Lambert, O., et al., "The Giant Bite of a New Raptorial Sperm Whale from the Miocene Epoch of Peru," *Nature* 466 (2010): 105–8.

33. Gill, B. C., et al., "Geochemical Evidence."

34. Benton, M. J., and R. J. Twitchett, "How to Kill (Almost) All Life: The End-Permian Extinction Event," *Trends in Ecology and Evolution* 18 (2003): 358–65.

35. Kidder, D. L., and T. R. Worsley, "Causes and Consequences of Extreme Permo-Triassic Warming to Globally Equable Climate and Relation to the Permo-Triassic Extinction and Recovery," *Palaeo* 203 (2004): 207–37.

36. Knoll, A. H., et al., "Paleophysiology and End-Permian Mass Extinction," *Earth and Planetary Science Letters* 256 (2007): 295–313.

Chapter 2: Food from the Sea

1. Landau, M., "Human Evolution As Narrative," *American Scientist* 72 (1984): 262–67.

2. Verhaegen, M., et al., "The Original Econiche of the Genus Homo: Open Plain or Waterside?" in S. I. Munoz, ed., *Ecology Research Progress* (New York: Nova Science Publishers, 2007), pp. 155–86.

3. Recently the term hominin has refined previous use of hominid. Hominin includes humans and all their ancestors, while hominid refers to all of the great apes (including us) and their ancestors.

4. Tishkoff, S. A., et al., "The Genetic Structure and History of Africans and African Americans," *Science* 324 (2009): 1035–44.

5. Marean, C. W., et al., "Early Human Use of Marine Resources and Pigment in South Africa During the Middle Pleistocene," *Nature* 449 (2007): 905–8.

6. Although Blombos was occupied for longer, it wasn't inhabited as early on as Pinnacle Point.

7. Brown, K. S., et al., "Fire as an Engineering Tool of Early Modern Humans," *Science* 325 (2009): 859–62.

8. Braun, D. R., et al., "Early Hominin Diet Included Diverse Terrestrial and Aquatic Animals 1.95 Ma in East Turkana, Kenya," *Proceedings of the National Academy of Sciences* 107 (2010): 10002–7.

9. For a long time it was thought that Neanderthals and modern humans overlapped in their habitation of Europe. But new dates for deposits seem to contradict that, pushing back the dates of the last Neanderthals to before our arrival. However, researchers from Spain suggest that their dates of thirty-two thousand years ago are robust. No doubt the debate will continue. http://news.sciencemag.org/sciencenow/2011/05/were-neandertals-and-modern-huma.html?ref=hp, accessed May 18, 2012; Stringer, C. B., et al., "Neanderthal Exploitation of Marine Mammals in Gibraltar," *Proceedings of the National Academy of Sciences* 105 (2008): 14319–24.

10. Clottes, J., and J. Courtin, *The Cave Beneath the Sea: Paleolithic Images at Cosquer* (New York: H. N. Abrams, New York, 1996).

11. Erlandson, J. M., "The Archaeology of Aquatic Adaptations: Paradigms for a New Millennium," *Journal of Archaeological Research* 9 (2001): 287–348; Erlandson, J. M., et al., "The Kelp Highway Hypothesis: Marine Ecology, the Coastal Migration Theory, and the Peopling of the Americas," *Journal of Island and Coastal Archaeology* 2 (2007): 161–74.

12. There is evidence for earlier sea journeys by other hominins. For example, *Homo erectus* made it to the island of Flores in Indonesia between eight hundred thousand and seven hundred thousand years ago, involving a couple of short sea journeys. Those journeys would never have required them to be out of sight of land, however. Similarly, Neanderthals probably crossed short stretches of sea in the Mediterranean before twenty-two thousand years ago (Erlandson 2001).

13. Hine, P., et al., "Antiquity of Stone-Walled Tidal Fish Traps on the Cape Coast, South Africa," *South African Archaeological Bulletin* 65 (2010): 35–44.

14. O'Connor, S., et al. "Pelagic Fishing at 42,000 Years before the Present and the Maritime Skills of Modern Humans," *Science* 334 : (2011): 1117–21.

15. Erlandson, J. M., et al., "Fishing up the Food Web?: 12,000 Years of Maritime Subsistence and Adaptive Adjustments on California's Channel Islands," *Pacific Science* 63 (2009): 711–24.

16. Geoff Bailey, University of York, personal communication.

17. Johannes, R. E., *Words of the Lagoon* (Berkeley: University of California Press, 1981).

18. Kvavadze, E., et al., "30,000-Year-Old Wild Flax Fibers," *Science* 325 (2009): 1359.

19. Soffer, O., "Recovering Perishable Technologies Through Use-Wear on Tools: Preliminary Evidence for Upper Palaeolithic Weaving and Net-making," *Current Anthropology* 45 (2004): 407–13.

20. Erlandson, J. M., and T. C. Rick, "Archaeology, Historical Ecology, and the Future of Ocean Ecosystems," in T. C. Rick and J. M. Erlandson, eds., *Human Impacts on Ancient Marine Ecosystems* (Berkeley: University of California Press, 2008), pp. 297–308.

21. Walters, I., "Fish Hooks: Evidence for Dual Social Systems in Southeastern Australia?" *Australian Archaeology* 27 (1988): 98–114.

22. Radcliffe, W., *Fishing from Earliest Times* (London: John Murray, 1921).

23. *Oppian's Halieuticks. Of the Nature of Fishes and Fishing of the Ancients*, in Mair, A. W., *Oppian, Colluthus and Tryphiodorus*, with an English translation (London: William Heinemann, 1928).

24. It is a testament to the popularity of Oppian's poem that fifty-eight contemporary copies survive. Bartley, A. N., *Stories from the Mountains, Stories from the Sea: The Digressions and Similes of Oppian's Halieutica and Cynegetica* (Gottingen, Germany: Vandenhoeck and Ruprecht, 2003).

25. Trakadas, A., "The Archaeological Evidence for Fish Processing in the Western Mediterranean," in T. Bekker-Nielsen, ed., *Ancient Fishing and Fish Processing in the Black Sea Region* (Copenhagen: Aarhus University Press, 2003), pp. 47–82.

26. *Oppian's Halieuticks.*

27. It might also be effective for catching large bottom dwellers like skates and rays.

28. Gosnell, M., *Ice: The Nature, the History, and the Uses of an Astonishing Substance,* (Chicago: Chicago University Press, 2007).

29. Radcliffe, *Fishing from the Earliest Times.*

30. Corcoran, T. H., "Roman Fish Sauces," *The Classical Journal* 58 (1963): 204–10.

31. Ibid.

32. Aquerreta, Y., et al., "Use of Exogenous Enzymes to Elaborate the Roman Fish Sauce 'Garum,' " *Journal of the Science of Food and Agriculture* 82 (2002): 107–12.

33. Curtis, R. I., "Source for Production and Trade of Greek and Roman Processed Fish," in Bekker-Nielsen, ed. *Ancient Fishing and Fish Processing*, pp. 31–46.

34. Ejstrud, B., "Size Matters: Estimating Trade of Wine, Oil and Fish-sauce from Amphorae in the First Century AD," in Bekker-Nielsen, ed., *Ancient Fishing and Fish Processing*, pp. 171–82.

35. Barrett, J. H., et al., "The Origins of Intensive Marine Fishing in Medieval Europe: The English Evidence," *Proceedings of the Royal Society B* 271 (2004): 2417–21; Barrett J. H. et al., "Dark Age Economics Revisited: The English Fish Bone Evidence AD 600–1600," *Antiquity* 78 (2004): 618–36.

36. Barrett, J. H., et al., "Detecting the Medieval Cod Trade: A New Method and First Results," *Journal of Archaeological Science* 35 (2008): 850–61.

37. Merwe, P. van der, ed., *Hooking, Drifting and Trawling: 500 Years of British Deep Sea Fishing* (London: National Maritime Museum, 1986).

38. Roberts, C. M., *The Unnatural History of the Sea* (Washington, DC: Island Press, 2007).

39. Duhamel du Monceau, M., *Traité Général des Pesches et Histoire des Poissons* (Paris: Saillant and Nyon, 1769).

40. Johannes, *Words of the Lagoon*.

41. Webster, G., "The Invention of the Kite," *The Kiteflyer* 98 (2004): 9–14.

Chapter 3: Fewer Fish in the Sea

1. Smith, J., *The Generall Historie of Virginia, New-England, and the Summer Isles* (London: I. D. and I. H. for Michael Sparkes, 1624; repr., Glasgow: J. MacLehose, 1907).

2. Thurstan, R. H. et al., "The Effects of 118 Years of Industrial Fishing on UK Bottom Trawl Fisheries," *Nature Communications* 1 (2010): 15; doi: 10.1038/ncomms1013.

3. We are very grateful to G .H. Engelhard for the calculations that made our analysis possible. Engelhard, G. H. "One Hundred and Twenty Years of Change in Fishing Power of English North Sea Trawlers," in A. Payne, J. Cotter, and T. Potter, eds., *Advances in Fisheries Science 50 years on from Beverton and Holt* (Oxford, UK: Blackwell Publishing, 2008), pp. 1–25.

4. Fish like menhaden were often caught in such vast numbers they were used as fertilizer, since most of the catch spoiled before it could be sold for human consumption and only a limited amount could be fed to available livestock. Brown Goode, G., *The Fisheries and Fishery Industries of the United States: Natural History of Useful Aquatic Animals* (Washington, DC: Government Printing Office, 1884), Sect. I.

5. Hoover, H., "The Food Armies of Liberty. The Winning Weapon: Food," *National Geographic* (September 1917): 187–212.

6. Collins, J. W., *Report on the Investigation of Fishing Grounds in the Gulf of Mexico* (Washington, DC: Government Printing Office, 1887).

7. Lewis Anspach, as quoted in *The Literary Gazette and Journal of Belles Lettres for the Year 1819* (London: William Pople, 1820).

8. Wallace, S., and B. Gisborne, *Basking Sharks: The Slaughter of BC's Gentle Giants* (Vancouver: New Star Books, 2006).

9. One vivid account from Ireland in 1744 describes the slaughter of a large group of porpoises: "Yesterday being a great spring tide, a vast army of porpusses came up Lough Foyle in pursuit of salmon. As they rolled by Londonderry, the sailors pursued them in their boats, and killed them all the way, drove them six miles farther up the lough, to the flats about Mount Gavelling. There a new chase began by our fishermen and country people, who stretched a net across the lough, and drove them up to the narrow passages of the Great Island, which lies a mile below this town; there they fell on them with guns, swords, hatchets, and all kinds of weapons, and made a terrible slaughter. There were killed here above one hundred and sixty, besides as many mortally wounded and carried off by the flood. Including those the men of Londonderry killed, there have at least fallen in this battle five hundred porpoises, generally weighing from 1,000 to 1,500 [pounds] weight, and very good oil. Some of them were full of young ones as big as calves; and some had from six to ten salmon in their stomachs. But we hope that since these grand devourers are destroyed, our fishing will hereafter flourish, and we are pretty well repaid by this oil for the damage they have done." From the *Post-Boy*, dated November 12, 1744, and quoted in T. de Voe, *The Market Assistant, containing a brief description of every article of human food sold in the public markets of the cities of New York, Boston, Philadelphia, and Brooklyn* (New York:

Orange, Judd and Company, 1866). The weights quoted seem too high for porpoises, suggesting they may have been dolphins.

10.   National Marine Fisheries Service, National Oceanic and Atmospheric Administration, "2010 Report to Congress. Status of U.S. Fisheries," Washington, DC, 2011; www.nmfs.noaa.gov/sfa/statusoffisheries/2010/2010_Report_to_Congress.pdf; accessed November 18, 2011.

11.   Continental shelves are the shallow areas adjacent to landmasses and are typically less than about 650 feet deep. Watling, L., and E. A. Norse, "Disturbance of the Seabed by Mobile Fishing Gear: A Comparison to Forest Clearcutting," *Conservation Biology* 12 (1998): 1180–97.

12.   Bertram, J. G., *The Harvest of the Sea* (London: John Murray, 1873).

13.   Bailey, D. M., et al., "Long-term Changes in Deep-water Fish Populations in the Northeast Atlantic: A Deeper Reaching Effect of Fisheries?" *Proceedings of the Royal Society B* (2009); doi: 10.1098/rspb.2009.0098.

14.   The study in which this figure first appeared (Myers, R. A., and B. Worm, "Rapid Worldwide Depletion of Predatory Fish Communities," *Nature* 423 [2003]: 280–83) has been hotly argued over since it came out and the figures questioned. Some critics say the wrong analysis was used and that all fish stocks will have collapsed by early in the twenty-second century rather than the mid-twenty-first. Others contend that catches are an inappropriate metric of the condition of fish stocks, given that they depend on fishing effort as well as how many fish are in the sea. In fact, the authors never suggested that all fish stocks would be gone by 2048, just that they might all have experienced collapse, defined as catches falling below 10 percent of their maximum. It is perfectly possible for some collapsed fisheries to have recovered by then, and for others to still produce fish, but in much lower quantities than they could at higher abundance. The picture has been revised in recent years (Worm, B., et al., "Rebuilding Global Fisheries," *Science* 325 [2009]: 578–85) to reflect efforts to recover fish stocks in the United States and elsewhere, and to account for differences when biomass rather than catches are used as measures of the condition of stocks (Branch, T. A., et al., "Contrasting Global Trends in Marine Fishery Status Obtained from Catches and from Stock Assessments," *Conservation Biology* [2011] doi: 10.1111/j.1523-1739.2011.01687.x), but the picture looks bleak at a global scale.

15.   Rick, T. C., and J. M. Erlandson, eds., *Human Impacts on Ancient Marine Ecosystems* (Berkeley: University of California Press, 2008).

16.   Swain, D. P., et al., "Evolutionary Response to Size-Selective Mortality in an Exploited Fish Population," *Proceedings of the Royal Society B* 274 (2007): 1015–22; Mollet, F. M., et al., "Fisheries-Induced Evolutionary Changes in Maturation Reaction Norms in North Sea Sole *Solea solea*," *Marine Ecology Progress Series* 351 (2007): 189–99.

17.   Hsieh C., et al., "Fishing Elevates Variability in the Abundance of Exploited Species," *Nature* 443 (2006): 859–62.

18.   Daniel Pauly coined the phrase fishing down the food web (Pauly, D., et al., "Fishing Down Marine Food Webs," *Science* 279 [1998]: 860–63). The phenomenon has been questioned by some scientists who suggest that what is happening is fishing through the food web (Essington, T. E., et al., "Fishing Through Marine Food Webs," *Proceedings of the National Academy of Sciences* 103 [2006]: 3171–75) and that species are simply added to the fishery over time as the big ones are still caught, or that the effect doesn't exist (Branch, T. A., et al., "The Trophic Fingerprint of Marine Fisheries," *Nature* 468 [2010]: 431–35). However you slice it, serial depletion of big fish is easy to see wherever there is a gradient of fishing pressure. Intensively exploited places lack the biggest

species and big individuals that less fished places have. Although the phenomenon is most often cited in relation to trophic level in the food web, in reality trophic level is just a correlate of what drives fishermen, which is profit. So fishers are really fishing down the value chain (Sethi, S. A., et al., "Global Fishery Development Patterns are Driven by Profit but Not Trophic Level," *Proceedings of the National Academy of Sciences* 107 [2010]: 12163–67).

19. Discovery of a new cod depot. Quoted in *Friends' Intelligencer* XVIII (Philadelphia: T. Ellwood Zell, 1890), pp. 618–19.

20. Beaufoy, H., "Third Report from the Committee Appointed to Inquire into the State of the British Fisheries, and into the Most Effectual Means for Their Improvement and Extension." Sixteenth Parliament of Great Britain, 2nd sess., (January 25, 1785–August 2, 1785), In *Reports from the Committees of the House of Commons 1715–1801, Miscellaneous Subjects Vol. 10: 1785–1801*(1803), pp. 18–189.

21. Thurstan, R. H., and C. M. Roberts, "Ecological Meltdown in the Firth of Clyde, Scotland: Two Centuries of Change in a Coastal Marine Ecosystem," *PLoS ONE* 5 (2010): e11767. doi:10.1371/journal.pone.0011767.

22. www.dunoon-observer.com/index.php/news/past-stories-covered-in-cowal-and -argyll/549-fishermens-leaders-slam-clyde-report; accessed December 27, 2011.

23. O'Leary, B., et al., "Fisheries Mismanagement," *Marine Pollution Bulletin* (2011) 62: 2642–48.

24. Fromentin J. M., "Lessons from the Past: Investigating Historical Data from Bluefin Tuna Fisheries," *Fish and Fisheries* 10 (2009): 197–216.

25. MacKenzie, B. R., et al., "Impending Collapse of Bluefin Tuna in the Northeast Atlantic and Mediterranean" *Conservation Letters* 2 (2009): 25–34.

Chapter 4: Winds and Currents

1. Lozier, M. S., "Deconstructing the Conveyor Belt," *Science* 328 (2010): 1507–11.

2. Had they been able to measure deeper, they would have found the temperature drop to just 37°F or 39°F.

3. Benjamin, Count of Rumford, "An Account of the Manner in Which Heat Is Propagated in Fluids, and Its General Consequences in the Economy of the Universe," *A Journal of Natural Philosophy, Chemistry and the Arts* 1 (London: William Nicholson, 1757).

4. Denny, M. W., *How the Ocean Works: An Introduction to Oceanography* (Princeton: Princeton University Press, 2008).

5. The total flow of world rivers is approximately 40,700 km³ per year. The Amazon accounts for 15 percent of this flow. The flow of the Atlantic Meridional Overturning Circulation, as the Atlantic arm of the Global Ocean Conveyor current is known, is about 490,000 km³ per year, or just one third of 1 percent of the volume of the sea. Willis, J. K., "Can In Situ Floats and Satellite Altimeters Detect Long-term Changes in Atlantic Ocean Overturning?" *Geophysical Research Letters* 37 (2010): art. L06602; Postel, S. L., et al., "Human Appropriation of Renewable Fresh Water," *Science* 271 (1996): 785–88.

6. Lenton T. M., et al., "Tipping Elements in the Earth's Climate System," *Proceedings of the National Academy of Science* 105 (2008): 1786–93.

7. Arrhenius won the Nobel Prize for Chemistry in 1903, not for the spectacular leap of logic he made in predicting global warming, but for his theory of the dissociation of ions in solution.

8. Planetary wobbles, called Milankovich cycles after their discoverer, are also critical to glaciations; http://en.wikipedia.org/wiki/Milankovitch_cycles; accessed May 22, 2011. Milankovitch, Milutin (1998) [1941]. *Canon of Insolation and the Ice Age Problem* (Belgrade: Zavod za Udzbenike i Nastavna Sredstva).

9. Arrhenius, S., *Worlds in the Making. The Evolution of the Universe* (London and New York: Harper & Brothers, 1908).

10. www.asi.org/adb/02/05/01/surface-temperature.html.

11. The value for 2007 was 388 ppm of CO_2. IPCC, *Climate Change 2007: Synthesis Report. Contribution of Working Groups I, II and III to the Fourth Assessment Report of the Intergovernmental Panel on Climate Change* [Core Writing Team, Pachauri, R. K., and A. Reisinger, (eds.)] (Geneva: IPCC, 2007).

12. Calculated over a time horizon of one hundred years; since methane reacts with other compounds over time, its global warming potential is high to begin with and declines as time passes. See IPCC Third Assessment Report, chapter 2: www.ipcc.ch/pdf/assessment-report/ar4/wg1/ar4-wg1-chapter2.pdf; accessed December 31, 2011.

13. There are also vast quantities of methane hydrates deep in the oceans. Past episodes of intense global warming, such as that at the great extinction at the end of the Permian period, led to their release.

14. One kg of seawater contains about 4.13 times the heat of 1 kg of air at the same temperature and pressure. A kilogram of air has a volume of 856.2 liters at a pressure of one atmosphere and 250C, compared to a volume of 1 liter for seawater under the same conditions. So the heat capacity of water is 3,536 times higher than air under these conditions.

15. Schubert, R., et al., *The Future Oceans—Warming Up, Rising High, Turning Sour* (Berlin: German Advisory Council on Global Change, 2006), p. 110.

16. Hoegh-Guldberg, O., and J. F. Bruno, "The Impact of Climate Change on the World's Marine Ecosystems," *Science* 328 (2010): 1523–28.

17. Schofield, O., et al., "How Do Polar Marine Ecosystems Respond to Rapid Climate Change?" *Science* 328 (2010): 1520–23.

18. www.timesonline.co.uk/tol/news/uk/article767459.ece; accessed September 26, 2011.

19. Kerr, R. A., "Is Battered Arctic Sea Ice Down for the Count?" *Science* 318 (2007): 33–34.

20. This notion of the Gulf Stream as the North Atlantic's winter warmer has recently been questioned (Seager, R., et al., "Is the Gulf Stream Responsible for Europe's Mild Winters?" *Quarterly Journal of the Royal Meteorological Society* 128 [2002]: 2563–86). The Rocky Mountains squeeze air into a thinner layer as it flows from the west and twists winds to the north in the process. As air flows east beyond the Rockies, the twist unwinds to the south, bending winds above the warm Sargasso Sea that surrounds Bermuda. There they pick up heat, which is carried northeast. Europe's mild climate is a result of ocean warmth, but the Gulf Stream carries little of that heat, or so the authors argue. However, this view is at odds with sea surface temperature maps built up from satellite measurements that show warm ocean waters pulled far into the northeastern Atlantic by the global ocean conveyor. Because of the much greater heat capacity of water than air, far more heat would be required to warm those waters than mere winds could transfer blowing from the tropics.

21. Rahmstorf, S., "Ocean Circulation and Climate During the Past 120,000 Years," *Nature* 419 (2002): 207–14.

22. Nesje, A., O. D. Svein, and J. Bakke, "Were Abrupt Late Glacial and Early-Holocene Climatic Changes in Northwest Europe Linked to Freshwater Outbursts to the North Atlantic and Arctic Oceans?" *The Holocene* 14 (2004): 299–310.

23. Sachs, J. P., and S. J. Lehman, "Subtropical North Atlantic Temperatures 60,000 to 30,000 Years Ago," *Science* 286 (1999): 756-59.

24. Rahmstorf, "Ocean Circulation."

25. Water at the surface veers at about 45° to the wind direction. Friction transfers wind energy from surface to deeper layers. If you take a pack of cards, place your finger in one corner, and fan them using a finger in the opposite corner, friction moves each card a little less than the one above until the cards do not move at all. In the same way, friction moves water less as you go deeper. However, a peculiarity of the Coriolis force is that as depth increases, water twists a little farther, so the angle of flow relative to the wind increases with depth, rather than lessens. The net flow, taking into account the whole water column, is at right angles to the wind direction.

26. Thermoclines are well developed and virtually permanent features of tropical seas, but they are poorly developed or absent from polar seas, where warming is less. In between they tend to be seasonal, forming in spring and breaking up in autumn when storms stir the sea.

27. Bakun, A., et al., "Greenhouse Gas, Upwelling-Favorable Winds, and the Future of Coastal Ocean Upwelling Ecosystems," *Global Change Biology* 16 (2010): 1213–28.

28. Bakun, A., and S. J. Weeks, "Greenhouse Gas Buildup, Sardines, Submarine Eruptions and the Possibility of Abrupt Degradation of Intense Marine Upwelling Ecosystems," *Ecology Letters* 7 (2004): 1015–23.

29. Ibid.

30. Bakun, A., "Global Climate Change and Intensification of Coastal Ocean Upwelling," *Science* 247 (1990): 198–201.

31. Ekau, W., et al., "Impacts of Hypoxia on the Structure and Processes in Pelagic Communities (Zooplankton, Macro-invertebrates and Fish)," *Biogeosciences* 7 (2010): 1669–99.

32. Daniel Pauly, whose breadth of understanding always amazes me, has come up with an elegant theory to explain much about why fish are built the way they are and do the things they do. He articulates the case for oxygen as a limiting constraint in Pauly, D., *Gasping Fish and Panting Squids: Oxygen, Temperature and the Growth of Water-Breathing Animals* (Hamburg, Germany: International Ecology Institute, 2010).

33. Bakun, A., "Afterword," in Pauly, D. ed. *Gasping Fish and Panting Squids*, pp. 137–40.

34. Field, J., "Jumbo Squid (*Dosidicus gigas*) Invasions in the Eastern Pacific Ocean," *CalCOFI Report* 49 (2008): 79–81.

35. Stramma, L., et al., "Expanding Oxygen-Minimum Zones in the Tropical Oceans," *Science* 320 (2008): 655–58.

36. Chan, F., et al., "Emergence of Anoxia in the California Current Large Marine Ecosystem," *Science* 319 (2008): 920.

37. www.sciencedaily.com/releases/2006/08/060812155855.htm; accessed May 22, 2011.

38. Zeidberg, L. D., and B. H. Robinson, "Invasive Range Expansion by the Humboldt Squid, *Dosidicus gigas*, in the Eastern North Pacific," *Proceedings of the National Academy of Sciences* 104 (2008): 12948–50.

Chapter 5: Life on the Move

1. de Lamarck, J.B., *Philosophie zoologique, ou exposition des considerations relatives à l'histoire naturelle des animaux. (Zoological Philosophy: An Exposition with Regard to the Natural History of Animals)*. Translated by Hugh Elliot (London: Macmillan, 1809, reprint edition 1914).

2. Roberts, C. M., et al., "Marine Biodiversity Hot Spots and Conservation Priorities for Tropical Reefs," *Science* (2002): 1280–84; Hawkins, J. P. et al., "The Threatened Status of Restricted Range Coral Reef Fish Species," *Animal Conservation* 3 (2000): 81–88.

3. Ushatinskaya, G. T., "Origin and Dispersal of the Earliest Brachiopods," *Paleontological Journal* 42 (2008): 776–91, doi 10.1134/S0031030108080029.

4. Sepkoski, J. J., "Biodiversity: Past, Present and Future," *Journal of Paleontolgy* 71 (1997): 533–39. Microbes, by contrast, seem almost immutable over vast stretches of geologic time.

5. Perry, A., et al., "Climate Change and Distribution Shifts in Marine Fishes" *Science* 308 (2005): 1912–15.

6. Brander, K., "Impacts of Climate Change on Marine Ecosystems and Fisheries," *Journal of the Marine Biological Association of India* 51 (2009): 1–13.

7. Beaugrand, G., et al., "Plankton Effect on Cod Recruitment in the North Sea," *Nature* 426 (2003): 661–64; Beaugrand, G. et al., "Reorganisation of North Atlantic Marine Copepod Biodiversity and Climate," *Science* 296 (2002): 1692–94.

8. Personal communication from John Pitchford, University of York.

9. Cheung, W. W. L., et al., "Large-Scale Redistribution of Maximum Fisheries Catch Potential in the Global Ocean under Climate Change," *Global Change Biology* 16 (2009): 24–35.

10. Sheppard, C. R. C., "Predicted Recurrences of Mass Coral Mortality in the Indian Ocean," *Nature* 425 (2003): 294–97.

11. Cheung, W. W. L., et al., "Application of Macroecological Theory to Predict Effects of Climate Change on Global Fisheries Potential," *Marine Ecology Progress Series* 365 (2008): 187–97.

12. Hamilton, L., et al., "Social Change, Ecology and Climate in Twentieth Century Greenland," *Climatic Change* 47 (2000): 193–211.

Chapter 6: Rising Tides

1. Countering this spurt in wetland creation, coastal construction and reclamation has reduced wetland area in some places.

2. With the caveat that solids like ice can take up more space than their warmer liquid phases.

3. Lambeck, K., et al., "Sea Level in Roman Time in the Central Mediterranean and Implications for Recent Change," *Earth and Planetary Science Letters* 224 (2004): 563–75.

4. Pilkey, O., and R. Young, *The Rising Sea* (Washington, DC. Island Press, 2009).

5. Nicholls, R. J., and A. Cazenave, "Sea-Level Rise and Its Impact on Coastal Zones," *Science* 328 (2010): 1517–20.

6. Jevrejeva, S., A. Grinsted, and J. C. Moore, "Anthropogenic Forcing Dominates Sea Level Rise Since 1850," *Geophysical Research Letters* 36 (2009): L20706; doi:10.1029/2009GL040216.

7. Fanos, A. M., "The Impact of Human Activities on the Erosion and Accretion of the Nile Delta Coast," *Journal of Coastal Research* 11 (1995): 821–33.

8. Bamber, J. L., et al., "Reassessment of the Potential Sea-level Rise from a Collapse of the West Antarctica Ice Sheet" (2009): doi: 10.1126/science.1169335. This study revises downward the widely reported six meter (about seven yards) rise in sea levels that melting of the West Antarctic ice shelf was expected to produce.

9. Weiss, J. L., et al., "Implications of Recent Sea Level Rise Science for Low-Elevation Areas in Coastal Cities of the Coterminous U.S.A.," *Climatic Change* 105 (2011): 635–45.

10. Blanchon, P., et al., "Rapid Sea-Level Rise and Reef Back-stepping at the Close of the Last Interglacial Highstand," *Nature* 458 (2009): 881–84.

11. Kopp, R. E., et al., "Probabilistic Assessment of Sea Level During the Last Interglacial Stage," *Nature* 462 (2009): 863–68.

12. Nicholls, R. J., et al., "Sea-Level Rise and Its Possible Impacts Given a 'Beyond 4°C World' in the Twenty-first Century," *Philosophical Transactions of the Royal Society A* 369 (2011): 161–81.

13. Jenkins, A., et al., "Observations beneath Pine Island Glacier in West Antarctica and Implications for Its Retreat," *Nature Geoscience* 3 (2010): 468–72; doi 10.1038/ngeo890.

14. Connor, S., "Vast Methane 'Plumes' Seen in Arctic Ocean as Sea Ice Retreats," *The Independent* (UK), December 13, 2011; www.independent.co.uk/news/science/vast-methane-plumes-seen-in-arctic-ocean-as-sea-ice-retreats-6276278.html.

15. UN-HABITAT, *State of the World's Cities 2008/2009–Harmonious Cities* (Nairobi, Kenya: United Nations Human Settlement Program, 2008); www.unhabitat.org/pmss/getPage.asp?page=bookView&book=2562; accessed August 18, 2011.

16. Kabat, P., et al., "Dutch Coasts in Transition," *Nature Geoscience* 2 (2009): 450–52.

17. Subsidence rates in the Maldives are around 0.15mm per year, which is very slow compared to subsidence rates in many deltas. Fürstenau, J., et al., "Submerged Reef Terraces of the Maldives, Indian Ocean," *Geo-Marine Letters* 30 (2010): 511–15.

18. www.parliament.uk/documents/post/postpn342.pdf; accessed May 22, 2011. Two other factors explain the difference: Scotland has fewer people and more hard coastal rock than England, so it has less to defend and less need for coastal defense.

19. Large dams are defined as those with walls higher than 15m. Syvitski, J. P. M., et al., "Impact of Humans on the Flux of Terrestrial Sediment to the Global Coastal Ocean," *Science* 308 (2005): 376–80.

20. Engelkemeir, R., et al., "Surface Deformation in Houston, Texas, Using GPS," *Tectonophysics* 490 (2010): 47: doi: 10.1016/j.tecto.2010.04.016.

21. Extremes will rise faster than the average if the extent of the variation around an average value is proportional to the value of the average.

22. Kerr, R. A., "Models Foresee More Intense Hurricanes in the Greenhouse," *Science* 327 (2010): 399.

23. Ibid.

24. Ericson, J. P., et al., "Effective Sea-Level Rise and Deltas: Causes of Change and Human Dimension Implications," *Global and Planetary Change* 50 (2006): 63–82.

25. Other effects of climate change, including higher temperatures and altered rainfall patterns, will also affect agricultural production.

26. Tanaka, N., et al., "Coastal Vegetation Structures and Their Function in Tsunami Protection: Experience of the Recent Indian Ocean Tsunami," *Landscape and Ecological Engineering* 3 (2007): 33–45.

27. Analyses of satellite images show that fourteen of the world's largest deltas lost 3 percent of their wetlands in the last fourteen years of the twentieth century. Coleman, J. M., O. K. Huh, and D. Braud, "Wetland Loss in World Deltas," *Journal of Coastal Research* 24 (2008): 1–14.

28. Alongi, D. M., "Present State and Future of the World's Mangrove Forests," *Environmental Conservation* 29 (2002): 331–49.

29. Waycott, M., et al. "Accelerating Loss of Sea Grasses across the Globe Threatens Coastal Ecosystems," *Proceedings of the National Academy of Sciences* 106 (2009): 12377–81.

30. Pilkey and Young, *The Rising Sea*.

31. Chatenoux, B., and P. Peduzzi, "Impacts from the 2004 Indian Ocean Tsunami: Analysing the Potential Protecting Role of Environmental Features," *Natural Hazards* 40 (2007): 289–304.

32. King, S. E., and J. N. Lester, "The Value of Salt Marsh as a Sea Defence," *Marine Pollution Bulletin* 30 (1995): 180–89.

Chapter 7: Corrosive Seas

1. W. Russell, quoted in Littell, E. *Littell's Living Age*, Vol. X (Boston: Waite, Pearce & Company, 1846).

2. Hall-Spencer, J., and E. Rauer, "Champagne Seas—Foretelling the Ocean's Future?" *Journal of Marine Education* 25 (2009): 11–12; Hall-Spencer, J. M. et al., "Volcanic Carbon Dioxide Vents Show Ecosystem Effects of Ocean Acidification," *Nature* 454 (2008): 96–99.

3. If you are familiar with the language of chemistry, the equations that represent the dissolution of carbon dioxide in the sea are:

$$CO_2 \text{ (carbon dioxide)} + H_2O \text{ (water)} \leftrightarrow H_2CO_3 \text{ (carbonic acid)}$$
$$H_2CO_3 \text{ (carbonic acid)} \leftrightarrow HCO_3^- \text{ (bicarbonate ion)} + H^+ \text{ (hydrogen ion)}$$
$$H^+ + CO_2^- \text{ (carbonate ion)} \leftrightarrow HCO_3^-$$

4. Forest clearance and wetland drainage liberate large quantities of stored carbon, most of which ends up in the atmosphere as carbon dioxide.

5. Secretariat of the Convention on Biological Diversity. *Scientific Synthesis of the Impacts of Ocean Acidification on Marine Biodiversity*, Montreal: Technical Series No. 46 (2009), 61 pages.

6. Ridgwell, A., and D. N. Schmidt, "Past Constraints on the Vulnerability of Marine Calcifiers to Massive Carbon Dioxide Release," *Nature Geoscience* (2010): doi: 10.1038/NGEO755.

7. This assumes business as usual in growth of carbon dioxide emissions.

8. Caldeira, K., and M. E. Wickett, "Anthropogenic Carbon and Ocean pH," *Nature* 425 (2003): 365.

9. De'ath, G., et al., "Declining Coral Calcification on the Great Barrier Reef," *Science* 323 (2009): 116–19.

10. Maier, C., et al., "Calcification Rates and the Effect of Ocean Acidification on Mediterranean Cold-Water Corals." *Proceedings of the Royal Society B* (2011): doi:10.1098/rspb.2011.1763.

11. The depth below which carbonate dissolves is shallower in the Atlantic than in the Pacific, because deep water in the Pacific is "older" (as you will remember, deep bottom water forms in the North and South Atlantic and flows around the world from there), so it has had more time to accumulate carbon dioxide from the breakdown of organic matter.

12. Veron, J. E. N., et al., "The Coral Reef Crisis: The Critical Importance of <350 ppm CO_2," *Marine Pollution Bulletin* 58 (2009): 1428–36. Some of my colleagues were pleased there was no agreement on emissions reduction in Copenhagen in 2010 because it left negotiating space for a tougher deal that could save coral reefs.

13. Fabricius, K. E., et al., "Losers and Winners in Coral Reefs Acclimatized to Elevated Carbon Dioxide Concentrations," *Nature Climate Change* (2011): doi: 10.1038/nclimate1122.

14. Deep waters are more acid because they contain higher levels of carbon dioxide as a result of respiration by the organisms that live there and less oxygen because there is no light for photosynthesis. Manzello, D. P., et al., "Poorly Cemented Coral Reefs of the Eastern Tropical Pacific: Possible Insights into Reef Development in a High-CO_2 World," *Proceedings of the National Academy of Sciences* 105 (2008): 10450–55.

15. Yamamoto-Kawai, M., et al.,"Aragonite Undersaturation in the Arctic Ocean: Effects of Ocean Acidification and Sea Ice Melt," *Science* 326 (2009): 1098–1100.

16. Tropical seas warmed about 9°F while those at the poles warmed up to 16°F. Deep bottom waters also warmed by 7 9°F. Zachos, J , et al , "Rapid Acidification of the Ocean During the Paleocene-Eocene Thermal Maximum," *Science* 308 (2005): 1611–15.

17. Kerr, R. A., "Ocean Acidification. Unprecedented, Unsettling," *Science* 328 (2010): 1500–1.

18. Carbon dioxide was removed by weathering of silicate rocks above sea level, as well as by the dissolution of carbonates in the sea.

19. Scheibner, C., and R. P. Speijer, "Decline of Coral Reefs During Late Paleocene to Early Eocene Global Warming" (2008); www.electronic-earth.net/3/19/2008/: doi: 10.5194/ee-3-19-2008. Charlie Veron thinks that there was a major reef coral extinction event as well, but the research has yet to be done to document it.

20. Zachos, J., et al., "Rapid Acidification of the Ocean."

21. There is a possibility that lack of oxygen in bottom waters killed these species rather than it being a direct effect of acidification. Zachos, J. C., et al., "An Early Cenozoic Perspective on Greenhouse Warming and Carbon Cycle Dynamics," *Nature* 451 (2008): 279–83.

22. Gibbs, S. J., et al., "Nannoplankton Extinction and Origination Across the Paleocene-Eocene Thermal Maximum," *Science* 314 (2006): 1770–73.

23. Although oxygen is fundamental to life, there is little chance we will run out of it quickly. Even if all photosynthesis were to cease tomorrow, it would take over a thousand years to draw down atmospheric oxygen to levels that would threaten human existence.

24. Fernandez, E., et al., "Production of Organic and Inorganic Carbon Within a Large-Scale Coccolithophore Bloom in the Northeast Atlantic Ocean," *Marine Ecology Progress Series* 97 (1993): 271–85.

25. Despite these encouraging results from lab experiments, a recent study has shown that calcification by coccolithophores in the sea is reduced by higher dissolved carbon dioxide levels. Whether this affects other aspects of their growth and oxygen production is not known. Hutchins, D. A., "Forecasting the Rain Ratio," *Nature* 476 (2011): 41–42.

26. Suttle, C. A., "Viruses in the Sea," *Nature* 437 (2005): 356–61. By weight, all the viruses in the sea are equivalent to seventy-five million adult blue whales (of which there are about ten thousand left today).

27. Shi, D., et al., "Effect of Ocean Acidification on Iron Availability to Marine Phytoplankton," *Science* 327 (2010): 676–79.

28. Dixson, D., et al., "Ocean Acidification Disrupts the Innate Ability of Fish to Detect Predator Olfactory Cues," *Ecology Letters* 13 (2010): 68–75.

29. Munday, P., et al., "Ocean Acidification Impairs Olfactory Discrimination and Hom-
     ing Ability of a Marine Fish," *Proceedings of the National Academy of Sciences* 106
     (2009): 1848–52.

30. Cigliano, M., et al., "Effects of Ocean Acidification on Invertebrate Settlement at
     Volcanic CO_2 Vents. *Marine Biology* 157 (2010): 2489–2502.

31. Kerr, "Ocean Acidification."

Chapter 8: Dead Zones and the World's Great Rivers

1.   Actually the flush toilet was a reinvention—the Romans had them, and there are
     claims in the archaeological literature for even more ancient examples.

2.   Another important source of nutrients to the sea is atmospheric deposition, especially
     of nitrogen.

3.   Beman, J. M., et al., "Agricultural Runoff Fuels Large Phytoplankton Blooms in
     Vulnerable Areas of the Ocean," *Nature* 434 (2005): 211–14.

4.   Diaz, R. J., and R. Rosenberg, "Spreading Dead Zones and Consequences for Marine
     Ecosystems," *Science* 321 (2008): 926–29.

5.   Rabalais, N. N., et al., "Sediments Tell the History of Eutrophication and Hypoxia
     in the Northern Gulf of Mexico," *Ecological Applications* 17 (2007): supp. t: S129-S143.

6.   http://CNN.com: edition.cnn.com/2008/TECH/science/08/18/dead.zone; accessed
     November 19, 2011. Dead zones are actually far from lifeless. In these places we have
     reproduced the conditions of ancient oceans and awakened the sulfur-loving micro-
     bial communities of old.

7.   Beman, "Agricultural Runoff Fuels."

8.   Fahlbusch, H., "Early Dams," *Proceedings of the Institution of Civil Engineers* 162
     (2009): 13–18; doi: 10.1680/ehh2009.162.1.13.

9.   Ibid.

10. Chen, C-T. A., "The Impact of Dams on Fisheries: Case of the Three Gorges Dam," in
     W. Steffen et al., eds., *Challenges of a Changing Earth* (Berlin: Springer, 2002), chap. 16.

11. Rabalais, "Sediments Tell the History."

12. Gwo-Ching, G., et al., "Reduction of Primary Production and Changing of Nutrient
     Ratio in the East China Sea: Effect of the Three Gorges Dam?" *Geophysical Research
     Letters* 33 (2006): doi:10.1029/2006GL025800.

13. Ryan, W. B. F., et al., "An Abrupt Drowning of the Black Sea Shelf," *Marine Geology*
     138 (1997): 119–26. A recent study suggests the flood was not as severe as this, with
     only a thirty-meter (about thirty-two yards) height difference between the Mediter-
     ranean and Black Sea lake, rather than eighty meters (about eighty-seven yards):
     Giosan, L., et al., "Was the Black Sea Catastrophically Flooded in the Early Holo-
     cene?" *Quaternary Science Reviews* 28 (2009): 1–6.

14. The Black Sea basin gets abundant freshwater inflow from rivers. This mixes with
     seawater to form a low-density, low-salinity layer, which flows out through the Bos-
     porus into the Eastern Mediterranean. Some of the outflowing water is replaced by a
     deeper return flow of high-salinity water from the Mediterranean, which sinks into
     the Black Sea basin. The two-layer nature of the Black Sea is thus maintained by
     salinity and temperature contrasts between deep and shallow water, which means that
     the surface water layer is much less dense and floats above the deep layer.

15. Savage, C., et al., "Effects of Land Use, Urbanization, and Climate Variability on
     Coastal Eutrophication in the Baltic Sea," *Limnology and Oceanography* 55 (2010):
     1033–46.

16. Conley, D. J., et al., "Long-term Changes and Impacts of Hypoxia in Danish Coastal Waters," *Ecological Applications* 17 (2007), supp.: S165–S184.

17. Jackson, J. B. C., et al., "Historical Overfishing and the Recent Collapse of Coastal Ecosystems," *Science* 293 (2001): 629–38.

18. Boesch, D. F., et al., *Coastal Dead Zones and Global Climate Change: Ramifications of Climate Change for Chesapeake Bay Hypoxia*. Pew Center on Global Climate Change and University of Maryland Center for Environmental Science (2007).

19. The causes of Florida red tides are less straightforward than a simple response to increased nutrient pollution, but this doubtless plays an important role in triggering and sustaining them. Alcock, F., *An Assessment of Florida Red Tide: Causes, Consequences and Management Strategies*. Technical Report 1190, Mote Marine Laboratory, Sarasota, FL (2007).

20. Kirkpatrick, B., et al., "Environmental Exposures to Florida Red Tides: Effects on Emergency Room Respiratory Diagnoses Admissions," *Harmful Algae* 5 (2006): 526–33; Kirkpatrick, B., et al., "Gastrointestinal Emergency Room Admissions and Florida Red Tide Blooms," *Harmful Algae* 9 (2010): 82–86. See also www.topcancernews .com/news/1797/1/Algal-toxin-commonly-inhaled-in-sea-spray-attacks-and -damages-DNA.

21. Knowler, D. J., et al., "An Open-access Model of Fisheries and Nutrient Enrichment in the Black Sea," *Marine Resource Economics* 16 (2002): 195–217.

22. Richardson, A. J., et al. "The Jellyfish Joyride: Causes, Consequences and Management Responses to a More Gelatinous Future," *Trends in Ecology and Evolution* 24 (2009): 312–22.

Chapter 9: Unwholesome Waters

1. Safina, C., "The 2010 Gulf of Mexico Oil Well Blowout; a Little Hindsight," *PLoS Biology* 9 (2011): e1001049. doi:10.1371/journal.pbio.1001049.

2. Crone, T. J., and M. Tolstoy, "Magnitude of the 2010 Gulf of Mexico Oil Leak," *Science* 330 (2010): 634.

3. This figure took me by surprise when I first saw it; the wellhead was gushing so fast at 68,000 barrels a day. But tankers are monumental these days, and it would take a long time to fill one at that rate. The rate and total amount of oil loss was estimated by experts in fluid dynamics who used footage of the leaking wellhead to measure the speed at which the oil gushed from it.

4. I am grateful to Nigel Haggan of the University of British Columbia for sharing this witticism!

5. Schrope, M., "The Lost Legacy of the Last Great Oil Spill," *Nature* 466 (2010): 305–6.

6. www.birdlife.org/news/news/2010/06/seabird-petition.html.

7. At the time of writing, October 2011, an area of 1,041 square miles around the Deepwater Horizon wellhead remained closed. http://sero.nmfs.noaa.gov/media/pdfs/2010/ Area%206%20and%207%20Press%20Release_FINAL.pdf; accessed October 29, 2011.

8. Safina, "The 2010 Gulf of Mexico Oil Well Blowout."

9. Jernelöv, A., "How to Defend Against Future Oil Spills," *Nature* 466 (2010): 182–3.

10. National Research Council, *Oil in the Sea III: Inputs, Fates, and Effects* (Washington, DC, 2003).

11. Wurl, O., and J. P. Obbard, "A Review of Pollutants in the Sea-Surface Microlayer (SML): A Unique Habitat for Marine Organisms," *Marine Pollution Bulletin* 48 (2004): 1016–30.

12. Quite a lot of pollutants are also carried long distances by wind alone to be deposited into the sea far from the sources.

13. Jernelöv, "How to Defend Against Future Oil Spills."

14. Kessler, J. D., et al., "A Persistent Oxygen Anomaly Reveals the Fate of Spilled Methane in the Deep Gulf of Mexico," *Science* 331 (2011): 312–15.

15. Wells, R.S., et al., "Integrating Life-History and Reproductive Success Data to Examine Potential Relationships with Organochlorine Compounds for Bottlenose Dolphins (*Tursiops truncatus*) in Sarasota Bay, Florida," *Science of the Total Environment* 349 (2005): 106–19.

16. Reddy, M. L., et al., "Opportunities for Using Navy Marine Mammals to Explore Associations Between Organochlorine Contaminants and Unfavourable Effects on Reproduction," *Science of the Total Environment* 274 (2001): 171–82.

17. Hoover, S. M., "Exposure to Persistent Organochlorines in Canadian Breast Milk: A Probabilistic Assessment," *Risk Analysis* 19 (1999): 527–45.

18. *Arctic Pollution 2009*. Report from the Arctic Monitoring and Assessment Programme (Oslo: AMAP, 2009), pp. xi, 83.

19. Porterfield, S. P., "Vulnerability of the Developing Brain to Thyroid Abnormalities: Environmental Insults to the Thyroid System," *Environmental Health Perspectives* 102 (1994): 125–30.

20. *Arctic Pollution,* 2009.

21. Sunderland, E. M., et al., "Mercury Sources, Distribution, and Bioavailability in the North Pacific Ocean: Insights from Data And Models," *Global Biogeochemical Cycles* 23 (2009): doi:10.1029/2008GB003425.

22. Anh-Thu, E. V., et al., "Temporal Increase in Organic Mercury in an Endangered Pelagic Seabird Assessed by Century-Old Museum Specimens," *Proceedings of the National Academy of Sciences* 108 (2011): 7466–71.

23. Frederick, P., and N. Jayasema, "Altered Pairing Behaviour and Reproductive Success in White Ibises Exposed to Environmentally Relevant Concentrations of Methylmercury," *Proceedings of the Royal Society B* (2010): doi: 10.1098/rspb.2010.2189.

24. Sunderland, E. M., "Mercury Exposure from Domestic and Imported Estuarine and Marine Fish in the U.S. Seafood Market," *Environmental Health Perspectives* 115 (2007): 235–42.

25. Lowenstein, J. H., "DNA Barcodes Reveal Species-Specific Mercury Levels in Tuna Sushi That Pose a Health Risk to Consumers," *Biology Letters* (2010): doi:10.1098/rsbl .2010.0156.

26. Tarbox, B. M., "Toxic Fish Counter" (2010); www.gotmercury.org.

27. Muir, D. C. G., and P. H. Howard, "Are There Other Persistent Organic Pollutants? A Challenge for Environmental Chemists," *Environmental Science and Technology* 40 (2006): 7157–66.

28. Law, R. J., et al., "Levels and Trends of Brominated Flame Retardants in the European Environment," *Chemosphere* 64 (2006): 187–208; Darnerud, P. O., "Toxic Effects of Brominated Flame Retardants in Man and in Wildlife," *Environment International* 29 (2003): 841–53.

29. Roze, E., et al., "Prenatal Exposure to Organohalogens, Including Brominated Flame Retardants, Influences Motor, Cognitive, and Behavioral Performance at School Age," *Environmental Health Perspectives* 117 (2009): 1953–58.

30. Arnold, K., et al., "Medicating the Environment: Impacts on Individuals and Populations," *Trends in Ecology and Evolution* (in press).

31. Heckmann, L.-H., et al., "Chronic Toxicity of Ibuprofen to *Daphnia magna:* Effects on Life History Traits and Population Dynamics," *Toxicology Letters* 172 (2007): 137–45.

32. Daigle, J. K., "Acute Responses of Freshwater and Marine Species to Ethinyl Estradiol and Fluoxetine," master's of science thesis, Louisiana State University (2010).

33. Hannah, W., and P. B. Thompson, "Nanotechnology, Risk and the Environment: A Review," *Journal of Environmental Monitoring* 10 (2008): 291–300.

34. Koehler, A., et al., "Effects of Nanoparticles in *Mytilus edulis* Gills and Hepato-pancreas—A New Threat to Marine Life?" *Marine Environmental Research* 66 (2008): 12–14.

35. Ylitalo, G. M., et al., "High Levels of Persistent Organic Pollutants Measured in Blubber of Island-Associated False Killer Whales (*Pseudorca crassidens*) around the Main Hawaiian Islands," *Marine Pollution Bulletin* 58 (2009): 1922–52.

Chapter 10: The Age of Plastic

1. Ebbesmeyer, C. C., and E. Scigliano, *Flotsametrics and the Floating World* (New York: HarperCollins, 2009).

2. Ballan, H., "Plastics and a Man Named Yarsley"; www.epsomandewellhistoryexplorer.org.uk/Yarsley.html; accessed November 3, 2011.

3. Yarsley, V. E., and E. G. Couzens, *Plastics* (Middlesex, UK: Penguin Books Ltd, 1941).

4. Thompson, R. C., et al., "Plastics, the Environment and Human Health: Current Consensus and Future Trends," *Philosophical Transactions of the Royal Society B* 364 (2009): 2153–66.

5. Yarsley, *Plastics.*

6. Hohn, Donovan, *Moby Duck* (New York: Viking, 2011).

7. Maury, M. F., *The Physical Geography of the Sea* (Edinburgh and New York: Thomas Nelson and Sons, 1883).

8. Purdy, J., *Memoir Descriptive and Explanatory, to Accompany the Charts of the Northern Atlantic Ocean* (London: R. H. Laurie, 1853).

9. Maury, *The Physical Geography of the Sea.*

10. Ebbesmeyer and Scigliano, *Flotsametrics.*

11. Moore, S. L. et al., "Composition and Distribution of Beach Debris in Orange County, California," 2001; Southern California Coastal Water Research project; ftp://www.sccwrp.org/pub/download/DOCUMENTS/AnnualReports/1999 AnnualReport/09_ar10.pdf; accessed May 20, 2011.

12. Barnes, D. K. A., "Remote Islands Reveal Rapid Rise of Southern Hemisphere Sea Debris," *The Scientific World Journal* 5 (2005): 915–21.

13. Barnes, D. K. A., et al., "Accumulation and Fragmentation of Plastic Debris in Global Environments," *Philosophical Transactions of the Royal Society B* 364 (2009): 1985–98.

14. Yamashita, R., and A. Tanimura, "Floating Plastic in the Kuroshio Current Area, Western North Pacific Ocean," *Marine Pollution Bulletin* 54 (2007): 464–88.

15. Gregory, M. R., "Environmental Implications of Plastic Debris in Marine Settings—Entanglement, Ingestion, Smothering, Hangers-on, Hitch-hiking and Alien Invasions," *Philosophical Transactions of the Royal Society B* 364 (2009): 2013–25.

16. Moore, C. J., et al., "A Comparison of Plastic and Plankton in the North Pacific Central Gyre," *Marine Pollution Bulletin* 42 (2001):1297–1300.

17. Lavender Law, K., "Plastic Accumulation in the North Atlantic Subtropical Gyre," *Science* 329 (2010): 1185–88.

18. Young, L. C., et al., "Bringing Home the Trash: Do Colony-Based Differences in Foraging Distribution Lead to Increased Plastic Ingestion in Laysan Albatrosses?" *PLoS ONE* 4 (2009): e7623; doi:10.1371/journal.pone.0007623.

19. As any frustrated golfer knows, golf balls sink in freshwater, but they float in higher density saltwater.

20. Ryan, P. G., et al., "Monitoring the Abundance of Plastic Debris in the Marine Environment," *Philsophical Transactions of the Royal Society B* 364 (2009): 1999–2012.

21. Mrosovsky, N., "Leatherback Turtles: The Menace of Plastic," *Marine Pollution Bulletin* 58 (2009): 287–89.

22. Plot, V., and J-Y. Georges, "Plastic Debris in a Nesting Leatherback Turtle in French Guiana," *Chelonian Conservation and Biology* 9 (2010): 27–70.

23. Teuten, E. L., et al., "Transport and Release of Chemicals from Plastics to the Environment and Wildlife," *Philosophical Transactions of the Royal Society B* 364 (2009): 2027–45.

24. Eriksson, C., and H. Burton, "Origins and Biological Accumulation of Small Plastic Particles in Fur Seals from Macquarie Island," *Ambio* 32 (2003): 380–84.

25. Stamper, M. A., et al., "Case Study: Morbidity in a Pygmy Sperm Whale (*Kogia breviceps*) Due to Ocean-Bourne Plastic," *Marine Mammal Science* 22 (2006): 719–22.

26. Tarpley, R. J., and S. Marwitz, "Plastic Debris Ingestion by Cetaceans along the Texas Coast: Two Case Reports," *Aquatic Mammals* 19 (2003): 93–98.

27. Jacobsen, J. K., et al., "Fatal Ingestion of Floating Net Debris by Two Sperm Whales (*Physeter macrocephalus*)," *Marine Pollution Bulletin* 60 (2010): 765–67.

28. Ryan, "Monitoring the Abundance of Plastic Debris."

29. American Chemical Society, "Hard Plastics Decompose in Oceans, Releasing Endocrine Disruptor BPA," *ScienceDaily*, March 24, 2010; www.sciencedaily.com/releases/2010/03/100323184607.htm; accessed May 20, 2011.

30. Mato, Y., et al., "Plastic Resin Pellets as a Transport Medium for Toxic Chemicals in the Marine Environment," *Environmental Science and Technology* 35 (2001): 318–24.

31. According to Captain Charles Moore, plastic debris from Central and North American coasts probably also arrives there via ocean highways joining the gyre from coastal inputs without having to go around.

32. Boerger, C. M., et al., "Plastic Ingestion by Planktivorous Fishes in the North Pacific Central Gyre," *Marine Pollution Bulletin* 60 (2010): 227–78.

33. Lavender Law, "Plastic Accumulation in the North Atlantic."

34. Ebbesmeyer, C. C., and E. Scigliano, *Flotsametrics.*

35. Ryan, P. G., et al., "Monitoring the Abundance of Plastic Debris."

Chapter 11: The Not So Silent World

1. Frisk, G. V., "Noiseonomics: The Relationship Between Ambient Noise Levels and Global Economic Trends"; http://pruac.apl.washington.edu/abstracts/Frisk.pdf; accessed February 14, 2011.

2. Denny, M. W., *Air and Water: The Biology and Physics of Life's Media* (Princeton: Princeton University Press, 1993).

3. Doug Anderson's favorite marine mammal call is that of the Weddell seal of Antarctica. You can hear a clip of this and many other marine mammals at Discovery of Sound in the Sea: www.dosits.org/audio/marinemammals/pinnipeds/weddellseal; accessed January 1, 2012.

4. The decibel scale provides a measure of sound levels, or sound pressure levels more precisely, that is referenced to the sensitivity of the human ear, where 0 decibels corresponds to something at the lowest limit of our hearing in air. All decibel measures must be referenced to a specific sound-pressure level, which is conventionally taken to be 20 µpa (micropascals) at one meter (about one yard) from the source in air and 1 µpa at one meter from the source underwater. What this means is that for a given level of sound pressure, the loudness underwater would be equivalent to hearing the sound one meter away, whereas the loudness of the same sound on land would be equivalent to hearing the sound twenty meters (about twenty-two yards) away.

5. BBC News: http://news.bbc.co.uk/1/hi/sci/tech/7641537.stm; accessed May 20, 2011.

6. Frantzis, A., "Does Acoustic Testing Strand Whales?" *Nature* 329 (1998): 29.

7. Not all military sonars produce sound within the hearing range of whales. Some use lower frequencies that may be perfectly safe. Cook, M. L. H., et al., "Beaked Whale Auditory Evoked Potential Hearing Measurements," *Journal of Comparative Physiology A* 192 (2006): 489–95.

8. Fernández, A., et al., " 'Gas and Fat Embolic Syndrome' Involving a Mass Stranding of Beaked Whales (Family *Ziphiidae*) Exposed to Anthropogenic Sonar Signals," *Veterinary Pathology* 42 (2005): 446l: doi: 10.1354/vp.42-4-446.

9. Tyack, P. L., et al., "Extreme Diving of Beaked Whales," *Journal of Experimental Biology* 209 (2006): 4238–53.

10. Nosengo, N., "The Neutrino and the Whale," *Nature* 462 (2009): 560–61.

11. Frankel, A. S., and C. W. Clark, "ATOC and Other Factors Affecting the Distribution and Abundance of Humpback Whales (*Megaptera novaeangliae*) off the North Shore of Kauai," *Marine Mammal Science* 18 (2002): 644–62.

12. Kroll, D. A., et al., "Only Male Fin Whales Sing Loud Songs," *Nature* 417 (2002): 809.

13. Researchers played dolphins the whistle sounds of others that had been synthesized electronically to ensure dolphins responded to the call rather than the sound of the voice of the caller. They still recognized one another as individuals. Janik, V. N., et al., "Signature Whistle Shape Conveys Identity Information to Bottlenose Dolphins," *Proceedings of the National Academy of Sciences* 103 (2006): 8293–97.

14. Buckstaff, K. C., "Effects of Watercraft Noise on the Acoustic Behaviour of Bottlenose Dolphins, *Tursiops truncatus*, in Sarasota Bay, Florida," *Marine Mammal Science* 20 (2004): 709–25.

15. Wysocki, L. E., et al., "Ship Noise and Cortisol Secretion in European Freshwater Fishes," *Biological Conservation* 128 (2006): 501–8.

16. Papoutsoglou, S.E. et al., "Effect of Mozart's Music (Romanze-Andante of 'Eine Kleine Nacht Musik,' sol major, K525) Stimulus on Common Carp (*Cyprinus carpio* L.) Physiology under Different Light Conditions," *Aquacultural Engineering* 36 (2007): 61–72.

17. Fay, R., "Soundscapes and the Sense of Hearing of Fishes," *Integrative Zoology* 4 (2009): 26–32.

18. Simpson, S. D., et al., "Homeward Sound," *Science* 308 (2005): 221.

19. Vermeij, M. J. A., et al., "Coral Larvae Move Towards Reef Sounds," *PLoS One* 5 (2010): e10660.

20. Lohse, D., B. Schmitz, and M. Versluis, "Snapping Shrimp Make Flashing Bubbles," *Nature* 413 (2001): 477–78.

21. Rowell, T., et al., "Use of Passive Acoustics to map Grouper Spawning Aggregations, with Emphasis on Red Hind, *Epinephelus guttatus*, off Western Puerto Rico"; presentation given at Gulf and Caribbean Fisheries Institute meeting, Puerto Rico, 2010;

www.gcfi.org/Conferences/63rd/GCFIBook_Of_AbstractsEngPDF.pdf; accessed May 22, 2011.

22. Vascocelos, R. O., et al., "Effects of Ship Noise on the Detectability of Communication Signals in the Lusitanian Toadfish," *Journal of Experimental Biology* 210 (2007): 2104–12.

23. Bass, A. H., et al., "Evolutionary Origins for Social Vocalization in a Vertebrate Hindbrain-Spinal Compartment," *Science* 321 (2008): 417–21.

24. Anderson, A., "Humming Fish Disturb the Peace," *New Scientist*, September 12, 1985, p. 64–65.

25. Parks, S. E., et al., "Individual Right Whales Call Louder in Increased Environmental Noise," *Biology Letters* (2010): doi: 10.1098/rsbl.2010.0451; *see also* article in Woods Hole Oceanographic Institution magazine, *Oceanus:* www.whoi.edu/oceanus/viewArticle.do?id=84868, accessed February 14, 2011.

26. Foote, A. D., et al., "Whale-Call Response to Masking Boat Noise," *Nature* 428 (2004): 910.

27. Slabbekoorn, H., and A. Den Boer-Visser, "Cities Change the Song of Birds," *Current Biology* 16 (2006): 2326–31.

28. Aguilar Soto, N. A., et al., "Does Intense Ship Noise Disrupt Foraging in Deep-Diving Cuvier's Beaked Whales (*Ziphius cavirostris*)?" *Marine Mammal Science* 22 (2006): 690–99.

29. Mooney, T. A., et al., "Sonar-Induced Temporary Hearing Loss in Dolphins," *Biology Letters* 5 (2009): 565–67.

30. A short-fin pilot whale also had profound hearing loss, while twelve animals from other species had normal hearing. Mann, D., et al., "Hearing Loss in Stranded Odontocete Dolphins and Whales." *PLoS ONE* 5 (2010): e13824: doi:10.1371/journal.pone.0013824.

31. Frisk, G. V., "Noiseonomics."

Chapter 12: Aliens, Invaders, and the Homogenization of Life

1. Coates, A. G., et al., "Closure of the Isthmus of Panama: The Near-Shore Marine Record of Costa Rica and Western Panama," *Geological Society of America Bulletin* 104 (1992): 814–28.

2. Brawley, S. H., et al., "Historical Invasions of the Intertidal Zone of Atlantic North America Associated with Distinctive Patterns of Trade and Emigration," *Proceedings of the National Academy of Sciences* 106 (2009): 8239–44.

3. There is a widely recounted tale, which turns out to be apocryphal, that lionfish were released to the Atlantic during Hurricane Andrew, in 1992, when storm surge overwhelmed Biscayne Bay aquarium. The source of the story has since retracted it. See *Science* magazine: http://news.sciencemag.org/scienceinsider/2010/04/mystery-of-the-lionfish-dont-bla.html#more; accessed November 17, 2011.

4. Albins, M. A., and M. A. Hixon, "Worst Case Scenario: Potential Long-Term Effects of Invasive Predatory Lionfish (*Pterois volitans*) on Atlantic and Caribbean Coral-Reef Communities," *Environmental Biology of Fishes* (2011): doi 10.1007/s10641-011-9795-1.

5. Mark Hixon was quoted in *The Times* (London), October 20, 2008; www.timesonline.co.uk/tol/news/environment/article4974396.ece; accessed May 22, 2011.

6. Bax, N., et al., "Marine Invasive Alien Species: A Threat to Global Biodiversity," *Marine Policy* 27 (2003): 313–23.

7. Carlton, J. T., "The Scale and Ecological Consequences of Biological Invasions in the World's Oceans," in Sandlund, O. T. et al., eds. *Invasive Species and Biodiversity Management* (Dordrecht, The Netherlands: Kluwer Academic Publishers, 1999), pp. 195–212.

8. Shiganova, T. A., and Y. V. Bulgakova, "Effects of Gelatinous Plankton on Black Sea and Sea of Azov Fish and Their Food Resources," *ICES Journal of Marine Science* 57 (2000): 641–48: doi:10.1006/jmsc.2000.0736.

9. Saltonstall, K., "Cryptic Invasion By A Non-Native Genotype of Phragmites australis into North America," *Proceedings of the National Academy of Sciences* 99 (2002): 2445–49.

10. Cohen, A. N., and J. T. Carlton, "Accelerating Invasion Rate in a Highly Invaded Estuary," *Science* 279 (1998): 555–58.

11. Personal communication from James Carlton, Williams College, Mystic, CT, USA.

12. Ray, G. L., "Invasive Marine and Estuarine Animals of California," ERDC/TN ANSRP-05-2, August 2005. Available from: http://el.erdc.usace.army.mil/elpubs/pdf/ansrp05-2.pdf; accessed January 1, 2012.

13. Meinesz, A., *Killer Algae* (Chicago: University of Chicago Press, 1999).

14. Musée Océanographique de Monaco.

15. Meinesz, A., *Killer Algae*.

16. Carlton, J. T., and L. Eldredge, "Marine Bioinvasions of Hawai'i: The Introduced and Cryptogenic Marine and Estuarine Animals and Plants of the Hawaiian Archipelago," *Bishop Museum Bulletin of Cultural and Environmental Studies* 4 (2009): 1–202.

17. Tan, K. S., and B. Morton, "The Invasive Caribbean Bivalve Mytilopsis sallei (*Dreissenidae*) Introduced to Singapore and Johor Bahru, Malaysia," *The Raffles Bulletin of Zoology* 54 (2006): 429–34.

18. Reid, P. C., et al., "A Biological Consequence of Reducing Arctic Ice Cover: Arrival of the Pacific Diatom Neodenticula seminae in the North Atlantic for the First Time in 800 000 Years," *Global Change Biology* 13 (2007): 1910–21.

19. Scheinin, A. P., et al., "Gray Whale (*Eschrichtius robustus*) in the Mediterranean Sea: Anomalous Event or Early Sign of Climate-Driven Distribution Change?" *Marine Biodiversity Records* (2011): doi:10.1017/S1755267211000042.

20. Lewis, P. N., et al., "Assisted Passage or Passive Drift: A Comparison of Alternative Transport Mechanisms for Non-indigenous Coastal Species into the Southern Ocean," *Antarctic Science* 17 (2005): 183–91.

21. Ricciardi, A., et al., "Should Biological Invasions Be Managed as Natural Disasters?" *BioScience* 61 (2011): 312–17.

22. Galil, B. S., "Loss or Gain? Invasive Aliens and Biodiversity in the Mediterranean Sea," *Marine Pollution Bulletin* 55 (200): 314–22.

23. One experimental test of this idea proved the case at a very small scale: Stachowicz, J. J., et al., "Species Diversity and Invasion Resistance in a Marine Ecosystem," *Science* 286 (1999): 1577–79.

24. Some scientists, like Williams College's James Carlton, think this conclusion has more to do with our ignorance of marine life than a real absence of extinctions.

25. Myers, J. H., et al., "Eradication Revisited: Dealing with Exotic Species," *Trends in Ecology and Evolution*, 15 (2000): 316–20.

26. Ecospeed, for example. Hydrex Underwater Technology: www.hydrex.be/ecospeed_hull_coating_system; accessed January 2, 2012.

Chapter 13: Pestilence and Plague

1. Nugues, M. M., and I. Nagelkerken, "Status of Aspergillosis and Sea Fan Populations in Curaçao Ten Years after the 1995 Caribbean Epizootic," *Revista de Biologia Tropical* 54 (2006): 153–60.

2. Darwin, C., *A Naturalist's Voyage* (London: John Murray, 1886).

3. Garrison, V. H., et al., "African and Asian Dust: From Desert Soils to Coral Reefs," *Bioscience* 53 (2003): 469–80. Others, like the University of North Carolina's John Bruno, doubt the dust explanation, pointing out that African dust has been falling for thousands of years without obvious harm. On the other hand, it may only have been recently that populations of animals like sea fans became sufficiently stressed to be susceptible to epidemics.

4. Patterson Sutherland, K., et al., "Human Sewage Identified as Likely Source of White Pox Disease of the Threatened Caribbean Elkhorn Coral, *Acropora palmata*," *Environmental Microbiology* 12 (2010): 122–31.

5. Gardner, T. A., et al., "Long-term Region-wide Declines in Caribbean Corals," *Science* 301 (2003): 958–60.

6. Arthur, K., et al., "The Exposure of Green Turtles (*Chelonia mydas*) to Tumour Promoting Compounds Produced by the Cyanobacterium *Lyngbya majuscula* and Their Potential Role in the Aetiology of Fibropapillomatosis," *Harmful Algae* 7 (2008): 114–25.

7. Ward, J. R., and K. D. Lafferty, "The Elusive Baseline of Marine Disease: Are Diseases in Ocean Ecosystems Increasing?" *PLoS Biology* 2 (2004): 542–47.

8. Lafferty, K. D., et al., "Reef Fishes Have Higher Parasite Richness at Unfished Palmyra Atoll Compared to Fished Kiritimati Island," *EcoHealth* 5 (2008): 338–45.

9. Thornton, Russell, *American Indian Holocaust and Survival: A Population History Since 1492* (Norman, OK: University of Oklahoma Press, 1990), pp. 26–32.

10. Lessios, H., "Mass Mortality of *Diadema antillarum* in the Caribbean: What Have We Learned?" *Annual Reviews in Ecology and Systematics* 19 (1988): 371–93.

11. Gulland, F. M. D., and A. J. Hall, "Is Marine Mammal Health Deteriorating? Trends in Global Reporting of Marine Mammal Disease," *EcoHealth* 4 (2007): 135–50.

12. Harvell, D., et al., "The Rising Tide of Ocean Diseases: Unsolved Problems and Research Priorities," *Frontiers in Ecology and Environment* 2 (2004): 375–82; *see also* BBC News: http://news.bbc.co.uk/1/hi/4729810.stm; accessed February 16, 2011.

13. Swart, R. L. de, et al., "Impaired Immunity in Harbour Seals (*Phoca vitulina*) Exposed to Bioaccumulated Environmental Contaminants: Review of a Long-term Feeding Study," *Environmental Health Perspectives* 104 (1996): 823–28.

14. Carey, C., "Infectious Disease and Worldwide Declines of Amphibian Populations, with Comments on Emerging Diseases in Coral Reef Organisms and in Humans," *Environmental Health Perspectives* 108 (2000): 143–50.

15. Nugues, M. M., "Impact of a Coral Disease Outbreak on Coral Communities in St. Lucia: What and How Much Has Been Lost?" Marine Ecology Progress Series 229 (2002): 61–71.

16. Watling L., and E. A. Norse, "Disturbance of the Seabed by Mobile Fishing Gear: A Comparison to Forest Clearcutting," *Conservation Biology* 12 (1998): 1180–97.

17. Ford, S. E., "Range Extension by the Oyster Parasite *Perkinsus marinus* into the Northeastern United States: Response to Climate Change?" *Journal of Shellfish Research* 15 (1996): 45–56.

18. Sokolow, S., "Effects of a Changing Climate on the Dynamics of Coral Infectious Disease: A Review of the Evidence," *Diseases of Aquatic Organisms* 87 (2009): 5–18.

19. Ibid. Extreme warmth of 31°C–35°C seems to damp down growth of pathogens.

20. Webster, N. S., "Sponge Disease: A Global Threat?" *Environmental Microbiology* 9 (2007): 1363–75.

21. Bruno, J. F., et al., "Nutrient Enrichment Can Increase the Severity of Coral Diseases," *Ecology Letters* 6 (2003): 1056–61.

22. Wood, C. L., et al., "Fishing out Marine Parasites? Impacts of Fishing on Rates of Parasitism in the Ocean," *Ecology Letters* 13 (2010): 761–75.

23. Raymundo, L. J., et al., "Functionally Diverse Reef-Fish Communities Ameliorate Coral Disease," *Proceedings of the National Academy of Sciences* 106 (2009): 17067–70. Lafferty, K. D., "Fishing for Lobsters Indirectly Increases Epidemics in Sea Urchins," *Ecological Applications* 14 (2004): 1566–73.

Chapter 14: Mare Incognitum

1. Cohen, J. E., "Human Population: The Next Half Century," *Science* 302 (2003): 1172–75.

2. UN Food and Agriculture Organization, "State of Fisheries and Aquaculture" (Rome: FAO, 1998).

3. Clarke, S., "Shark Product Trade in Hong Kong and Mainland China and Implementation of Cites Shark Listings," *TRAFFIC East Asia, Hong Kong, China* (2004).

4. Montgomery, D. R., *Dirt: The Erosion of Civilizations* (Berkeley: University of California Press, 2007).

5. Richardson, A. J., et al., "The Jellyfish Joyride: Causes, Consequences and Management Responses to a More Gelatinous Future," *Trends in Ecology and Evolution* 24 (2009): 312–22.

6. Arai, M. N., "Predation on Pelagic Coelenterates: A Review," *Proceedings of the Marine Biological Association of the UK* 85 (2005): 523–36.

7. Jackson, J. B. C., "Ecological Extinction and Evolution in the Brave New Ocean," *Proceedings of the National Academy of Sciences* 105 (2008): 11458–65.

8. Boyce, D. G., et al., "Global Phytoplankton Decline over the Past Century," *Nature* 466 (2010): 591–96. The 40 percent decline in phytoplankton productivity since 1950 claimed in this study has been challenged as an artifact of differences in sampling methods over time. The authors have rejected this criticism, although it seems well founded. This is an area where more research is needed for clarity to emerge. *See also* Stramma, L., et al., "Expanding Oxygen-Minimum Zones in the Tropical Oceans," *Science* 320 (2008): 655–58.

9. Roopnarine, P., et al., "The Whale Pump: Marine Mammals Enhance Primary Productivity in a Coastal Basin," *PLoS ONE* (2010); e13255 doi: 10.1371/journal.pone .0013255; http://harvardmagazine.com/2011/05/why-whales; accessed May 23, 2011.

10. Iglesias-Rodriguez, M. D., et al., "Phytoplankton Calcification in a High CO_2 World," *Science* 320 (2008): 336–40.

11. Riebesell, U., et al., "Reduced Calcification of Marine Plankton in Response to Increased Atmospheric CO_2," *Nature* 407 (2000): 364–67.

12. Moy, A. D., et al., "Reduced Calcification in Modern Southern Ocean Planktonic Foraminifera," *Nature Geoscience* doi: 10.1038/NGEO460 (2009).

13. Kolbert, E., "Acid Oceans," *National Geographic* 219 (April 2011): 100–21.

14. Muir, J., *My First Summer in the Sierra* (Boston and New York: Houghton Mifflin, 1911).

15. Ward, P., and R. M. Myers, "Shifts in Open Ocean Fish Communities Coinciding with the Commencement of Commercial Fishing," *Ecology* 86 (2005): 835–47.

16. Myers, R. A., et al., "Cascading Effects of the Loss of Apex Predatory Sharks from a Coastal Ocean," *Science* 315 (2007): 1846–50.

17. Airoldi, L., et al., "The Gray Zone: Relationships Between Habitat Loss and Marine Biodiversity and Their Implications for Conservation," *Journal of Experimental Marine Biology and Ecology* 366 (2008): 8–15.

18. Nugues, M. M., "Algal Contact as a Trigger for Coral Disease," *Ecology Letters* 7 (2004): 919–23.
19. McIntyre, P. B., et al. "Fish Extinctions Alter Nutrient Recycling in Tropical Freshwaters." *Proceedings of the National Academy of Sciences* 104 (2007): 4461–66; Schindler, D. E., "Fish Extinctions and Ecosystem Functioning in Tropical Ecosystems," *Proceedings of the National Academy of Sciences* 104 (2007): 5707–8.

Chapter 15: Ecosystems at Your Service

1. Seaman, O., "The Uses of the Ocean," H. Watson, ed., *Ode to the Sea: Poems to Celebrate Britain's Maritime History* (London: National Trust Books, 2011): pp 162–63.
2. Bellamy, J. C., *The Housekeeper's Guide to the Fish-market for Each Month of the Year, and an Account of the Fish and Fisheries of Devon and Cornwall* (London: Longman, Brown, Green and Longmans, 1843).
3. Shephard, S., et al., "Benthivorous Fish May Go Hungry on Trawled Seabed," *Proceedings of the Royal Society B* (2011); doi: 10.1098/rspb.2010.2713.
4. Thurstan, R. H., et al., "Effects of 118 Years of Industrial Fishing on UK Bottom Trawl Fisheries," *Nature Communications* 1 (2010):15, doi: 10.1038/ncomms1013.
5. Department of Health, "The Burden of Asthma in Washington State: 2008 Update," DOH Publication 345-240; Rev 05/2009.
6. Millennium Ecosystem Assessment, *Ecosystems and Human Well-being: Synthesis* (Washington DC: Island Press, 2005).
7. Wackernagel, M., "Tracking the Ecological Overshoot of the Human Economy," *Proceedings of the National Academy of Sciences* 99 (2002): 9266–71.
8. Stern, N., *The Stern Review on the Economics of Climate Change* (UK Government, HM Treasury, 2006); www.webcitation.org/5nCeyEYJ.
9. Myers, N., "Environmental Refugees," *Population and Environment* 19 (1997): 167–82.
10. Diamond, J. *Collapse: How Societies Choose to Fail or Survive* (London: Penguin Books, 2005).
11. Ibid.

Chapter 16: Farming the Sea

1. Herubel, M. A., *Sea Fisheries: Their Treasures and Toilers* (London: T. Fisher Unwin, 1912).
2. To see this trend we have to first correct for systematic overreporting of Chinese catches. In a centrally planned economy it seems it is better to cook the books than not meet production targets. We also have to take out Peruvian anchovy catches, which make up the largest single species landed in good years. Anchovy abundance jumps up and down in response to the El Niño–Southern Oscillation climate swings that alter the strength of upwelling in the eastern Pacific and therefore dictate the productivity of these seas. Watson, R., and D. Pauly, "Systematic Distortions in World Fisheries Catch Trends," *Nature* 414 (2001): 534–36.
3. Thurstan, R., and C. M. Roberts, "Health Recommendations and Global Fish Availability: Are There Enough Fish to Go Around?" (in press). The figures represent fish available after processing losses (inedible bits like heads, tails, bones and shells). Of course, these calculations are only a mind game, because fish are not distributed evenly, and not all of the global fish catch is eaten directly. For countries like the Philippines and Sierra Leone, fish provide most of the dietary protein. For others, fish

are hardly eaten at all; Brazil jumps to mind as a place where restaurants seem to sell almost nothing that isn't red and meaty.

4. Food and Agriculture Organization of the United Nations, *The State of World Fisheries and Aquaculture 2010* (Rome: FAO, 2010).

5. Costa-Pierce, B. A., "Aquaculture in Ancient Hawaii," *Bioscience* 37 (1987): 320–31.

6. Higginbotham, J., *Piscinae: Artificial Fishponds in Roman Italy* (Chapel Hill and London: University of North Carolina Press, 1997).

7. Pliny the Elder, *Naturalis Historiae,* Book 9 (AD 79), ch. LIV.

8. Higginbotham, J., *Piscinae.*

9. Costa-Pierce, B. A., "Aquaculture in Ancient Hawaii."

10. *An Encyclopedia of New Zealand, 1966*; www.teara.govt.nz/en/1966/fish-introduced -freshwater/1; accessed January 2, 2012.

11. Herubel, M. A., *Sea Fisheries.*

12. Ibid.

13. There has recently been renewed interest in restocking and enhancing wild fisheries using hatchery-reared animals, but evidence for success is still limited. Bell, J. D., et al., "Restocking and Stock Enhancement of Coastal Fisheries: Potential, Problems and Progress," *Fisheries Research* 80 (2006): 1–8.

14. Williams, J., *Clam Gardens: Aboriginal Mariculture On Canada's West Coast* (Vancouver: New Star Books Ltd., 2006).

15. Naylor, R. N., et al., "Feeding Aquaculture in an Era of Finite Resources," *Proceedings of the National Academy of Science* 106 (2009): 15103–10.

16. Ibid.

17. Lobo, A. S., et al., "Commercializing Bycatch Can Push a Fishery Beyond Economic Extinction," *Conservation Letters* 3 (2010): 277–85.

18. Tacon, A. G. J., and M. Metian, "Fishing for Feed or Fishing for Food: Increasing Global Competition for Small Pelagic Forage Fish," *Ambio* 38 (2009): 294–302.

19. Brashares, J. S., et al., "Bushmeat Hunting, Wildlife Declines, and Fish Supply in West Africa," *Science* 306 (2004): 1180–83.

20. Schiermeier, Q., "Ecologists Fear Antarctic Krill Crisis," *Nature* 467 (2010): 15.

21. Krkošek, M., et al., "Epizootics of Wild Fish Induced by Farm Fish," *Proceedings of the National Academy of Sciences* 103 (2006): 15506–10.

22. Murray, A. G., "Shipping and the Spread of Infectious Salmon Anemia in Scottish Aquaculture," *Emerging Infectious Diseases* 8 (2002): 1–5.

23. Cabello, F. C., "Heavy Use of Prophylactic Antibiotics in Aquaculture: A Growing Problem for Human and Animal Health and for the Environment," *Environmental Microbiology* 8 (2006): 1137–44.

24. Sapkota, A., et al., "Aquaculture Practices and Potential Human Health Risks: Current Knowledge and Future Priorities," *Environment International* 34 (2008): 1215–26.

25. MacGarvin, M., *Scotland's Secret? Aquaculture, Nutrient Pollution, Eutrophication and Toxic Algal Blooms*, (Aberfeldy, Scotland: World Wildlife Federation, 2000); www .wwf.org.uk/fileLibrary/pdf/secret.pdf; accessed May 24, 2011.

26. Biao, X., and Y. Kaijin, "Shrimp Farming in China: Operating Characteristics, Environmental Impact and Perspectives," *Ocean and Coastal Management* 50 (2007): 538–50.

27. Primavera, J. H., "Overcoming the Impacts of Aquaculture on the Coastal Zone," *Ocean and Coastal Management* 49 (2006): 531–45.

28. Jing, G., "China's Bohai Sea Drowns in Discharged Waste," *Caixin Times*, September 9, 2011; http://english.caixin.cn/2011-09-14/100304938.html; accessed October 28,

2011; *see also* Wang, B., et al., "Water Quality in Marginal Seas off China in the Last Two Decades," *International Journal of Oceanography*, (2011); doi:10.1155/2011/731828.

29. Marris, E., "Transgenic Fish Go Large," *Nature* 467 (2010): 259.
30. Cao, L., et al., "Environmental Impact of Aquaculture and Countermeasures to Aquaculture Pollution in China," *Environmental Science and Pollution Research* 14 (2007): 452–62.
31. Azad, A.K., "Coastal Aquaculture Development in Bangladesh: Unsustainable and Sustainable Experiences," *Environmental Management* 44 (2009): 800–9.
32. Cressey, D., "Future Fish," *Nature* 458 (2009): 398–400.

Chapter 17: The Great Cleanup

1. Piskaln, C. H., et al., "Resuspension of Bottom Sediment by Bottom Trawling in the Gulf of Maine and Potential Geochemical Consequences," *Conservation Biology* 12 (1998): 1223–29.
2. Stockholm Convention: chm.pops.int/Countries/StatusofRatification/tabid/252/language/en-US/Default.aspx; accessed January 2, 2012. The number of countries is a rough total since it often changes.
3. Jones, O. A. H., et al., "Questioning the Excessive Use of Advanced Treatment to Remove Organic Micropollutants from Wastewater," *Environmental Science and Technology* 41 (2007): 5085–89.
4. Harvey, F., "EU Unveils Plans to Pay Fishermen to Catch Plastic," Guardian; www.guardian.co.uk/environment/2011/may/04/eu-fishermen-catch-plastic; accessed January 2, 2012.
5. Thompson, R. C., et al., "Plastics, the Environment and Human Health: Current Consensus and Future Trends," *Philosophical Transactions of the Royal Society B* 364 (2009): 2153–66.
6. O'Brine, T., and R. C. Thompson, "Degradation of Plastic Carrier Bags in the Marine Environment," *Marine Pollution Bulletin*, 60 (2010): 2279–83.
7. Marine Conservation Society UK: www.mcsuk.org/downloads/pollution/beachwatch/latest2011/Methods%20&%20Results%20BW10.pdf; accessed December 28, 2011.
8. Page, B., et al., "Entanglement of Australian Sea Lions and New Zealand Fur Seals in Lost Fishing Gear and Other Marine Debris Before and After Government and Industry Attempts to Reduce the Problem," *Marine Pollution Bulletin* 49 (2004): 33–42; Arnould, J. P. Y., and J. P. Croxall, "Trends in Entanglement of Antarctic Fur Seals (*Arctocephalus gazella*) in Man-made Debris at South Georgia," *Marine Pollution Bulletin* 30 (1995): 707–12.
9. Mee, L. D., et al., "Restoring the Black Sea in Times of Uncertainty," *Oceanography* 18 (2005): 100–11.
10. Conley, D. J., et al., "Tackling Hypoxia in the Baltic Sea: Is Engineering a Solution?" *Environmental Science and Technology* 43 (2009): 3407–11.
11. Weaver, D.E, "Environmental Impacts of Bottom Trawling Suspended Solids Generation. Report for the United Anglers of Southern California" (2008); web.me.com/deweaver/bottom_trawling/Links_to_Docs_files/Bottom_Trawling3.pdf; accessed May 24, 2011; *see also*: Palanques et al., "Impact of Bottom Trawling on Water Turbidity and Muddy Sediment of an Unfished Continental Shelf," *Limnology and Oceanography* 46 (2001): 1100–10.

12. Kunzig, R., "Seven Billion," *National Geographic* (January 2011), p. 42–63. In 1900, there were an estimated 1.6 billion people on planet Earth. In 2011, the total reached 7 billion.

Chapter 18: Can We Cool Our Warming World?

1. Keith, D. W., "Why Capture CO_2 from the Atmosphere?" *Science* 325 (2009): 1654–55.
2. Bollmann, M., et al., *World Ocean Review. Living With the Ocean* (Hamburg: Maribus GmbH, 2010).
3. Lindeboom, H. J., et al., "Short-term Ecological Effects of an Offshore Wind Farm In the Dutch Coastal Zone: A Compilation," *Environmental Research Letters* (2011); doi:10.1088/1748-9326/6/3/035101.
4. EurekAlert Press Release, "World Needs Climate Emergency Backup Plan, Says Expert," November 7, 2008: www.eurekalert.org/pub_releases/2008-11/ci-wnc110708 .php; accessed March 21, 2011.
5. We will also have to deal with sources of methane. However, as methane has a much shorter life in the atmosphere than carbon dioxide (a half-life of about seven years as compared to about a century for carbon dioxide), direct methane removal will not be necessary—only reduced emissions.
6. Jones, I. S. F., and H. E. Young, "The Potential of the Ocean for the Management of Global Warming," *International Journal of Global Warming* 1 (2009): 43–56.
7. Strong, A. L., et al., "Ocean Fertilization: Science, Policy and Commerce," *Oceanography* 22 (2009): 236–61.
8. "Carbon Sequestration," *Science* 325 (2009): 1644–45.
9. Haszeldine, R. S., "Carbon Capture and Storage: How Green Can Black Be?" *Science* 325 (2009): 1647–52.
10. Rasch, P. J., et al., "Geoengineering by Cloud Seeding: Influence on Sea Ice and Climate System," *Environmental Research Letters* 4 (2009): 1–8.
11. Keith, D. W., et al., "Research on Global Sunblock Needed Now," *Nature* 463 (2010): 426–27.
12. Turner, W. R., et al., "A Force to Fight Global Warming," *Nature* 462 (2009): 278–79.
13. Nellemann, C., et al., eds., *Blue Carbon. A Rapid Response Assessment* (GRID-Arendal, Norway: United Nations Environment Programme, 2009); www.grida.no.
14. These figures seem high to me and may not stand up to long-term scrutiny. But they are the best we have at the moment.

Chapter 19: A New Deal for the Oceans

1. Hueber, A., "Fake Sea Grass Could Boost Fish Numbers," *New Zealand Herald*, January 30, 2011.
2. Outdoor Alabama, official Web site of the Alabama Department of Conservation and Natural Resources; www.outdooralabama.com/fishing/saltwater/where/artificial-reefs/ reefhist.cfm; accessed January 2, 2012.
3. All U.S. states have regulations that require contaminant chemicals like engine oil to be removed before dumping.
4. Community of Arran Seabed Trust, UK: www.arrancoast.com.
5. Howarth, L. M., et al., "Complex Habitat Boosts Scallop Recruitment in a Fully Protected Marine Reserve," *Marine Biology* (2011); doi 10.1007/s00227-011-1690-y.

6. Alcala, A. C., and G. R. Russ, "A Direct Test of the Effects of Protective Management on Abundance and Yield of Tropical Marine Resources," *Journal du Conseil Internationale pour L'Exploration de la Mer*, 46 (1990): 40–47; Russ, G. R., and Alcala, A. C., "Marine Reserves: Rates and Patterns of Recovery and Decline of Large Predatory Fish," *Ecological Applications* 6 (1996): 947–61; Russ, G. R., and A. C. Alcala, "Enhanced Biodiversity Beyond Marine Reserve Boundaries: The Cup Spillith Over," *Ecological Applications* 21 (2011): 241–50.

7. García-Charton J., et al., "Effectiveness of European Atlanto-Mediterranean MPAs: Do They Accomplish the Expected Effects on Populations, Communities and Ecosystems?" *Journal for Nature Conservation* 16,(2008): 193–221; Ault, J. S., et al., "Building Sustainable Fisheries in Florida's Coral Reef Ecosystem: Positive Signs in the Dry Tortugas," *Bulletin of Marine Science* 78 (2006): 633–54.

8. Here is a way you can visualize how marine reserves help to steady the variability of fish stocks. Think of habitats in the sea as being like a sponge that is capable of holding only so much life. Imagine that sponge is held beneath a tap that rains new life upon it to replace the life that dribbles away through mortality by natural and human causes. The tap sometimes flows faster, sometimes more slowly, depending on how favorable conditions are for survival of young animals and plants. Fishing reduces reproduction, so the tap merely trickles and the sponge is only occasionally saturated. Creating marine reserves turns the tap on more forcefully. Although its flow still varies, the sponge is kept full.

9. Thurstan, R. H., S. Brockington, and C. M. Roberts, "The Effects of 118 Years of Industrial Fishing on UK Bottom Trawl Fisheries," *Nature Communications* 1 (2010): 15; doi: 10.1038/ncomms1013.

10. *Fishing News*, April 15, 2011.

11. Mascia, M. B., et al., "Impacts of Marine Protected Areas on Fishing Communities," *Conservation Biology* 24 (2010): 1424–29.

12. Goñi, R., et al., "Spillover from Six Western Mediterranean Marine Protected Areas: Evidence from Artisanal Fisheries," *Marine Ecology Progress Series* 366 (2008): 159–74.

13. Murawksi, S. A., et al., "Effort Distribution and Catch Patterns Adjacent to Temperate MPAs," *ICES Journal of Marine Science* 62 (2005): 1150–67.

14. Hoskin, M. G., et al., "Variable Population Responses by Large Decapods Crustaceans to the Establishment of a Temperate Marine No-take Zone," *Canadian Journal of Fisheries and Aquatic Sciences* 68 (2011): 185–200.

15. PISCO, *The Science of Marine Reserves*, European edition (PISCO Consortium, 2011); www.piscoweb.org. Lester, S. E., et al., "Biological Effects Within No-take Marine Reserves: A Global Synthesis," *Marine Ecology Progress Series* 384 (2009): 33–46.

16. Svedäng, H., "Long-term Impact of Different Fishing Methods on the Ecosystem in the Kattegat and Öresund," (European Parliament, 2010); IP/B/PECH/IC/2010_24. Available from Common Fisheries Policy ReformWatch: www.cfp-reformwatch.eu/pdf/kattegatt_oresund_trawl.pdf; accessed January 2, 2012.

17. Schindler, D. W., et al., "Population Diversity and the Portfolio Effect in an Exploited Species," *Nature* 465 (2010): 609–12.

18. Kean, S., "The Secret Lives of Ocean Fish," *Science* 327 (2010): 264.

19. Unpublished research by myself and colleagues Julie Hawkins, Gemma Aitken, and John Bainbridge, Environment Department, University of York.

Chapter 20: Life Renewed

1. *The Sunken Billions: The Economic Justification for Fisheries Reform* (Washington DC: The World Bank and Rome: FAO, 2009).
2. Froese, R., and A. Proelß, "Rebuilding Fish Stocks No Later Than 2015: Will Europe Meet the Deadline?" *Fish and Fisheries* (2010); doi: 10.1111/j.1467-2979.2009.00349.x.
3. Worm, B., et al., "Rebuilding Global Fisheries," *Science* 325 (2009): 578–85.
4. O'Leary, B., et al., "Fisheries Mismanagement," *Marine Pollution Bulletin* 62 (2011): 2642–48.
5. Prince Charles made this comment in a speech at St. James' Palace, London, in March 2011.
6. Bromley, D. W., "Abdicating Responsibility. The Deceits of Fisheries Policy," *Fisheries* 34 (2009): 280–90.
7. Essington, T. E., "Ecological Indicators Display Reduced Variation in North American Catch Share Fisheries," *Proceedings of the National Academy of Sciences* 107 (2010): 754–59.
8. The Seth Macinko quote is from a talk I heard him give in Arran, Scotland, in 2010.
9. Schrope, M., "What's the Catch?" *Nature* 465 (2010): 540–42.
10. Fox, C. J., "West Coast Fishery Trials of a Twin Rigged *Nephrops* Trawl Incorporating a Large Mesh Topsheet for Reducing Commercial Gadoid Species Bycatch," *Scottish Industry Science Partnership Report*, no. 03/10 (Marine Scotland, 2010).
11. Wooldridge, S. A., and T. J. Done, "Improved Water Quality Can Ameliorate Effects of Climate Change on Corals," *Ecological Applications* 19 (2009): 1492–99.
12. Gibson, L., and N. S. Sodhi, "Habitats at Risk: A Step Forward, a Step Back," *Science* 331 (2010): 1137.
13. Stern, N., "The Economics of Climate Change," *The Stern Review* (London: The Cabinet Office, 2007).
14. Mörner, N., et al., "New Perspectives for the Future of the Maldives," *Global and Planetary Change* 40 (2004): 177–82.

Chapter 21: Saving the Giants of the Sea

1. Swimmer, S., et al., "Sustainable Fishing Gear: The Case of Modified Circle Hooks in a Costa Rican Longline Fishery," *Marine Biology* 158 (2010): 757–67. The authors tested a modified hook that would reduce these horrific bycatch rates. The new hooks reduced the number of turtles caught per mahi-mahi from 2.3 to "just" 1.7, but the authors concluded they would not be acceptable to the fishing industry, because they also reduced capture rates of mahi-mahi by 15 percent.
2. Edwards, E. F., "Fishery Effects on Dolphins Targeted by Tuna Purse-seiners in the Eastern Tropical Pacific Ocean," *International Journal of Comparative Psychology* 20 (2007): 217–27; *see also* Cramer, K. L., et al., "Declines in Reproductive Output in Two Dolphin Populations Depleted by the Yellowfin Tuna Purse-seine Fishery," *Marine Ecology Progress Series* 369 (2008): 273–85.
3. The best option for guilt-free tuna is "pole and line" caught. Look out for this on the label. Anything that doesn't mention pole and line will probably have been caught with longlines or purse seines, so many other animals will have been killed to put your fish in the can.
4. See Tagging of Pacific Predators, www.topp.org, for exciting examples of how tagging can reveal the enigmatic lives of ocean wanderers.

5. Fujiwara, M., and H. Caswell, "Demography of the Endangered North Atlantic Right Whale," *Nature* 414 (2001): 537–41.

6. United Nations: www.un.org/depts/los/convention_agreements/convention_overview_convention.htm; accessed January 2, 2012.

7. Personal communication from Guy Stevens, Maldivian Manta Ray Project.

8. This effort involved many people. My team comprised Beth O'Leary, Rachel Brown, Melanie O'Rourke, Andrew Davies, and Tina Molodtsova. Of the rest, particular credit should go to the German delegation to OSPAR under the direction of Henning von Nordheim, together with Jeff Ardron and Tim Packeiser, WWF International, and the delegations of The Netherlands and Portugal, all ably assisted by David Johnson and his team within the OSPAR secretariat.

9. O'Leary, B. C., et al., "The First Network of Marine Protected Areas (MPAs) in the High Seas: The Process, the Challenges and Where Next," *Marine Policy* 36 (2012): 598–605.

10. Iceland tried to scupper the North Atlantic marine protected areas but fortunately were persuaded against it this time. I find it ironic that a country that knows more about environmental degradation than almost any other should wish the same upon the world oceans. In truth, I think the majority of Icelanders don't want that. From what I have heard it comes down more to the prejudices of their international negotiators, whose worldview was forged in the 1960s' fog of cod wars and unilateral enlargement of sovereign waters.

11. Haeckel, E., *Kunstformen der Natur* (Leipzig, Germany: Bibliographisches Institut, 1899).

Chapter 22: Preparing for the Worst

1. Barnosky, A. D., et al., "Has the World's Sixth Mass Extinction Already Arrived?" *Nature* 471 (2011): 51–57.

2. Lutz, W., et al., "The End of World Population Growth," *Nature* 412 (2001): 543–45.

3. Butchart, S. H. M. et al., "Global Biodiversity: Indicators of Recent Declines," *Science* 328 (2010): 1164–68.

4. The Chagossian people were forcibly removed from their homeland by the UK government in the 1960s and 1970s to make way for the U.S. military base.

5. Sheppard, C. R. C., et al., "Archipelago-wide Coral Recovery Patterns Since 1998 in the Chagos Archipelago, Central Indian Ocean," *Marine Ecology Progress Series* 362 (2008): 109–17.

6. In late 2011, the Australian government launched a consultation on a proposal to turn the whole Coral Sea into a Marine Protected Area, half of which would be protected from all exploitation. Australian Government Department of Sustainability, Environment, Water, Population and Communities: www.environment.gov.au/coasts/mbp/coralsea/consultation/index.html; accessed January 3, 2012.

7. United Nations Environment Program—World Conservation Monitoring Centre: www.unep-wcmc.org/medialibrary/2011/06/23/b3b09e87/Gough%20and%20Inaccessible%20Islands.pdf.

8. Roberts, C. M., et al., "Marine Biodiversity Hot Spots and Conservation Priorities for Tropical Reefs," *Science* 295 (2002): 1280–84; Tittensor, D. P. et al., "Global Patterns and Predictors of Marine Biodiversity Across Taxa," *Nature* 466 (2010): 1098–1101; Trebilco, R., et al., "Mapping Species Richness and Human Impact Drivers to

Inform Global Pelagic Conservation Prioritisation," *Biological Conservation* 144 (2011): 1758–66.

9. Fuller, R. A., et al., "Replacing Underperforming Protected Areas Achieves Better Conservation Outcomes," *Nature* 466 (2010): 365–67. To be fair to the authors they did mention in passing that their cost-benefit metric might be replaced by another that measured amenity value, but this seemed very much an afterthought.

10. In other words, to levels required to achieve the maximum long-term yield from a stock, i.e., maximum sustainable yield.

11. Roberts, C. M., et al., "Guidance on the Size and Spacing of Marine Protected Areas in England (Commissioned Report, NECR037)," (Peterborough, UK: Natural England, 2008).

12. Munday, P. L., et al., "Climate Change and Coral Reef Connectivity," *Coral Reefs* 28 (2008): 379–95.

13. Alward, G. L., *The Sea Fisheries of Great Britain and Ireland* (Grimsby, UK: Albert Gait, 1932).

Acknowledgments

This is a book that I hadn't planned to write. My first, *The Unnatural History of the Sea*, took in a thousand years of history and had the world as its stage. It took five years to research and write, so by the time it was complete I thought I would try something a little less daunting. My agent, Patrick Walsh, and Will Goodlad, my soon-to-be editor at Penguin UK, had other ideas. Over a splendid lunch, they persuaded me to go even bigger: the story of the oceans from the beginning of the world to their possible futures a hundred years from now. I am not sure how or why I agreed, but I am glad I did, even though it has taken another five years! I am grateful to both of you for the boldness of your vision and your faith in me to take it on.

I had the great good fortune to have Joy de Menil from Viking as my editor. Joy has a gift of seeing what is essential and showing an author how to say it with elegance and clarity. She suggested I cut many things from my draft, some of which I thought at first were indispensable. If you have made it this far, you will be glad she wielded her red pen so skillfully.

In researching this book I visited subjects I scarcely even knew existed at the outset. I have been privileged to have so many excellent guides who have been generous with their advice and time. I am indebted to the following for their comments on draft chapters, suggestions for improvements, and for helping me understand: Rebecca Atkins, Geoff Bailey, Andrew Bakun, Bryce Beukers-Stewart, Dee Boersma, Alistair Boxall, Ian Boyd, Sally Brown, John Bruno, Ken Caldeira, Jim Carlton, William Cheung, Daniel Conley, Sarika

Cullis-Suzuki, Curtis Ebbesmeyer, Jon Erlandson, Rainer Froese, Jason Hall-Spencer, Julie Hawkins, Mark Hixon, Leigh Howarth, Joanie Kleypas, Dane Klinger, Andrew Knoll, Kevin Lafferty, James McCarthy, Carol Milner, Charles Moore, Roz Naylor, Maggy Nugues, Beth O'Leary, Daniel Pauly, Howard Peters, Jurgenne Primavera, Yasmin Primavera, Nancy Rabalais, Judith Sealey, John Shepherd, Steven Simpson, Albert Tacon, Ruth Thurstan, Carol Turley, Peter Tyack, Charlie Veron, and Oliver Wurl. These folks spared me from many errors. Those that remain are my own. I have been lucky to be able to try out many of the ideas in this book on dozens of enthusiastic and very bright students. They tested the limits of my understanding and forced me to improve my explanations when they were found wanting. I am also grateful for the kindness of the following colleagues who let me use their images in this book: Marco Octavio Aburto-Oropeza, Bryce Beukers-Stewart, Nancy Boucha, Victor Hugo Casillas Romo, Aaron J. Cavosie, Sarika Cullis-Suzuki, Katharina Fabricius, Dave Harasti (who is a superb underwater photographer and whose pictures can be seen at www.daveharasti.com), Elizabeth Gates, Ryan Goehrung, Karin Malmstrom, Guy Marcovaldi, Curtis Marean, Loren McClenchan, Stephen McGowan, Tim Pusack, Link Roberts, William Rodriguez Schepis (Instituto EcoFaxina), Steve Spring, Guy Stevens (The Manta Trust: www.mantatrust.org), Silke Stukenbrock, Bob Talbot, Glen Tepke, John Valley, Kyle Van Houten, Cynthia Vanderlip, and Reuven Walder. Thanks also to Devin Harvey of SeaWeb for his great help in sourcing images.

I have my wife, Julie, and two wonderful daughters to thank for patience, understanding, encouragement, and love throughout the course of writing this book. I am fortunate indeed.

Index